Huajin Tang, Kay Chen Tan and Zhang Yi

Neural Networks: Computational Models and Applications

# Studies in Computational Intelligence, Volume 53

Editor-in-chief
Prof. Janusz Kacprzyk
Systems Research Institute
Polish Academy of Sciences
ul. Newelska 6
01-447 Warsaw
Poland
*E-mail:* kacprzyk@ibspan.waw.pl

Further volumes of this series
can be found on our homepage:
springer.com

Vol. 31. Ajith Abraham, Crina Grosan, Vitorino Ramos (Eds.)
*Stigmergic Optimization,* 2006
ISBN 978-3-540-34689-0

Vol. 32. Akira Hirose
*Complex-Valued Neural Networks,* 2006
ISBN 978-3-540-33456-9

Vol. 33. Martin Pelikan, Kumara Sastry, Erick Cantú-Paz (Eds.)
*Scalable Optimization via Probabilistic Modeling,* 2006
ISBN 978-3-540-34953-2

Vol. 34. Ajith Abraham, Crina Grosan, Vitorino Ramos (Eds.)
*Swarm Intelligence in Data Mining,* 2006
ISBN 978-3-540-34955-6

Vol. 35. Ke Chen, Lipo Wang (Eds.)
*Trends in Neural Computation,* 2007
ISBN 978-3-540-36121-3

Vol. 36. Ildar Batyrshin, Janusz Kacprzyk, Leonid Sheremetor, Lotfi A. Zadeh (Eds.)
*Preception-based Data Mining and Decision Making in Economics and Finance,* 2006
ISBN 978-3-540-36244-9

Vol. 37. Jie Lu, Da Ruan, Guangquan Zhang (Eds.)
*E-Service Intelligence,* 2007
ISBN 978-3-540-37015-4

Vol. 38. Art Lew, Holger Mauch
*Dynamic Programming,* 2007
ISBN 978-3-540-37013-0

Vol. 39. Gregory Levitin (Ed.)
*Computational Intelligence in Reliability Engineering,* 2007
ISBN 978-3-540-37367-4

Vol. 40. Gregory Levitin (Ed.)
*Computational Intelligence in Reliability Engineering,* 2007
ISBN 978-3-540-37371-1

Vol. 41. Mukesh Khare, S.M. Shiva Nagendra (Eds.)
*Artificial Neural Networks in Vehicular Pollution Modelling,* 2007
ISBN 978-3-540-37417-6

Vol. 42. Bernd J. Krämer, Wolfgang A. Halang (Eds.)
*Contributions to Ubiquitous Computing,* 2007
ISBN 978-3-540-44909-6

Vol. 43. Fabrice Guillet, Howard J. Hamilton (Eds.)
*Quality Measures in Data Mining,* 2007
ISBN 978-3-540-44911-9

Vol. 44. Nadia Nedjah, Luiza de Macedo Mourelle, Mario Neto Borges, Nival Nunes de Almeida (Eds.)
*Intelligent Educational Machines,* 2007
ISBN 978-3-540-44920-1

Vol. 45. Vladimir G. Ivancevic, Tijana T. Ivancevic
*Neuro-Fuzzy Associative Machinery for Comprehensive Brain and Cognition Modeling,* 2007
ISBN 978-3-540-47463-0

Vol. 46. Valentina Zharkova, Lakhmi C. Jain
*Artificial Intelligence in Recognition and Classification of Astrophysical and Medical Images,* 2007
ISBN 978-3-540-47511-8

Vol. 47. S. Sumathi, S. Esakkirajan
*Fundamentals of Relational Database Management Systems,* 2007
ISBN 978-3-540-48397-7

Vol. 48. H. Yoshida (Ed.)
*Advanced Computational Intelligence Paradigms in Healthcare,* 2007
ISBN 978-3-540-47523-1

Vol. 49. Keshav P. Dahal, Kay Chen Tan, Peter I. Cowling (Eds.)
*Evolutionary Scheduling,* 2007
ISBN 978-3-540-48582-7

Vol. 50. Nadia Nedjah, Leandro dos Santos Coelho, Luiza de Macedo Mourelle (Eds.)
*Mobile Robots: The Evolutionary Approach,* 2007
ISBN 978-3-540-49719-6

Vol. 51. Shengxiang Yang, Yew Soon Ong, Yaochu Jin Honda (Eds.)
*Evolutionary Computation in Dynamic and Uncertain Environment,* 2007
ISBN 978-3-540-49772-1

Vol. 52. Abraham Kandel, Horst Bunke, Mark Last (Eds.)
*Applied Graph Theory in Computer Vision and Pattern Recognition,* 2007
ISBN 978-3-540-68019-2

Vol. 53. Huajin Tang, Kay Chen Tan, Zhang Yi
*Neural Networks: Computational Models and Applications,* 2007
ISBN 978-3-540-69225-6

Huajin Tang
Kay Chen Tan
Zhang Yi

# Neural Networks: Computational Models and Applications

With 103 Figures and 27 Tables

Springer

Dr. Huajin Tang
Queensland Brain Institute
University of Queensland
QLD 4072 Australia
E-mail: huajin.tang@ieee.org

Prof. Kay Chen Tan
Department of Electrical
and Computer Engineering
National University of Singapore
4 Engineering Drive 3
Singapore 117576
E-mail: eletankc@nus.edu.sq

Prof. Zhang Yi
Computational Intelligence Laboratory
School of Computer Science
and Engineering
University of Electronic Science
and Technology of China
Chengdu 610054 P.R. China
E-mail: zhangyi@uestc.edu.cn

Library of Congress Control Number: 2006939344

ISSN print edition: 1860-949X
ISSN electronic edition: 1860-9503
ISBN-10  3-540-69225-8 Springer Berlin Heidelberg New York
ISBN-13  978-3-540-69225-6 Springer Berlin Heidelberg New York

This work is subject to copyright. All rights are reserved, whether the whole or part of the material is concerned, specifically the rights of translation, reprinting, reuse of illustrations, recitation, broadcasting, reproduction on microfilm or in any other way, and storage in data banks. Duplication of this publication or parts thereof is permitted only under the provisions of the German Copyright Law of September 9, 1965, in its current version, and permission for use must always be obtained from Springer-Verlag. Violations are liable to prosecution under the German Copyright Law.

Springer is a part of Springer Science+Business Media
springer.com
© Springer-Verlag Berlin Heidelberg 2007

The use of general descriptive names, registered names, trademarks, etc. in this publication does not imply, even in the absence of a specific statement, that such names are exempt from the relevant protective laws and regulations and therefore free for general use.

Cover design: deblik, Berlin
Typesetting by the authors using a Springer LaTeX macro package
Printed on acid-free paper    SPIN: 11544272    89/SPi    5 4 3 2 1 0

To our family, for their loves and supports.
Huajin Tang
Kay Chen Tan
Zhang Yi

# Preface

Artificial neural networks, or simply called neural networks, refer to the various mathematical models of human brain functions such as perception, computation and memory. It is a fascinating scientific challenge of our time to understand how the human brain works. Modeling neural networks facilitates us in investigating the information processing occurred in brain in a mathematical or computational manner. On this manifold, the complexity and capability of modeled neural networks rely on our present understanding of biological neural systems. On the other hand, neural networks provide efficient computation methods in making intelligent machines in multidisciplinary fields, e.g., computational intelligence, robotics and computer vision.

In the past two decades, research in neural networks has witnessed a great deal of accomplishments in the areas of theoretical analysis, mathematical modeling and engineering application. This book does not intend to cover all the advances in all aspects, and for it is formidable even in attempting to do so. Significant efforts are devoted to present the recent discoveries that the authors have made mainly in feedforward neural networks, recurrent networks and bio-inspired recurrent network studies. The covered topics include learning algorithm, neuro-dynamics analysis, optimization and sensory information processing, etc. In writing, the authors especially hope the book to be helpful for the readers getting familiar with general methodologies of research in the neural network areas, and to inspire new ideas in the concerned topics.

We received significant support and assistance from the Department of Electrical and Computer Engineering, and we are especially grateful to the colleagues in the Control & Simulation Lab and Zhang Yi's CI lab, where many simulation experiments and meaningful work were carried out.

We are grateful to a lot of people who helped us directly or indirectly. We are extremely grateful to the following professors for their valuable suggestions, continuous encouragements and constructive criticisms: Ben M. Chen, Jian-Xin Xu, National University of Singapore; Weinian Zhang, Sichuan University, China; C.S. Poon, Massachusetts Institute of Technology; Xin Yao, University of Birmingham, UK; Hussein A. Abbass, University of New South

Wales, Australian Defence Force Academy Campus. We are very thankful to Eujin Teoh for his careful reading of the draft.

Huajin Tang would like to thank Geoffrey Goodhill, University of Queensland, for his valuable helps. Zhang Yi would like to thank Hong Qu, Jian Chen Lv and Lei Zhang, University of Electronic Science and Technology of China, for their useful discussions.

Australia            *Huajin Tang*
Singapore          *Kay Chen Tan*
China                *Zhang Yi*
May 2006

# Contents

List of Figures .................................................. XV

List of Tables ................................................... XXI

1 **Introduction** .................................................. 1
   1.1 Backgrounds ................................................ 1
      1.1.1 Feed-forward Neural Networks ....................... 1
      1.1.2 Recurrent Networks with Saturating Transfer
           Functions ........................................ 2
      1.1.3 Recurrent Networks with Nonsaturating Transfer
           Functions ........................................ 4
   1.2 Scopes .................................................... 5
   1.3 Organization .............................................. 6

2 **Feedforward Neural Networks and Training Methods** ....... 9
   2.1 Introduction .............................................. 9
   2.2 Error Back-propagation Algorithm .......................... 9
   2.3 Optimal Weight Initialization Method ...................... 12
   2.4 The Optimization-Layer-by-Layer Algorithm ................ 14
      2.4.1 Optimization of the Output Layer .................... 15
      2.4.2 Optimization of the Hidden Layer .................... 16
   2.5 Modified Error Back-Propagation Method .................. 18

3 **New Dynamical Optimal Learning for Linear Multilayer
FNN** ............................................................ 23
   3.1 Introduction .............................................. 23
   3.2 Preliminaries ............................................. 24
   3.3 The Dynamical Optimal Learning .......................... 25
   3.4 Simulation Results ........................................ 28
      3.4.1 Function Mapping ................................... 28
      3.4.2 Pattern Recognition ................................. 30

## Contents

- 3.5 Discussions .... 33
- 3.6 Conclusion .... 33

**4 Fundamentals of Dynamic Systems** .... 35
- 4.1 Linear Systems and State Space .... 35
  - 4.1.1 Linear Systems in $\mathbf{R}^2$ .... 35
  - 4.1.2 Linear Systems in $\mathbf{R}^n$ .... 38
- 4.2 Nonlinear Systems .... 41
- 4.3 Stability, Convergence and Bounded-ness .... 41
- 4.4 Analysis of Neuro-dynamics .... 46
- 4.5 Limit Sets, Attractors and Limit Cycles .... 51

**5 Various Computational Models and Applications** .... 57
- 5.1 RNNs as a Linear and Nonlinear Programming Solver .... 57
  - 5.1.1 Recurrent Neural Networks .... 58
  - 5.1.2 Comparison with Genetic Algorithms .... 59
- 5.2 RNN Models for Extracting Eigenvectors .... 66
- 5.3 A Discrete-Time Winner-Takes-All Network .... 68
- 5.4 A Winner-Takes-All Network with LT Neurons .... 70
- 5.5 Competitive-Layer Model for Feature Binding and Segmentation .... 74
- 5.6 A Neural Model of Contour Integration .... 76
- 5.7 Scene Segmentation Based on Temporal Correlation .... 77

**6 Convergence Analysis of Discrete Time RNNs for Linear Variational Inequality Problem** .... 81
- 6.1 Introduction .... 81
- 6.2 Preliminaries .... 82
- 6.3 Convergence Analysis: $A$ is a Positive Semidefinite Matrix .... 83
- 6.4 Convergence Analysis: $A$ is a Positive Definite Matrix .... 85
- 6.5 Discussions and Simulations .... 87
- 6.6 Conclusions .... 96

**7 Parameter Settings of Hopfield Networks Applied to Traveling Salesman Problems** .... 99
- 7.1 Introduction .... 99
- 7.2 TSP Mapping and CHN Model .... 100
- 7.3 The Enhanced Lyapunov Function for Mapping TSP .... 102
- 7.4 Stability Based Analysis for Network's Activities .... 104
- 7.5 Suppression of Spurious States .... 105
- 7.6 Setting of Parameters .... 112
- 7.7 Simulation Results and Discussions .... 112
- 7.8 Conclusion .... 115

## 8 Competitive Model for Combinatorial Optimization Problems ... 117
   8.1 Introduction ... 117
   8.2 Columnar Competitive Model ... 118
   8.3 Convergence of Competitive Model and Full Valid Solutions ... 120
   8.4 Simulated Annealing Applied to Competitive Model ... 123
   8.5 Simulation Results ... 124
   8.6 Conclusion ... 128

## 9 Competitive Neural Networks for Image Segmentation ... 129
   9.1 Introduction ... 129
   9.2 Neural Networks Based Image Segmentation ... 130
   9.3 Competitive Model of Neural Networks ... 131
   9.4 Dynamical Stability Analysis ... 132
   9.5 Simulated Annealing Applied to Competitive Model ... 134
   9.6 Local Minima Escape Algorithm Applied to Competitive Model ... 135
   9.7 Simulation Results ... 137
      9.7.1 Error-Correcting ... 137
      9.7.2 Image Segmentation ... 140
   9.8 Conclusion ... 143

## 10 Columnar Competitive Model for Solving Multi-Traveling Salesman Problem ... 145
   10.1 Introduction ... 145
   10.2 The MTSP Problem ... 146
   10.3 MTSP Mapping and CCM Model ... 148
   10.4 Valid Solutions and Convergence Analysis of CCM for MTSP . 151
      10.4.1 Parameters Settings for the CCM Applied to MTSP ... 152
      10.4.2 Dynamical Stability Analysis ... 153
   10.5 Simulation Results ... 156
   10.6 Conclusions ... 159

## 11 Improving Local Minima of Columnar Competitive Model for TSPs ... 161
   11.1 Introduction ... 161
   11.2 Performance Analysis for CCM ... 162
   11.3 An Improving for Columnar Competitive Model ... 165
      11.3.1 Some Preliminary Knowledge ... 165
      11.3.2 A Modified Neural Representation for CCM ... 167
      11.3.3 The Improvement for Columnar Competitive Model ... 168
   11.4 Simulation Results ... 171
   11.5 Conclusions ... 174

## 12 A New Algorithm for Finding the Shortest Paths Using PCNN ........................................................... 177
12.1 Introduction .................................................... 177
12.2 PCNNs Neuron Model ........................................ 178
12.3 The Multi-Output Model of Pulse Coupled Neural Networks (MPCNNs) ..................................................... 180
    12.3.1 The Design of MPCNNs ................................. 180
    12.3.2 Performance Analysis of the Travelling of Autowaves in MPCNNs ........................................... 181
12.4 The Algorithm for Solving the Shortest Path Problems using MPCNNs ....................................................... 183
12.5 Simulation Results ............................................. 184
12.6 Conclusions ..................................................... 188

## 13 Qualitative Analysis for Neural Networks with LT Transfer Functions ............................................................ 191
13.1 Introduction .................................................... 191
13.2 Equilibria and Their Properties ................................ 192
13.3 Coexistence of Multiple Equilibria ............................. 197
13.4 Boundedness and Global Attractivity .......................... 199
13.5 Simulation Examples ........................................... 203
13.6 Conclusion ...................................................... 206

## 14 Analysis of Cyclic Dynamics for Networks of Linear Threshold Neurons ........................................ 211
14.1 Introduction .................................................... 211
14.2 Preliminaries .................................................... 212
14.3 Geometrical Properties of Equilibria ........................... 213
14.4 Neural States in $D_1$ and $D_2$ ................................. 214
    14.4.1 Phase Analysis for Center Type Equilibrium in $D_1$ .... 214
    14.4.2 Phase Analysis in $D_2$ .................................. 215
    14.4.3 Neural States Computed in Temporal Domain ........ 217
14.5 Rotated Vector Fields .......................................... 217
14.6 Existence and Boundary of Periodic Orbits ................... 219
14.7 Winner-take-all Network ....................................... 226
14.8 Examples and Discussions ..................................... 229
    14.8.1 Nondivergence Arising from A Limit Cycle ............ 229
    14.8.2 An Example of WTA Network ........................ 229
    14.8.3 Periodic Orbits of Center Type ........................ 229
14.9 Conclusion ...................................................... 231

## 15 LT Network Dynamics and Analog Associative Memory ... 235
15.1 Introduction .................................................... 235
15.2 Linear Threshold Neurons ...................................... 236
15.3 LT Network Dynamics (Revisited) ............................. 238

15.4 Analog Associative Memory ............................. 243
  15.4.1 Methodology........................................ 243
  15.4.2 Design Method..................................... 244
  15.4.3 Strategies of Measures and Interpretation............. 247
15.5 Simulation Results....................................... 247
  15.5.1 Small-Scale Example............................... 248
  15.5.2 Single Stored Images............................... 249
  15.5.3 Multiple Stored Images ............................ 251
15.6 Discussion ............................................. 252
  15.6.1 Performance Metrics ............................... 252
  15.6.2 Competition and Stability .......................... 252
  15.6.3 Sparsity and Nonlinear Dynamics ................... 253
15.7 Conclusion ............................................. 255

# 16 Output Convergence Analysis for Delayed RNN with Time Varying Inputs ................................. 259
16.1 Introduction ............................................ 259
16.2 Preliminaries........................................... 261
16.3 Convergence Analysis .................................. 264
16.4 Simulation Results...................................... 274
16.5 Conclusion ............................................. 276

# 17 Background Neural Networks with Uniform Firing Rate and Background Input ..................................... 279
17.1 Introduction ............................................ 279
17.2 Preliminaries........................................... 280
17.3 Nondivergence and Global Attractivity ..................... 282
17.4 Complete Stability...................................... 283
17.5 Discussion ............................................. 285
17.6 Simulation............................................. 286
17.7 Conclusions............................................ 286

**References** ................................................. 289

# List of Figures

2.1 Feed-forward network with one hidden layer and a scalar output (top panel) and its linearized network structure (bottom panel). .................................................. 14
2.2 Linear and error portion of the nodes in the hidden layer (left panel) and the output layer (right panel). .................... 19

3.1 A general FNN model........................................ 25
3.2 DOL error performance for function mapping with 100 inputs: 85 epochs .................................................... 30
3.3 SBP error performance for function mapping with 100 inputs: 3303 epochs .................................................. 31
3.4 The character patterns for the pattern recognition problem .... 31
3.5 DOL error performance for the characters recognition: 523 epochs ...................................................... 32
3.6 SBP error performance for the characters recognition: 2875 epochs ...................................................... 32
3.7 The dynamical optimal learning rate for the function mapping problem ..................................................... 34

4.1 The phase portraits of stable nodes at the origin. (a) $\lambda = \mu$, (b) $\lambda < \mu$, (c) $\lambda < 0$. ........................................... 36
4.2 A saddle at the origin. ....................................... 37
4.3 The phase portraits of a stable focus at the origin. (a) $b > 0$, (b) $b < 0$. .................................................... 37
4.4 The phase portraits of a center at the origin. (a) $b > 0$, (b) $b < 0$. .................................................... 38
4.5 Representative trajectories illustrating Lyapunov stability. ...... 42
4.6 A limit cycle is contained in an annulus (gray region) with boundaries r1 and r2. The inner boundary encircles an unstable equilibrium point A. ................................. 53

## List of Figures

4.7 A finite number of limit cycles exist in the annular region: Two stable limit cycles (L1 and L3) enclose an unstable limit cycle (L2)..... 54

4.8 Wilson-Cowan network oscillator. Excitatory connections are shown by arrows and inhibitory by solid circles. The simplified network on the right is mathematically equivalent to that on the left by symmetry if all E→E connections are identical..... 54

4.9 Limit cycle of the Wilson-Cowan oscillator. In the above panel an asymptotic limit cycle is plotted in the phase plane along with two isoclines (upper panel). The lower panel plots $E(t)$ (solid line) and $I(t)$ (dashed line) as function of time. Adapted from Wilson (1999)..... 56

5.1 Function optimization using genetic algorithm..... 62
5.2 Evolution of best solution and average solution..... 62
5.3 State trajectories of Chua's network with $s = 10$ (Example 1)..... 65
5.4 State trajectories of Chua's network with $s = 100$ (Example 1)..... 65
5.5 State trajectories of Chua's network with $s = 100$ (Example 2)..... 66
5.6 Estimation of an eigenvector corresponding to the largest eigenvalue of the matrix $A$. Curtesy of Yi and Tan (2004)..... 68
5.7 The discrete-time WTA network (5.15)..... 69
5.8 The convergence of the network states $x_i$..... 70
5.9 The convergence of the outputs of the network..... 71
5.10 The neural states in 50 seconds. Initial point is randomly generated and the neuron $x_3$ becomes the winner. $w_i = 2, \tau = 0.5, h = (1, 2, 3)^\top$..... 72
5.11 The neural states in 50 seconds. Initial point is randomly generated. Differing from the previous figure, the neuron $x_2$ becomes the winner for the same set of parameters..... 73
5.12 The phase portrait in the $x_1 - L$ plane. 20 initial points are randomly generated. It is shown that there exist two steady states..... 73

6.1 The function $r(\alpha)$..... 86
6.2 Convergence of each component of one trajectory..... 90
6.3 Spatial convergence of the network with 50 trajectories..... 90
6.4 Convergence of random initial 30 points in Example 2..... 91
6.5 Spatial convergence of the network with 300 trajectories..... 92
6.6 Convergence of the network (5) in Example 3..... 93
6.7 Spatial representation of the trajectories in the plane $(x_1, x_2)$ in Example 3..... 94
6.8 Spatial representation of the trajectories in the plane $(x_1, x_3)$ in Example 3..... 94

| | | |
|---|---|---|
| 6.9 | Spatial representation of the trajectories in the plane $(x_2, x_3)$ in Example 3. | 95 |
| 6.10 | Spatial representation of solution trajectories in 3-D space in Example 3. | 96 |
| 6.11 | Solution trajectories of each components in Example 4. | 97 |
| 7.1 | Vertex point, edge point and interior point. | 106 |
| 7.2 | Optimum tour state | 114 |
| 7.3 | Near-Optimum tour state | 114 |
| 7.4 | Comparison of average tour length between the modified formulation (new setting) and H-T formulation (old setting). | 116 |
| 7.5 | Comparison of minimal tour length between the modified formulation (new setting) and H-T formulation (old setting). | 116 |
| 8.1 | Optimum tour generated by CCM | 125 |
| 8.2 | Near-optimum tour generated by CCM. | 126 |
| 9.1 | Corrupted Image | 137 |
| 9.2 | Wrongly classified image | 138 |
| 9.3 | Correct classifications for different variances | 138 |
| 9.4 | Original Lena image | 141 |
| 9.5 | Lena image segmented with 3 classes. | 141 |
| 9.6 | Lena image segmented with 6 classes. | 142 |
| 9.7 | Energies of the network for different classes | 142 |
| 10.1 | Valid tours of various problems. Subplots a, b, c and d are valid tours of problems 1, 2, 3, 4, respectively. | 147 |
| 10.2 | The left is the feasible solution tour. The right is the tour after adding the virtual city. F is the virtual city | 149 |
| 10.3 | The city network. | 157 |
| 10.4 | The road example | 158 |
| 10.5 | The convergence of the energy function with different parameters. The left panel $A = 300, B = 0.05, A - Bn^2 < 2MAX - MINX$, the middle panel $A = 400, B = 0.05, A - Bn^2 \doteq 2MAX - MIN$, the right panel $A = 400, B = 0.01, A - Bn^2 > 2MAX - MIN$ | 159 |
| 11.1 | The example. 1 is the start city. | 164 |
| 11.2 | Block diagram of the improved CCM | 170 |
| 11.3 | Tour length histograms for 10-city problems 1, 2, and 3 produced by CCM (top panels) and Improved CCM (bottom panels). | 174 |
| 11.4 | Average iteration number histograms for problems 1, 2, and 3 produced by CCM and the improved CCM | 175 |
| 11.5 | The tours found by CCM and the improved CCM | 175 |

## List of Figures

12.1 PCNNs Neuron Model .................................... 179
12.2 The Neuron's Model of MPCNNs.......................... 180
12.3 A symmetric weighted graph. ............................ 185
12.4 The finding processing of the instance from "start" node 1 to others. ................................................ 186
12.5 A symmetric weighted graph randomly generated with $N = 100$ in a square of edge length $D = 50$.............. 188
12.6 (a), (b) and (c) show the shortest paths form differently specified "start" nodes to all others ......................... 189

13.1 Equilibria distribution in $w_{11}, w_{22}$ plane. .................. 200
13.2 Center and stable node coexisting in quadrants I and III of network (13.18) ........................................ 204
13.3 Three equilibria coexisting in quadrants I, II and IV .......... 205
13.4 Four equilibria coexisting in four quadrants ................. 206
13.5 Global attractivity of the network (14.43) .................. 206
13.6 Projection on $(x_1, x_2)$ phase plane ........................ 207
13.7 Projection on $(x_2, x_3)$ phase plane ........................ 207
13.8 Projection on $(x_1, x_3)$ phase plane ........................ 208

14.1 Rotated vector fields prescribed by (14.24). (a) is under the conditions $w_{12} < 0, 0 < k_1 < k_2$, and (b) $w_{12} < 0, k_1 < 0 < k_2$. If $w_{12} > 0$, the arrows from $L_1$ and $L_2$ are reversed and $p < s < 0$ for both (a) and (b). ........................... 218
14.2 The vector fields described by the oscillation prerequisites. The trajectories in $D_1$ and $D_2$ ($\Gamma_R$ and $\Gamma_L$, respectively) forced by the vector fields are intuitively illustrated. .......... 220
14.3 Phase portrait for $0 < w_{22} < 1$. ........................... 221
14.4 Phase portrait for $w_{22} < 0$. .............................. 222
14.5 Phase portrait for $w_{22} = 0$. .............................. 222
14.6 Phase portrait for $w_{22} > 1$. .............................. 223
14.7 Nondivergence arising from a stable limit cycle. (a) The limit cycle in the phase plane. Each trajectory from its interior or its outside approaches it. It is obviously an attractor. (b) The oscillations of the neural states. One of the neural states endlessly switches between stable mode and unstable mode. An Euler algorithm is employed for the simulations with time step 0.01. ....................................... 232
14.8 Cyclic dynamics of the WTA network with 6 excitatory neurons. $\tau = 2, \theta_i = 2$ for all $i$, $h = (1, 1.5, 2, 2.5, 3, 3.5)^T$. The trajectory of each neural state is simulated for 50 seconds. The dashed curve shows the state of the global inhibitory neuron $L$. ................................................. 233

14.9 A periodic orbit constructed in $x_6 - L$ plane. It shows the trajectories starting from five random points approaching the periodic orbit eventually. ................................... 233

14.10 Periodic orbits of center type. The trajectories with three different initial points $(0, 0.5), (0, 1)$ and $(0, 1.2)$ approach a periodic orbit that crosses the boundary of $D_1$. The trajectory starting from $(0, 1)$ constructs an outer boundary $\Gamma = \widehat{pqr} \cup \overline{rp}$ such that all periodic orbits lie in its interior. .... 234

15.1 LT activation function with gain $k = 1$, threshold $\theta = 0$, relating the neural activity output to the induced local field. .... 237
15.2 Original and retrieved patterns with stable dynamics. ......... 249
15.3 Illustration of convergent individual neuron activity. .......... 250
15.4 Collage of the 4, $32 \times 32$, 256 gray-level images. ............. 251
15.5 Lena: $SNR$ and $MaxW^+$ with $\alpha$ in increment of 0.0025. ....... 252
15.6 Brain: $SNR$ (solid line) and $MaxW^+$ with $\alpha$ in increment of 0.005. .................................................. 253
15.7 Lena: $\alpha = 0.32, \beta = 0.0045, \omega = -0.6, SNR = 5.6306$; zero mean Gaussian noise with 10% variance. .................... 254
15.8 Brain: $\alpha = 0.43, \beta = 0.0045, \omega = -0.6, SNR = 113.8802$; zero mean Gaussian noise with 10% variance. .................... 255
15.9 Strawberry: $\alpha = 0.24, \beta = 0.0045, \omega = 0.6, SNR = 1.8689$; 50% Salt-&-Pepper noise. ................................. 256
15.10 Men: $\alpha = 0.24, \beta = 0.0045, \omega = 0.6, SNR = 1.8689$; 50% Salt-&-Pepper noise. ...................................... 257

16.1 Output convergence of network (16.22). The four output trajectories are generated by four state trajectories starting from $(3, 3, 3)^T$, $(3, -3, 0)^T$, $(-3, 0, -3)^T$ and $(-3, 3, 0)^T$, respectively. Clearly, the four output trajectories converge to $(-0.7641, -0.1502, 0.1167)^T$. ............................ 275
16.2 Convergence of output of network (16.23) ................... 276

17.1 Complete convergence of the network (17.1) in condition (a) in Example 1. .......................................... 287
17.2 Complete convergence of the network (17.1) in condition (a) in Example 1. .......................................... 287

# List of Tables

3.1 Improvement ratio of DOL over SBP for 100 inputs with $mse = 0.001$ .................................... 29
3.2 Improvement ratio of DOL over SBP for 9 patterns with $mse = 0.001$ .................................... 33

7.1 Performance of the parameter settings obtained from Talaván .. 113
7.2 Performance of the new parameter settings ................. 113

8.1 The performance of original Hopfield model for the 24-city example ................................................. 125
8.2 The performance of CCM for the 24-city example ........... 125
8.3 The performance of CCM with SA for the 24-city example..... 127
8.4 The performance of CCM for the 48-city example ........... 127
8.5 The performance of CCM with SA for various city sizes ....... 127

9.1 The performance of the segmentation model (Cheng et al., 1996) 139
9.2 The performance of the proposed model .................... 140
9.3 Image Segmentation using only WTA ....................... 141
9.4 Image Segmentation with SA or LME ....................... 143

10.1 The resulting paths ...................................... 158
10.2 The computational result of CCM for MTSP with tabu search method .................................................. 159

11.1 The probability of the convergence to every valid solution ..... 165
11.2 City Coordinates for The Three 10-City Problems ............ 172
11.3 The Performance of CCM Applied to 10-City TSP............ 172
11.4 The Performance of Improved CCM Applied to 10-City TSP... 173
11.5 Comparison performance for 10-city problems ............... 173

12.1 Results of MPCNN for the graph in Fig. 12.3 ................ 186
12.2 Simulation Results with the different value of $V_\theta$ .............. 187

12.3 Simulation Results with the different value of $V_\theta$ .............. 187

13.1 Properties and distributions of the equilibria ................. 194

14.1 Properties and distributions of the equilibria ................. 214

15.1 Nomenclature ............................................. 245

17.1 The relationship of the existence and stability of equilibrium point to $\Delta$. ................................................ 286

# 1

# Introduction

Typically, models of neural networks are divided into two categories in terms of signal transmission manner: feed-forward neural networks and recurrent neural networks. They are built up using different frameworks, which give rise to different fields of applications.

## 1.1 Backgrounds

### 1.1.1 Feed-forward Neural Networks

Feed-forward neural network (FNN), also referred to as multilayer perceptrons (MLPs), has drawn great interests over the last two decades for its distinction as a universal function approximator (Funahashi, 1989; Scalero and Tepedelenlioglu, 1992; Ergezinger and Thomsen, 1995; Yu et al., 2002). As an important intelligent computation method, FNN has been applied to a wide range of applications, including curve fitting, pattern classification and nonlinear system identification and so on (Vemuri, 1995).

FNN features a supervised training with a highly popular algorithm known as the error back-propagation algorithm. In the standard back-propagation (SBP) algorithm, the learning of a FNN is composed of two passes: in the forward pass, the input signal propagates through the network in a forward direction, on a layer-by-layer basis with the weights fixed; in the backward pass, the error signal is propagated in a backward manner. The weights are adjusted based on an error-correction rule. Although it has been successfully used in many real world applications, SBP suffers from two infamous shortcomings, i.e., slow learning speed and sensitivity to parameters. Many iterations are required to train small networks, even for a simple problem. The sensitivity to learning parameters, initial states and perturbations was analyzed in (Yeung and Sun, 2002). Behind such drawbacks the learning rate plays a key role in affecting the learning performance and it has to be chosen carefully. If the learning rate is large, the network may exhibit chaotic behavior so learning might not succeed, while a very small learning rate will result in slow convergence, which is also not desirable. The chaotic phenomena was studied from a dynamical system point of view (Bertels et al., 2001) which reported that when the learning rate falls in some unsuitable range, it may result in chaotic behaviors in the network learning, and for non-chaotic learning rates the network converges faster than for chaotic ones.

Since the shortcomings of the SBP algorithm limit the practical use of FNN, a significant amount of research has been carried out to improve the training performance and to better select the training parameters. A modified back-propagation algorithm was derived by minimizing the mean-squared error with respect to the inputs summation, instead of minimizing with respect to weights like SBP, but its convergence heavily depended on the magnitude of the initial weights. An accelerated learning algorithm OLL (Ergezinger and Thomsen, 1995) was presented based on a linearization of the nonlinear processing nodes and optimizing cost functions layer by layer. Slow learning was attributed to the effect of unlearning and a localizing learning algorithm was developed to reduce unlearning (Weaver and Polycarpou, 2001). Bearing in mind that the derivative of the activation has a large value when the outputs of the neurons in the active region, a method to determine optimal initial weights was put forward in (Yam and Chow, 2001). This method was able to prevent the network from getting stuck in the early stage of training, thus increasing the training speed.

Existing approaches have improved the learning performance in terms of the reduction of iteration numbers, however, none of them dealt with dynamical adaption of the learning rate for different parameters and training phases, which certainly contributes to the sensitivity of such algorithms. An optimal learning rate for a given two layers' neural network was derived in the work of (Wang et al., 2001), but a two-layer neural network has very limited generalization ability. Finding a suitable learning rate is a very experimental technique, since for the multilayer FNN with squashing sigmoid functions, it is difficult to deduce an optimal learning rate and even impossible to predetermine the value of such a parameter for different problems and different initial parameters. Indeed, the optimal learning rate keeps changing along with the training iterations. Finding a dynamical optimal learning algorithm being able to reduce the sensitivity and improve learning motivate developing a new and efficient learning algorithm for multilayer FNN.

### 1.1.2 Recurrent Networks with Saturating Transfer Functions

Unlike feed-forward neural networks, recurrent neural networks (RNN) are described by a system of differential equations that define the exact evolution of the model dynamics as a function of time. The system is characterized by a large number of coupling constants represented by the strengths of individual junctions, and it is believed that the computational power is the result of the collective dynamics of the system. Two prominent computation models with saturating transfer functions, the Hopfield network and cellular neural network, have stimulated a great deal of research efforts over the past two decades because of their great potential of applications in associative memory, optimization and intelligent computation (Hopfield, 1984; Hopfield and Tank, 1985; Tank and Hopfield, 1986; Bouzerdoum and Pattison, 1993; Maa and Shanblatt, 1992; Zak et al., 1995b; Tan et al., 2004; Yi et al., 2004).

As a nonlinear dynamical system, intrinsically, the stability is of primary interest in the analysis and applications of recurrent networks, where the Lyapunov stability theory is a fundamental tool and widely used for analyzing nonlinear systems (Grossberg, 1988; Vidyasagar, 1992; Yi et al., 1999; Qiao et al., 2003). Based on the Lyapunov method, the conditions of global exponential stability of a continuous-time RNN were established and applied to bound-constrained nonlinear differentiable optimization problems (Liang and Wang, 2000). A discrete-time recurrent network solving strictly convex quadratic optimization problems with bound constraints was analyzed and stability conditions were presented (Pérez-Ilzarbe, 1998). Compared with its continuous-time counterpart, the discrete-time model has its advantages in digital implementation. However, there is lack of more general stability conditions for the discrete-time network in the previous work (Pérez-Ilzarbe, 1998), which deserves further investigation.

Solving NP-hard optimization problems, especially the traveling salesman problem (TSP) using recurrent networks has become an active topic since the seminal work of (Hopfield and Tank, 1985) showed that the Hopfield network could give near optimal solutions for the TSP. In the Hopfield network, the combinatorial optimization problem is converted into a continuous optimization problem that minimizes an energy function calculated by a weighted sum of constraints and an objective function. The method, nevertheless, faces a number of disadvantages. Firstly, the nature of the energy function causes infeasible solutions to occur most of the time. Secondly, several penalty parameters need to be fixed before running the network, while it is nontrivial to optimally set these parameters. Besides, low computational efficiency, especially for large scale problems, is also a restriction.

It has been a continuing research effort to improve the performance of the Hopfield network (Aiyer et al., 1990; Abe, 1993; Peng et al., 1993; Papageorgiou et al., 1998; Talaván and Yáñez, 2002a). The authors in (Aiyer et al., 1990) analyzed the dynamic behavior of a Hopfield network based on the eigenvalues of connection matrix and discussed the parameter settings for TSP. By assuming a piecewise linear activation function and by virtue of studying the energy of the vertex at a unit hypercube, a set of convergence and suppression conditions were obtained (Abe, 1993). A local minima escape (LME) algorithm was presented to improve the local minima by combining the network disturbing technique with the Hopfield network's local minima searching property (Peng et al., 1993).

Most recently, a parameter setting rule was presented by analyzing the dynamical stability conditions of the energy function (Talaván and Yáñez,

2002a), which shows promising results compared with previous work, though much effort has to be paid to suppress the invalid solutions and increase convergence speed. To achieve such objectives, incorporating the winner-take-all (WTA) learning mechanism (Cheng et al., 1996; Yi et al., 2000) is one of the more promising approaches.

### 1.1.3 Recurrent Networks with Nonsaturating Transfer Functions

In recent years, the linear threshold (LT) network which underlies the behavior of visual cortical neurons has attracted extensive interests of scientists as the growing literature illustrates (Hartline and Ratliff, 1958; von der Malsburg, 1973; Douglas et al., 1995; Ben-Yishai et al., 1995; Salinas and Abbott, 1996; Adorjan et al., 1999; Bauer et al., 1999; Hahnloser, 1998; Hahnloser et al., 2000; Wersing et al., 2001a; Yi et al., 2003). Differing from the Hopfield type network, the LT network possesses nonsaturating transfer functions of neurons, which is believed to be more biologically plausible and has more profound implications in the neurodynamics. For example, the network may exhibit multistability and chaotic phenomena, which will probably give birth to new discoveries and insights in associative memory and sensory information processing (Xie et al., 2002).

The LT network has been observed to exhibit one important property, i.e., multistability, which allows the networks to possess multiple steady states coexisting under certain synaptic weights and external inputs. The multistability endows the LT networks with distinguished application potentials in decision, digital selection and analogue amplification (Hahnloser et al., 2000). It was proved that local inhibition is sufficient to achieve nondivergence of LT networks (Wersing et al., 2001b). Most recently, several aspects of LT dynamics were studied and the conditions were established for boundedness, global attractivity and complete convergence (Yi et al., 2003). Nearly all the previous research efforts were devoted to stability analysis, thus the cyclic dynamics has yet been elucidated in a systematic manner. In the work of (Hahnloser, 1998), periodic oscillations were observed in a multistable WTA network when slowing down the global inhibition. He reported that the epileptic network switches endlessly between stable and unstable partitions and eventually the state trajectory approaches a limit cycle (periodic oscillation) which was shown by computer simulations. It was suggested that the appearance of periodic orbits in linear threshold networks was related to the existence of complex conjugate eigenvalues with positive real parts. However, there was lack of theoretical proof about the existence of limit cycles. It also remains unclear what factors will affect the amplitude of the oscillations.

Studying recurrent dynamics is also of crucial concern in the realm of modeling the visual cortex, since recurrent neural dynamics is a basic computational substrate for cortical processing. Physiological and psychophysical data suggest that the visual cortex implements preattentive computations such as contour enchancement, texture segmentation and figure-ground segregation (Kapadia et al., 1995; Gallant et al., 1995; Knierim and van Essen, 1992). Various models have addressed particular components of the cortical computation (Grossberg and Mingolla, 1985; Zucker et al., 1989; Yen and Finkel, 1998). A fully functional and dynamically well-behaved model has been proposed to achieve the designed cortical computations (Li and Dayan, 1999; Li, 2001). The LEGION model uses the mechanism of oscillation to perform figure-ground segmentation (Wang and Terman, 1995; Wang and Terman, 1997; Wang, 1999; Chen and Wang, 2002). The CLM model, formulated by the LT network, realizes an energy-based approach to feature binding and texture segmentation and has been successfully applied to segmentation of real-world images (Ontrup and Ritter, 1998; Wersing et al., 1997; Wersing and Ritter, 1999). Dynamic binding in a neural network is of great interest for the vision research, a variety of models have been addressed using different binding approaches, such as temporal coding and spatial coding (Hummel and Biederman, 1992; Feldman and Ballard, 1982; Williamson, 1996). Understanding the complex, recurrent and nonlinear dynamics underlying the computation is essential to explore its power as well as for computational design.

These facts have provided substantial motivations for the extensive investigations of neural networks, both in dynamics analysis and applications.

## 1.2 Scopes

One focus of this book lies in the improvement of training algorithms for feed-forward neural networks by analyzing the mean-squared error function from the perspective of dynamical stability. The dynamical learning method is able to adaptively and optimally set the value of learning rate, hence the elimination of sensitivity of FNN networks with a fixed learning rate can be expected, as well as the reduction of convergence iterations and time.

Another emphasis is on the neurodynamics. The dynamics of the recurrent networks with saturating and nonsaturating transfer functions are analyzed extensively. New theoretical results on the nondivergence, stability and cyclic dynamics are established, which facilitate the applications of the recurrent networks in optimizations and sensory information segmentation. As an important application of the attractor networks, the analog associative memory of the LT network is also investigated. It shows that the LT network can successfully retrieve gray level images.

A special focus is on developing a competitive network incorporating winner-take-all mechanism. The competitive network deals with the constraints in optimization problems in an elegant way, so it has attractive advantages both in suppressing invalid solutions and in increasing convergence speed. The latter is a great concern when solving large scale problems. Probabilistic optimization methods, such as simulated annealing and local minima escape, are also applicable to the competitive network, which can further improve the solution quality.

The significance of this book falls into two basic grounds. Above all, the book will serve the purpose of exploring the computational models of neural networks, and promoting our understanding of the functions of biological neural systems such as computation, perception and memory. Secondly, the theories and methods in this book can provide meaningful techniques for developing real-world applications.

## 1.3 Organization

The first chapter motivates the issue of dynamics analysis as a crucial step to understand the collective computation property of neural systems and describes the scope and contributions of the book.

The second chapter describes the typical learning algorithm of feedforward networks and several prominent modified algorithms among existing approaches. Chapter 3 presents a new dynamical optimal training algorithm for feed-forward neural networks. The new training method aims to avoid the serious drawback of the standard feed-forward neural network's training algorithm, i.e., sensitivity to initial parameters and different problems.

Chapter 4 introduces the fundamentals of mathematical analysis for linear and nonlinear systems, which underlie the analysis of neuro-dynamics.

Chapter 5 is devoted to various computational models based on recurrent neural networks and winner-take-all networks. Some useful applications, such as linear and nonlinear programming, extracting eigenvalues, feature binding and segmentation are introduced. In Chapter 6, a class of discrete-time recurrent networks is discussed and is applied to the typical nonlinear optimization problems. The global exponential stability condition is established which ensures the network globally convergent to the unique optimum.

Chapters 7 and 8 are focused on the neural networks applied to combinatorial optimization problems, where the issue of parameter settings of Hopfield networks, and new competitive model is presented respectively. Subsequently, the competitive model is extended as an algorithm for image segmentation in Chapter 9. In Chapter 10, the model is proposed to solve the multi traveling salesman problems. Chapter 11 is focused on studying the local minima problem of the competitive network and an improvement strategy is provided. In Chapter 12, a new algorithm for finding the shortest path based on the pulsed coupled networks is proposed.

The next consecutive Chapters (13-15) are devoted to a prominent biologically motivated model, i.e., the recurrent network with linear threshold (LT) neurons. In Chapter 13 qualitative analysis is given regarding the geometrical properties of equilibria and the global attractivity. Chapter 14 analyzes one of important dynamic behaviors of the LT networks, periodic oscillation. Conditions for the existence of periodic orbits are established. Chapter 15 presents new conditions which ensure roundedness and stability for nonsymmetric and symmetric LT networks. As an important application, the analog associative memory is exploited in terms of storing gray images. The stability results are used to design such an associative memory network.

Chapters 16 and 17 are more focused on approaches of studying the dynamical properties of recurrent neural networks: delayed networks with time varying inputs and background neural networks with uniform firing rate and background input.

# 2

# Feedforward Neural Networks and Training Methods

## 2.1 Introduction

Multilayer feedforward neural networks (FNN), or equivalently referred to as multilayer perceptrons (MLP), have a layered structure and process information flow in feedforward manner: an input layer consisting of sensory nodes, one or more hidden layers of computational nodes, and an output layer that calculates the outputs of the network. By virtue of their universal function approximation property, multilayer FNNs play a fundamental role in neural computation, as they have been widely applied in many different areas including pattern recognition, image processing, intelligent control, time series prediction, etc. From the universal approximation theorem, a feedforward network of a single hidden layer is sufficient to compute a uniform approximation for a given training set and its desired outputs, hence this chapter is restricted to discuss the single hidden layer FNN, unless otherwise specified.

## 2.2 Error Back-propagation Algorithm

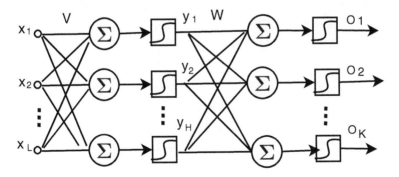

The error signal at the output node $k, k = 1, \cdots, K$, is defined by

$$e_k^p = d_k^p - o_k^p,$$

where the subscript $p$ denotes the given pattern, $p = 1, \cdots, P$. The mean squared error (given the $p$-th pattern) is written as

$$E = \frac{1}{K} \sum_{k=1}^{K} \frac{1}{2} (e_k^p)^2 = \frac{1}{K} \sum_{k=1}^{K} \frac{1}{2} (d_k^p - o_k^p)^2. \tag{2.1}$$

In the bach mode training, the weights are updated after all the training patterns are fed into the inputs and thus the cost function becomes:

$$E = \frac{1}{PK} \sum_{p=1}^{P} \sum_{k=1}^{K} \frac{1}{2} (d_k^p - o_k^p)^2. \tag{2.2}$$

For the output layer, compute the derivative of $E$ with respect to (w.r.t.) the weights $w_{kj}$:

$$\frac{\partial E}{\partial w_{kj}} = \frac{\partial E}{\partial e_k} \cdot \frac{\partial e_k}{\partial o_k} \cdot \frac{\partial o_k}{\partial \alpha_k} \cdot \frac{\partial \alpha_k}{\partial w_{kj}}$$
$$= e_k \cdot (-1) \cdot \varphi_k'(\alpha_k) \cdot y_j$$

Similarly, we compute the partial derivative w.r.t. the hidden layer weights $v_{ji}$:

$$\frac{\partial E}{\partial v_{ji}} = \sum_k \frac{\partial E}{\partial e_k} \cdot \frac{\partial e_k}{\partial \alpha_k} \cdot \frac{\partial \alpha_k}{\partial y_j} \cdot \frac{\partial y_j}{\partial \alpha_j} \cdot \frac{\partial \alpha_j}{v_{ji}}$$
$$= \sum_k e_k \cdot (-1) \cdot \varphi_k'(\alpha_k) w_{kj} \cdot \varphi_j'(\alpha_j) \cdot x_i$$
$$= -x_i \cdot \varphi_j'(\alpha_j) \cdot \sum_k e_k \varphi_k'(\alpha_k) w_{kj}$$

The functions $\varphi_j, \varphi_k$ are called *activation functions* which are continuously differentiable. The activation functions commonly used in feed-forward neural networks are described below:

1. *Logistic function.* Its general form is defined by

$$y = \varphi(x) = \frac{1}{1 + \exp(-ax)}, \quad a > 0, x \in \mathbf{R}. \tag{2.3}$$

The output value lies in the range $0 \leq y \leq 1$. Its derivative is computed as

$$\varphi'(x) = ay(1 - y). \tag{2.4}$$

2. *Hyperbolic tangent function.* This type of activation functions takes the following form
$$y = \frac{1 - \exp(-2x)}{1 + exp(-2x)}. \tag{2.5}$$
The output lies in the range $-1 \le y \le 1$. Its derivative takes the simple form
$$\varphi'(x) = (1+y)(1-y). \tag{2.6}$$
3. *Linear function.* It is simply defined as
$$y = \varphi(x) = x. \tag{2.7}$$

The kind of activation functions to be used is dependent on the applications. The former two types of activation functions are called *sigmoidal nonlinearity*. The hidden layer and output layer can take either form 1 or 2. In particular, for function approximation problem, the output layer often employs the linear activation function.

The optimization of the error function over the weights $w_{kj}$ and $v_{ji}$ is typically taking the *steepest descent algorithm*, that is, the successive adjustments applied to the weights are in the direction of steepest descent (a direction opposite to the gradient):

$$\begin{aligned}\Delta w_{kj} &= -\eta \frac{\partial E}{\partial w_{kj}} \\ &= \eta e_k \cdot \varphi'_k(\alpha_k) \cdot y_j \end{aligned} \tag{2.8a}$$

$$\begin{aligned}\Delta v_{ji} &= -\eta \frac{\partial E}{\partial v_{ji}} \\ &= \eta x_i \cdot \varphi'_j(\alpha_j) \cdot \sum_k e_k \varphi'_k(\alpha_k) w_{kj} \end{aligned} \tag{2.8b}$$

where $\eta$ is a positive constant and called the *learning rate*. The steepest descent method has a zig-zag problem when approaching the minimum, and in order to remedy this drawback, a momentum term is often added into the above equations:

$$\Delta w_{kj}(t) = -\eta \frac{\partial E(t)}{\partial w_{kj}} + \beta \Delta w_{kj}(t-1) \tag{2.9a}$$

$$\Delta v_{ji}(t) = -\eta \frac{\partial E(t)}{\partial v_{ji}} + \beta \Delta v_{ji}(t-1) \tag{2.9b}$$

where $t$ denotes the iteration number, and $\beta$ is a positive constant between 0 and 1 called the *momentum constant*.

From different starting points, there are different strategies used to improve the training performance, for example, optimal weight initialization (Yam and Chow, 2001).

## 2.3 Optimal Weight Initialization Method

Consider a single hidden-layer feedforward network, and let $\mathbf{x} = (x_1, x_2, \cdots, x_L)$ be the input vector, $\mathbf{h} = (h_1, h_2, \cdots, h_M)$ and $\mathbf{o} = (o_1, o_2, \cdots, o_N)$ be the output vectors of the hidden layer and output layer respectively. The weight matrix between the input and hidden layers is an $L \times M$ matrix denoted by $W = (w_{ij})$. Similarly, the weight matrix between the hidden and output layers is an $M \times N$ matrix denoted by $V = (v_{ij})$. Training patterns are indicated by the superscript as $\mathbf{x}^s, s = 1, \cdots, p$

Suppose the weights $w_{ji}$ and $v_{ji}$ are independent, identically distributed uniform random variables satisfying

$$w_{ji} \in (-w_{\max}, w_{\max}), \quad v_{ji} \in (-v_{\max}, v_{\max}).$$

The weighted sum $a_j$ to the $j$th hidden neuron is calculated by

$$a_j = w_{j0} + \sum_{i=1}^{L} w_{ji} x_i^s. \tag{2.10}$$

where $w_{j0}$ is the threshold value of the $j$th hidden node.

The typical sigmoid function with the output value between 0 and 1 is employed by the hidden layer:

$$f(\theta) = \frac{1}{1 + \exp(-\theta)}. \tag{2.11}$$

The derivative is equal to $f'(\theta) = f(\theta)(1 - f(\theta))$.

Thus, the output of the $j$th hidden node is

$$h_j = f(a_j) = \frac{1}{1 + \exp(-a_j)}. \tag{2.12}$$

The *active region* is assumed to be the region where the derivative of the sigmoid function is larger than one-twentieth of the maximum derivative, i.e., the region that

$$|a_j| \leq 4.36. \tag{2.13}$$

If $|a_j| > 4.36$, the derivative becomes very small and the region is called the *saturation region*. Within this region, the change of weight is small even when there is a substantial error such that the training process slows down.

Define a hyperplane in the input space:

$$P(a_j) = w_{j0} + \sum_{i=1}^{L} w_{ji} x_i^s - a_j. \tag{2.14}$$

For a fixed $a_j$, the orientation and position of $P(a_j)$ are determined by the values of the weights and threshold. For a particular set of $w_{ji}$, the hyperplanes

## 2.3 Optimal Weight Initialization Method

$P(-4.36)$ and $P(4.36)$ are the boundaries between the active region and the saturation region, and $P(0)$ is the hyperplane where the derivative of the activation function has a maximum value.

The distance between the two hyperplanes $P(-4.36)$ and $P(4.36)$ is given by

$$d^{in} = \frac{8.72}{\sqrt{\sum_{i=1}^{L} w_{ji}^2}},$$

and the maximum possible distance $D^{in}$ between two points of the input space is given by

$$D^{in} = \sqrt{\sum_{i=1}^{L} (\max(x_i) - \min(x_i))^2}. \qquad (2.15)$$

In order to ensure the outputs of the hidden neurons are within the active region, it is required that $d^{in} \geq D^{in}$. In the following, the equality is chosen. Then we have

$$\sqrt{\sum_{i=1}^{L} w_{ji}^2} = \frac{8.72}{D^{in}}.$$

The left hand, the length of the weight vector can be approximated by use of the second moment of the weights $E(W^2) = \frac{w_{max}^2}{3}$, such that $\sqrt{\sum_{i=1}^{L} w_{ji}^2} = \sqrt{L \times E(W^2)}$. Therefore, the maximal magnitude of weights is given by

$$w_{max} = \frac{8.72}{D^{in}} \sqrt{\frac{3}{L}}. \qquad (2.16)$$

The threshold is adjusted to move the hyperplane $P(0)$ to cross the center of the input space $C^{in} = (\frac{\max(x_1)+\min(x_1)}{2}, \dots, \frac{\max(x_L)+\min(x_L)}{2})$, and thus the threshold is given as

$$w_{j0} = -\sum_{i=1}^{L} c_i^{in} w_{ji}. \qquad (2.17)$$

The weights $V = (v_{ji})$ is evaluated in a similar manner. The distance between two hyperplanes enclosing the active region is given by

$$d^{hid} = \frac{8.72}{\sqrt{\sum_{i=1}^{M} v_{ji}^2}}.$$

Since the outputs of the hidden layer are limited between zero and one, the maximal possible distance between two points is

$$D^{hid} = \sqrt{M}.$$

By setting $d^{hid} = D^{hid}$, the maximal magnitude of $v_{ji}$ is calculated by

$$v_{\max} = \frac{15.10}{M}. \tag{2.18}$$

Since the center of the hyperspace of the hidden outputs is $(0.5, \cdots, 0.5)^\top$, the threshold $v_{j0}$ is given as

$$v_{j0} = -0.5 \sum_{i=1}^{M} v_{ji}. \tag{2.19}$$

## 2.4 The Optimization-Layer-by-Layer Algorithm

Consider a network with one hidden layer and a scalar output node. The activation functions for the hidden layer are sigmoidal function, and the linear functions for the output layer (see Fig. 2.1).

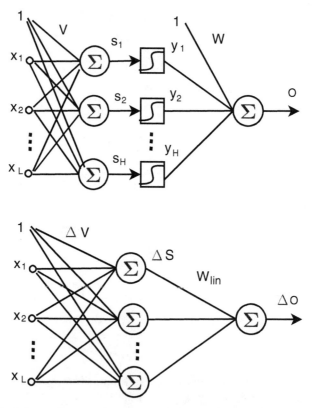

**Fig. 2.1.** Feed-forward network with one hidden layer and a scalar output (top panel) and its linearized network structure (bottom panel).

## 2.4 The Optimization-Layer-by-Layer Algorithm

The network defines a continuous nonlinear mapping $G : \mathbf{R}^L \to \mathbf{R}$ from an input vector $\mathbf{x}^p \in \mathbf{R}^L$ to a scalar output $o^p \in \mathbf{R}$:

$$o^p = \sum_{i=0}^{M} w_m y_m^p = w_0 + \sum_{h=1}^{H} w_m f(s_h^p)$$

$$= w_0 + \sum_{h=1}^{H} w_h f\left(\sum_{l=0}^{L} v_{hl} x_l^p\right) \qquad (2.20)$$

The cost function to be minimized is the mean squared error:

$$E(V, \mathbf{w}) = \frac{1}{P}\sum_{p=1}^{P} E^p = \frac{1}{P}\sum_{p=1}^{P} \frac{1}{2}(d^p - o^p)^2. \qquad (2.21)$$

The basic ideas of the optimization-layer-by-layer (OLL) algorithm are as follows (Ergezinger and Thomsen, 1995):

(1) The weights in each layer are modified dependent on each other but separately from all other layers.
(2) In this case the optimization of the output layer is a linear problem.
(3) The optimization of the hidden-layer weights is reduced to a linear problem by linearization of the sigmoidal activation function.

### 2.4.1 Optimization of the Output Layer

Recall that the affine output is simply the weighted sum:

$$o^p = \sum_{h=0}^{H} w_h y_h^p = \mathbf{w}^\top \mathbf{y}^p. \qquad (2.22)$$

The optimization of the output layer is formulated as

$$E(\mathbf{w}^*) = \min_{\mathbf{w}} E(\mathbf{w}) = \min_{\mathbf{w}} \frac{1}{P}\sum_{p=1}^{P} \frac{1}{2}(d^p - \mathbf{w}^\top \mathbf{y}^p)^2.$$

Let the gradient of $E(\mathbf{w})$ with respect to $\mathbf{w}$ equal to zero:

$$\nabla_{\mathbf{w}} E(\mathbf{w}) = \frac{1}{P}\sum_{p=1}^{P}(\mathbf{w}^\top \mathbf{y}^p - d^p)\mathbf{y}^p = 0.$$

Thus by solving the above linear equations, the optimal weight vector $\mathbf{w}^*$ is obtained

$$\mathbf{w}^* = A^{-1} b \qquad (2.23)$$

where the matrix $A$ and the vector $b$ are given by:

$$A = (a_{ij}); \quad a_{ij} = \sum_{p=1}^{P} y_i^p y_j^p; \quad i,j = 0, \cdots, H$$

$$b = (b_i); \quad b_i = \sum_{p=1}^{P} d^p y_i^p; \quad i = 0, \cdots, H$$

Note that $A$ is symmetric and positive definite.

### 2.4.2 Optimization of the Hidden Layer

The optimization of the hidden layer is first done through a linearization procedure. Assuming there is a change $\Delta v_{hl}$, which results in a change $\Delta s_h$ in the net input to the hidden node as:

$$s_{new,h}^p = \sum_{l=0}^{L} v_{new,hl} x_l^p$$

$$= \sum_{l=0}^{L} v_{old,hl} x_l^p + \sum_{l=0}^{L} \Delta v_{hl} x_l^p$$

$$= s_{old,h}^p + \Delta s_h^p$$

Next, we need to compute effects of $\Delta s_h$ on the network output $o$ (see Fig. 2.1, bottom panel). The Taylor series expansion is used to do the linearization of the sigmoidal activation function, consequently

$$y_{new,h}^p = \begin{cases} f(s_{old,h}) + \frac{\partial f(s_{old,h}^p)}{\partial s_{old,h}^p} \Delta s_h^p & \text{for } h = 1, \cdots, H \\ y_{old,h}^p = 1 & \text{for } h = 0. \end{cases}$$

Thus, the change in the output can be approximated as

$$o_{new}^p = \sum_{h=0}^{H} w_h y_{new,h}^p$$

$$\approx \sum_{h=0}^{H} w_h y_{old,h}^p + \sum_{h=1}^{H} f'(s_{old,h}^p) w_h \Delta s_h^p$$

$$= o_{old}^p + \Delta o^p.$$

The purpose of optimizing the weight matrix $V$ of the hidden layer is to further reduce the mean squared error

$$E = \frac{1}{P} \sum_{p=1}^{P} \frac{1}{2} (d^p - o_{old}^p)^2 = \frac{1}{P} \sum_{p=1}^{P} \frac{1}{2} (e_{old}^p)^2.$$

## 2.4 The Optimization-Layer-by-Layer Algorithm

measured for the original network (see Fig. 2.1, top panel) with optimized output-layer weight $\mathbf{w}^*$.

For the linearized network, to find the optimal $\Delta V$ a new cost function is defined as

$$E_{lin} = \frac{1}{P}\sum_{p=1}^{P} \frac{1}{2}(e_{old}^p - \Delta o^p)^2. \qquad (2.24)$$

In the linearized network the activation function of the hidden layer were approximated by a first order Taylor series expansion, while the approximation error can be estimated by the quadratic term

$$\epsilon_h^p = \frac{1}{2} f''(s_{old}^h)(\Delta s_h^p)^2.$$

Next, a penalty term estimating the quality of the linear approximation by averaging the absolute values $|\epsilon_h^p v_h|$ over all hidden nodes and training patterns is introduced:

$$E_{pen} = \frac{1}{PH} \sum_{p=1}^{P} \sum_{h=1}^{H} |\epsilon_h^p v_h|. \qquad (2.25)$$

Therefore, the modified cost function for the optimization of the hidden layer is given by

$$E_{hid} = E_{lin} + \mu E_{pen}. \qquad (2.26)$$

Now we compute the partial derivatives of $E_{hid}$ over the weight changes $\Delta v_{hl}$ separately. First, we have

$$\frac{\partial E_{lin}}{\partial \Delta v_{hl}} = \frac{1}{P}\sum_{p=1}^{P} \frac{\partial E_{lin}^p}{\partial \Delta w_{hl}}$$

$$= \frac{1}{P}\sum_{p=1}^{P} (\Delta o^p - e_{old}^p) \cdot \frac{\partial \Delta o^k}{\partial \Delta w_{hl}}$$

$$= \frac{1}{P}\sum_{p=1}^{P}\left[\left(\sum_{j=1}^{H} w_{lin,j}^p \sum_{i=0}^{L} x_i^p \Delta v_{ji}\right) - e_{old}^p\right]$$

$$= \sum_{i=0}^{L}\sum_{j=1}^{H} \Delta v_{ji} \cdot \frac{1}{P}\sum_{p=1}^{P} w_{lin,h}^p w_{lin,j}^p x_l^p x_i^p - \frac{1}{P}\sum_{p=1}^{P} e_{old}^p x_l^p w_{lin,h}^p.$$

The partial derivatives of the penalty term are computed:

$$\frac{\partial E_{pen}}{\partial \Delta v_{hl}} = \frac{1}{PH}\sum_{p=1}^{P}\sum_{j=1}^{H} |w_j f''(s_{old,j}^p))| \cdot \Delta s_j^p \cdot \frac{\partial \Delta s_j^p}{\partial \Delta v_{hl}}$$

$$= \frac{1}{PH}\sum_{i=0}^{L} \Delta v_{hi}|w_h| \cdot \sum_{p=1}^{P} |f''(s_{old,h}^p)| x_i^p x_l^p. \qquad (2.27)$$

To find the optimal weight changes $\Delta V^*$, all partial derivatives are set to zero, then

$$\frac{\partial E_{hid}}{\partial \Delta v_{hl}} = 0 = \sum_{i=0}^{L}\sum_{j=1}^{H} a_{hl,ji}\Delta v_{ji} + \frac{\mu}{H}\sum_{i=0}^{L} c_{hli}\Delta v_{hi} - \tilde{b}_{hl} \qquad (2.28)$$

where the following abbreviations are used:

$$a_{hl,ji} = \sum_{p=1}^{P} w_{lin,h}^p x_l^p \cdot w_{lin,j}^p x_i^p,$$

$$\tilde{b}_{hl} = \sum_{p=1}^{P} e_{old}^p \cdot w_{lin,h}^p \cdot x_l^p,$$

$$c_{hl,i} = |w_h| \cdot \sum_{p=1}^{P} |f''(s_{old,h}^p)| \cdot x_i^p x_l^p.$$

Let

$$\tilde{a}_{hl,ji} = \begin{cases} a_{hl,ji}, & \text{if } j \neq h \\ a_{hl,ji} + \frac{\mu}{H}c_{hl,i}, & \text{if } j = h \end{cases}$$

the above equation can be reduced to the matrix form

$$\tilde{A} \cdot \Delta V^* = \tilde{b}, \qquad (2.29)$$

where $\tilde{A} = (\tilde{a}_{hl,ji})$, $\Delta V^* = (\Delta v_{10}^*, \cdots, \Delta v_{1L}^*, \cdots, \Delta v_{H0}^*, \cdots, \Delta v_{HL}^*)^\top$. The matrix $\tilde{A}$ is symmetric and positive definite. Thus

$$\Delta V^* = \tilde{A}^{-1}\tilde{b}.$$

## 2.5 Modified Error Back-Propagation Method

In (Scalero and Tepedelenlioglu, 1992), a modified form of the back-propagation algorithm was developed. The basic idea is (see Fig. 2.2): if all inputs (**x** or **y**) to the nodes, and summations ($s_h$ or $s_k$) were specified, the problem would be reduced to a linear problem, i.e., a system of linear equations that relates the weights (**v** or **w**) to the summations and the inputs.

The problem is to minimize a mean-squared error function with respect to the net-input (weight summation):

$$E = \sum_{p=1}^{P}(\bar{d}^p - s^p)^2, \qquad (2.30)$$

where $\bar{d}^p$ and $s^p$ are the desired and actual net-input for the $p$th training pattern. Clearly, for the output layer, $\bar{d}^p = f^{-1}(o^p)$.

## 2.5 Modified Error Back-Propagation Method

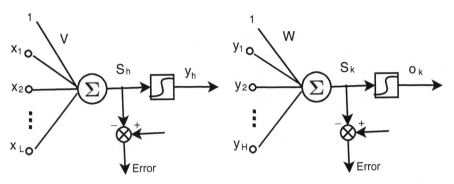

**Fig. 2.2.** Linear and error portion of the nodes in the hidden layer (left panel) and the output layer (right panel).

Minimizing $E$ with respect to a weight $w_n$ produces

$$\frac{\partial E}{\partial w_n} = 0 = -2\sum_{p=1}^{P}(\bar{d}^p - s^p)\frac{\partial s^p}{\partial w_n}$$

$$= -2\sum_{p=1}^{P}(\bar{d}^p - \sum_{i=0}^{N} w_i x_i^p)x_n^p.$$

Thus,

$$\sum_{p=1}^{P} \bar{d}^p x_n^p = \sum_{p=1}^{P}\sum_{i=0}^{N} w_i x_i^p x_n^p \quad (n=0,\cdots,N). \tag{2.31}$$

Define $R = \sum_{p=1}^{P} \mathbf{x}^p \mathbf{x}^{p\top}$ and $\rho = \sum_{p=1}^{P} \bar{d}^p \mathbf{x}^p$, the above equation can be expressed in the matrix form:

$$\rho = R\mathbf{w}. \tag{2.32}$$

Consider the iterative manner, then

$$R(t) = \sum_{k=1}^{t} \mathbf{x}(k)\mathbf{x}(k)^\top \tag{2.33a}$$

$$\rho(t) = \sum_{k=1}^{t} \bar{d}(k)\mathbf{x}(k), \tag{2.33b}$$

where $t$ is the number of the present iteration number. Thus equation (2.32) becomes

$$\mathbf{w}(t) = R^{-1}(t)\rho(t). \tag{2.34}$$

Equation (2.33b) can be further written in the recursive form as

$$R(t) = bR(t-1) + \mathbf{x}(t)\mathbf{x}(t)^\top \qquad (2.35a)$$
$$\rho(t) = b\rho(t-1) + \bar{d}(t)\mathbf{x}(t) \qquad (2.35b)$$

where $b$ is a forgetting factor.

The solution of (2.34) can be computed by using the Kalman filter method. Firstly compute the Kalman gain vector:

$$k_j(t) = \frac{R_j^{-1}(t-1)\mathbf{x}_{j-1}(t)}{b_j + \mathbf{x}_{j-1}^\top(t)R_j^{-1}(t-1)\mathbf{x}_{j-1}(t)} \qquad (2.36)$$

where the subscript $j$ indicates the layer index. For a one-hidden layer network, $j = 1$ indicates the hidden layer and $j = 2$ indicates the output layer. The recursive equation for the inverse matrix is computed by

$$R_j^{-1}(t) = \left[R_j^{-1}(t-1) - k_j(t)x_{j-1}^\top(t)R_j^{-1}(t-1)\right]b_j^{-1}. \qquad (2.37)$$

It needs to initialize $R^{-1}(0)$. Finally, the update equations for the weight vectors of the output layer and hidden layers are respectively,

$$\mathbf{w}_{jk}(t) = \mathbf{w}_{jk}(t-1) + k_j(t)(\bar{d}_k - s_{jk}), \text{ for } j = 2 \qquad (2.38a)$$
$$\mathbf{w}_{jk}(t) = \mathbf{w}_{jk}(t-1) + k_j(t)e_{jk}\mu_j, \text{ for } j = 1. \qquad (2.38b)$$

The algorithm is summarized in the following.

1. Initialize all weights randomly. Initialize the inverse matrix $R^{-1}$.
2. Randomly select an input/ouput pair to feed into the network.
3. For each layer $j = 1, 2$, calculate the summation

$$s_{jk} = \sum_{i=0}^{N}(x_{j-1,i}w_{jki})$$

and the node output

$$x_{jk} = f(s_{jk}) = \frac{1 - \exp(-as_{jk})}{1 + \exp(-as_{jk})}$$

for every node $k$. $N$ is the number of inputs to a node, and $a$ is the sigmoid slope constant.

4. Invoke the Kalman filter equations. For each layer $j = 1, 2$, calculate the Kalman gain

$$k_j = \frac{R_j^{-1}\mathbf{x}_{j-1}}{b_j + \mathbf{x}_{j-1}^\top R_j^{-1}\mathbf{x}_{j-1}}$$

and update the inverse matrix

$$R_j^{-1} = [R_j^{-1} - k_j\mathbf{x}_{j-1}^\top R_j^{-1}]b_j^{-1}.$$

## 2.5 Modified Error Back-Propagation Method

5. Back-propagate error signals. Compute the derivative of $f(s_{jk})$ as

$$f'(s_{jk}) = \frac{2a\exp(-as_{jk})}{[1+\exp(-as_{jk})]^2}.$$

Then calculate the error signals in the output layer, $j = 2$, by evaluating

$$e_{2k} = f'(s_{2k})(d_k - o_k).$$

For the hidden layer, $j = 1$, the error signals are evaluated by

$$e_{1k} = f'(s_{1k}) \sum_i e_{2,i} w_{2,i,k}.$$

6. Find the desired summation. Calculate the desired summation using the inverse function

$$\bar{d}_k = \frac{1}{a} \ln \frac{1+d_k}{1-d_k}.$$

7. Update the weights using equations (2.38).
8. Test if the stop criterion is reached. If not, go back to step 2.

# 3

# New Dynamical Optimal Learning for Linear Multilayer FNN

## 3.1 Introduction

[1]Generalization ability and learning stability are fundamental issues for multilayer feedforward neural networks (FNN) (Haykin, 1999). There are many research works that explore the learning of multilayer FNN with squashing activation function as well as the applications of FNN to function approximation and pattern recognition (Lippmann, 1987; Duda et al., 2001). However, there are very few studies on the learning issue of FNN with linear activation function in the hidden layer. On the other hand, network learning parameters, for example, the learning rate, are key factors in the training of FNN. The use of error back propagation algorithm is not always successful because of its sensitivity to learning parameters, initial state, and perturbations (Yeung and Sun, 2002; Bertels et al., 2001). When the learning rate falls within an unsuitable range, it will result in chaotic behaviors in the network learning, and for non-chaotic learning rates the network converges faster than for the chaotic ones (Bertels et al., 2001). Many research works are concerned with the improvement of training performance and the better selection of training parameters (Wang et al., 2001; Abid et al., 2001). Evolutionary algorithms have also been used to train neural networks (Yao and Liu, 1997; Yao, 1999). The optimal learning rate for a given two-layer neural network was derived in (Wang et al., 2001), but there is limited learning capability for the two-layer neural network. For conventional multilayer FNN with squashing sigmoid functions (Zurada, 1995), it is neither easy to deduce an optimal learning rate nor to practically determine the learning rate in a priori. In fact, the optimal learning rate always changes during the training process which makes it very difficult to be set optimally.

---

[1] Reuse of the materials of "New dynamical optimal learning for linear multilayer FNN", 15(6), 2004, 1562–1568, IEEE Trans. on Neural Networks, with permission from IEEE.

H. Tang et al.: *New Dynamical Optimal Learning for Linear Multilayer FNN*, Studies in Computational Intelligence (SCI) **53**, 23–34 (2007)
www.springerlink.com © Springer-Verlag Berlin Heidelberg 2007

In this chapter, a new dynamical optimal learning algorithm for three-layer linear neural networks is presented and it is proved that this learning is stable in the sense of Lyapunov. The proposed learning scheme reduces the sensitivity to perturbation of initial learning parameters or unsuccessful learning associated with standard back propagation (SBP) algorithm. The proposed learning algorithm is validated upon the problems of function mapping and pattern recognition, and the results are compared with the SBP algorithm.

## 3.2 Preliminaries

**Definition 3.1.** $\phi(x)$ *is a sigmoid function if it is a bounded, continuous and increasing function.*

It follows directly from the above definition, for $u_1, u_2 > 0$, the ramp function

$$\phi(x) = \begin{cases} -u_1, & \text{if } x \in (-\infty, -u_1] \\ x, & \text{if } x \in (-u_1, u_2) \\ u_2, & \text{if } x \in [u_2, \infty) \end{cases}$$

is a sigmoid function. If $u_1$ and $u_2$ are assumed to be large enough, i.e., $\phi(x)$ does not go beyond its linear region, the sigmoid type function can be treated as a linear function. This assumption is reasonable since the inputs to the neural network in our model are bounded. Here, networks with such ramp-type transfer functions are called as *linear neural networks*.

**Lemma 3.2.** *(Funahashi, 1989) Let $\phi(x)$ be a sigmoid function and $\Omega$ be a compact set in $R^n$. Given any continuous function $f : R^n \to R$ on $\Omega$, and for an arbitrary $\epsilon > 0$, there exists an integer $N$ and real constants $c_i, \theta_i, w_{ij}$, $i = 1, \cdots, N$, $j = 1, \cdots n$, such that*

$$\bar{f}(x_1, \cdots, x_n) = \sum_{i=1}^{N} c_i \phi(\sum_{j=1}^{n} w_{ij} x_j - \theta_i) \tag{3.1}$$

*satisfies* $\max_{x \in \Omega} \|f(x_1, \cdots, x_n) - \bar{f}(x_1, \cdots, x_n)\|_2 < \epsilon.$

Based on this Lemma, a three-layer FNN can be formulated where the hidden layer transfer functions employ $\phi(x)$ and the transfer functions for the input and output layers are linear.

Notations: All vectors are column vectors. A superscript such as in $d_p^k$ refers to a specific output vector's component. The network input, hidden-layer output, output-layer output and desired output is represented by matrix notations $X = [x_1 \ x_2 \cdots x_P] \in R^{L \times P}$, $Y = [y_1 \ y_2 \cdots y_P] \in R^{H \times P}$, $O = [o_1 \ o_2 \cdots o_P] \in R^{K \times P}$, $D = [d_1 \ d_2 \cdots d_P] \in R^{K \times P}$, respectively. $L$, $K$, $P$ denotes the number of input-layer neurons, output-layer neurons and patterns, respectively. $W$ and $V$ represents the hidden-output layer and input-hidden layer weights matrix, respectively. Subscript $t$ denotes the index of the iterative learning epochs.

The objective of the network training is to minimize an error function $J$,

$$J = \frac{1}{2PK} \sum_{p=1}^{P} \sum_{k=1}^{K} (o_p^k - d_p^k)^2 \qquad (3.2)$$

In the steepest (gradient) descent algorithm (Zurada, 1995), the updates of weights are computed by $W_{t+1} = W_t - \beta \frac{\partial J}{\partial W_t}$ and $V_{t+1} = V_t - \beta \frac{\partial J}{\partial V_t}$.

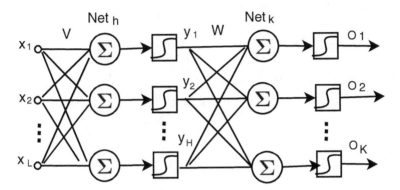

**Fig. 3.1.** A general FNN model

## 3.3 The Dynamical Optimal Learning

Consider a three-layer (one hidden layer) FNN as shown in Figure 3.1, the network error matrix is defined by the error between actual outputs and desired outputs,

$$E_t = O_t - D = W_t Y_t - D = W_t V_t X - D. \qquad (3.3)$$

It follows from (3.2) that

$$J_t = \frac{1}{2PK} Tr(E_t E_t^T) \qquad (3.4)$$

where $Tr$ is the matrix trace operator. So $W_{t+1}$ and $V_{t+1}$ are updated by,

$$W_{t+1} = W_t - \beta_t \frac{\partial J}{\partial W_t} = W_t - \beta_t \frac{1}{PK} E_t Y_t^T \qquad (3.5)$$

$$V_{t+1} = V_t - \beta_t \frac{\partial J}{\partial V_t} = V_t - \beta_t \frac{1}{PK} W_t^T E_t X^T \qquad (3.6)$$

From equations (3.3)-(3.6), it holds that

$$E_{t+1}E_{t+1}^T$$
$$= (W_{t+1}Y_{t+1} - D)(W_{t+1}Y_{t+1} - D)^T$$
$$= [(W_t - \frac{\beta_t}{PK}E_tY_t^T)(V_t - \frac{\beta_t}{PK}W_t^TE_tX^T)X - D]$$
$$[(W_t - \frac{\beta_t}{PK}E_tY_t^T)(V_t - \frac{\beta_t}{PK}W_t^TE_tX^T)X - D]^T$$
$$= [E_t - \frac{\beta_t}{PK}(W_tW_t^TE_tX^TX + E_tY_t^TV_tX) + \frac{\beta_t^2}{(PK)^2}E_tY_t^TW_t^TE_tX^TX]$$
$$[E_t - \frac{\beta_t}{PK}(W_tW_t^TE_tX^TX + E_tY_t^TV_tX) + \frac{\beta_t^2}{(PK)^2}E_tY_t^TW_t^TE_tX^TX]^T$$
$$= E_tE_t^T - \frac{\beta_t}{PK}[E_t(W_tW_t^TE_tX^TX)^T + E_t(E_tY_t^TV_tX)^T + W_tW_t^TE_tX^TXE_t^T$$
$$+ E_tY_t^TV_tXE_t^T] + \frac{\beta_t^2}{(PK)^2}[E_t(E_tY_t^TW_t^TE_tX^TX)^T + E_tY_t^TW_t^TE_tX^TXE_t^T$$
$$+ E_tY_t^TV_tX(W_tW_t^TE_tX^TX)^T + W_tW_t^TE_tX^TX(E_tY_t^TV_tX)^T$$
$$+ E_tY_t^TV_tX(E_tY_t^TV_tX)^T + W_tW_t^TE_tX^TX(W_tW_t^TE_tX^TX)^T]$$
$$- \frac{\beta_t^3}{(PK)^3}[W_tW_t^TE_tX^TX(E_tY_t^TW_t^TE_tX^TX)^T + E_tY_t^TV_tX(E_tY_t^TW_t^TE_tX^TX)^T$$
$$+ E_tY_t^TW_t^TE_tX^TX(W_tW_t^TE_tX^TX)^T + E_tY_t^TW_t^TE_tX^TX(E_tY_t^TV_tX)^T]$$
$$+ \frac{\beta_t^4}{(PK)^4}E_tY_t^TW_t^TE_tX^TX(E_tY_t^TW_t^TE_tX^TX)^T \tag{3.7}$$

Omitting subscript $t$ for simplicity, the following is derived

$$J_{t+1} - J_t = \frac{1}{2PK}(A\beta^4 + B\beta^3 + C\beta^2 + M\beta) \equiv \frac{1}{2PK}g(\beta), \tag{3.8}$$

where

$$A = \frac{1}{(PK)^4}Tr[EY^TW^TEX^TX(EY^TW^TEX^TX)^T]$$
$$B = -\frac{1}{(PK)^3}Tr[WW^TEX^TX(EY^TW^TEX^TX)^T + EY^TVX(EY^TW^TEX^TX)^T$$
$$+ EY^TW^TEX^TX(WW^TEX^TX)^T + EY^TW^TEX^TX(EY^TVX)^T]$$
$$C = \frac{1}{(PK)^2}Tr[E(EY^TW^TEX^TX)^T + WW^TEX^TX(EY^TVX)^T$$
$$+ EY^TVX(WW^TEX^TX)^T + EY^TVX(EY^TVX)^T$$
$$+ WW^TEX^TX(WW^TEX^TX)^T + EY^TW^TEX^TXE^T]$$
$$M = -\frac{1}{PK}Tr[E(WW^TEX^TX)^T + E(EY^TVX)^T + WW^TEX^TXE^T + EY^TVXE^T]$$

## 3.3 The Dynamical Optimal Learning

*Remark 3.3.* It can be observed that $A > 0$ is always assured. According to the algebra theorem (Scherk, 2000), $g(\beta)$ has two or four real roots including one constant zero root, and there exist one or two intervals in $\beta$ coordinate satisfying $g(\beta) < 0$. It needs to find the minimum value of $g(\beta)$ resulting in the largest reduction of $J$ at each epoch by solving the derivative of $g(\beta)$.

Differentiating $g(\beta)$ with respect to $\beta$ yields

$$\frac{\partial g}{\partial \beta} = 4A\beta^3 + 3B\beta^2 + 2C\beta + M$$
$$= 4A(\beta^3 + a\beta^2 + b\beta + c) \equiv 4A \cdot h(\beta), \quad (3.9)$$

where $a = \frac{3B}{4A}, b = \frac{2C}{4A}, c = \frac{M}{4A}$. Therefore the problem of finding the optimal learning rates is reduced to solving a real cubic equation. Below the results obtained from the general algebra theorem (Chapters 15 and 16 in (Scherk, 2000)) are introduced.

**Lemma 3.4.** *(Scherk, 2000) For a general real cubic equation, $h(x) = x^3 + ax^2 + bx + c$, let $D = discrim(h,x) = -27c^2 + 18cab + a^2b^2 - 4a^3c - 4b^3$, where $discrim(h,x)$ denotes the discriminant of $h(x)$. Then*

1. *if $D < 0$, $h(x)$ has one real root;*
2. *if $D \geq 0$, $h(x)$ has three real roots;*
    a) *$D > 0$, $h(x)$ has three different real roots;*
    b) *$D = 0, 6b - 2a^2 \neq 0$, $h(x)$ has one single root and one multiple root;*
    c) *$D = 0, 6b - 2a^2 = 0$, $h(x)$ has one root of three multiplicities.*

It needs to compute the roots of $h(\beta)$ (two of which are conjugated),

$$\beta_1 = -\frac{1}{3}a + \frac{1}{6}(36ab - 108c - 8a^3 + 12\omega_1)^{\frac{1}{3}} - \frac{6\omega_2}{(36ab - 108c - 8a^3 + 12\omega_1)^{\frac{1}{3}}},$$

$$\beta_{2,3} = -\frac{1}{3}a - \frac{1}{12}(36ab - 108c - 8a^3 + 12\omega_1)^{\frac{1}{3}} + \frac{3\omega_2}{(36ab - 108c - 8a^3 + 12\omega_1)^{\frac{1}{3}}}$$
$$\pm \frac{1}{2}I\sqrt{3}\{\frac{1}{6}(36ab - 108c - 8a^3 + 12\omega_1)^{\frac{1}{3}} + \frac{6\omega_2}{(36ab - 108c - 8a^3 + 12\omega_1)^{\frac{1}{3}}}\},$$

where

$$\omega_1 = \sqrt{81c^2 - 54cab - 3a^2b^2 + 12a^3c + 12b^3},$$
$$\omega_2 = \frac{1}{3}b - \frac{1}{9}a^2.$$

**Theorem 3.5.** *For a given polynomial $g(\beta)$, if*

$$\beta_{opt} = \{\beta_i | g(\beta_i) = min\,(g(\beta_1), g(\beta_2), g(\beta_3)), i \in \{1,2,3\}\} \quad (3.10)$$

*where $\beta_i$ are the real roots of $\partial g / \partial \beta$, then $\beta_{opt}$ is the optimal learning rate and this learning process is stable in the sense of Lyapunov.*

*Proof.* Firstly it needs to find the stable learning range of $\beta_t$. To do so, the Lyapunov function is constructed as $V_t = J_t^2$ and the difference of $V_t$ is $\Delta V_t = J_{t+1}^2 - J_t^2$. It is known that if $\Delta V_t < 0$, the dynamical system is guaranteed to be stable (Vidyasagar, 1992). For $\Delta V_t < 0$, it can be derived that $J_{t+1} - J_t < 0$. Here the batch mode training is considered where the input matrix $X$ remains the same during the whole training process. According to (3.8), it only needs to know the range of $\beta$ satisfying $A\beta^4 + B\beta^3 + C\beta^2 + M\beta < 0$. Since $\beta_1$, $\beta_2$, $\beta_3$ are the zero roots of the derivative of $g(\beta)$, one of these roots (i.e., $\beta_{opt}$) must result in the minimum of $g(\beta)$ according to the algebra theorem. Obviously, the minimum value of $g(\beta)$ is less than zero which results in the largest reduction of $J_t$ at each step of the learning process.

*Remark 3.6.* If the discriminant $D$ of $h(\beta)$ is less than zero, then the optimal learning rate is equal to the only one real root of $h(\beta)$. If $D > 0$, the optimal learning rate is given by equation (3.10).

The dynamical optimal learning (DOL) algorithm is described below,

**Algorithm 3.1 Dynamical Optimal Learning Algorithm**

1. *Set initial weights $W_t$ and $V_t$, which are random values usually in the range of $[-1, 1]$;*
2. *Compute error function $E_t$. If $E_t < \epsilon$ (a desired accuracy) then the training stops, else go to the next step;*
3. *Compute $A$, $B$, $C$, $M$, and $\beta_1$, $\beta_2$, $\beta_3$, $D$;*
4. *If $D < 0$, then $\beta_{opt}$ is the real root; if $D > 0$, then compute $\beta_{opt}$ by equation (3.10);*
5. *Update the weights $W_t$ and $V_t$ by equations (3.5) and (3.6), respectively; then go to step 2.*

## 3.4 Simulation Results

In this section, the proposed DOL algorithm is validated upon both benchmark and classical problems. Furthermore, its performance is compared with the SBP training algorithm without adding momentum term, considering that both the SBP and DOL algorithms are originally derived from the steepest descent method.

### 3.4.1 Function Mapping

Due to the multilayer FNN's good generalization and nonlinear mapping abilities, it has been widely used as a general function approximator or time series predictor (Yam and Chow, 2001), (Ergezinger and Thomsen, 1995). In this section, a one-hidden layer FNN with linear activation functions is designed and the proposed dynamical optimal learning algorithm is employed to train the network so as to fulfill a class of high dimensional nonlinear function approximation. In this problem, eight dimensional input vectors $x^k$ are mapped into three dimensional output vectors $d^k$ by the following function: $f : R^8 \to R^3$; $d^k = f(x^k)$, $x^k \in R^8$; $d^k \in R^3$ with

$$d_1 = \frac{x_1x_2 + x_3x_4 + x_5x_6 + x_7x_8}{4}$$

$$d_2 = \frac{x_1 + x_2 + x_3 + x_4 + x_5 + x_6 + x_7 + x_8}{8}$$

$$d_3 = \frac{1}{2}\sqrt{4 - x_1x_2 + x_3x_4 + x_5x_6 + x_7x_8}$$

These training sets contain 50 and 100 input-output vectors $[x^k, d^k]$, respectively. The components of the input vectors are random numbers uniformly distributed over the interval of $[0, 1]$. A bias $-1$ was added to the input layer as a dummy neuron. The hidden layer consists of 14 neurons and results in a $9 - 14 - 3$ neural network.

For the network with SBP, its transfer functions of hidden layer and output layer are based on bipolar sigmoid function. Since it is difficult to determine the optimal value of SBP learning rate, many experiments over a wide range of different initial conditions were performed. It was observed that the network obtains the fastest convergence on average by setting the learning rate $lr = 0.5$, beyond which its learning becomes unstable, while $lr < 0.1$ leads to very slow convergence. The results also show that when the range of initial weights was increased to $[-50, 50]$, the SBP learning could easily become unstable. In contrast, the proposed DOL algorithm was found to be robust to different initial states. For different initial weights randomly located in the range of $[-1, 1]$, the network with SBP and DOL was run 10 times respectively, and the improvement ratios of DOL over SBP given in Table 3.1.

**Table 3.1.** Improvement ratio of DOL over SBP for 100 inputs with $mse = 0.001$

| Index | 1 | 2 | 3 | 4 | 5 | 6 | 7 | 8 | 9 |
|---|---|---|---|---|---|---|---|---|---|
| SBP($lr = 0.5$), Iterations | 1780 | 2629 | 4180 | 2951 | 2692 | 2226 | 3838 | 2396 | 3261 |
| cpu time(sec) | 19.69 | 28.58 | 43.36 | 31.14 | 28.5 | 23.83 | 39.09 | 24.64 | 33.42 |
| DOL, Iterations | 78 | 72 | 108 | 111 | 141 | 110 | 121 | 113 | 116 |
| cpu time(sec) | 0.50 | 0.47 | 0.58 | 0.59 | 0.69 | 0.61 | 0.63 | 0.61 | 0.63 |

Average improvement ratio: iteration ratio = 25.9, time ratio = 49.3.

It was observed that the learning epochs of SBP increase significantly as the number of input patterns was increased from 50 to 100. However, the learning epochs of the proposed DOL algorithm were not increased apparently, which show that the DOL algorithm is less sensitive to different inputs. The typical learning procedures of DOL and SBP are depicted in Fig. 3.2 and Fig. 3.3, respectively.

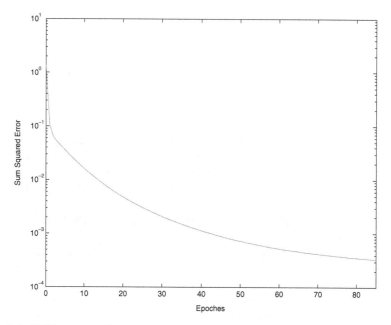

**Fig. 3.2.** DOL error performance for function mapping with 100 inputs: 85 epochs

### 3.4.2 Pattern Recognition

In this section, the proposed dynamical optimal learning algorithm is applied to recognize different characters presented to the inputs of a neural network (Scalero and Tepedelenlioglu, 1992). The characters' pixels are set to a level of 0.1 or 0.9, which represent the analog values rather than digital binary values. The output is likewise treated as analog. The $7 \times 7$ pixel patterns in Figure 3.4 are the inputs to a three-layer feedforward network with 10 nodes in the hidden layer. The desired output vector is in accordance with a 2, 3, or 4 binary string, which depends on the number of patterns presented to train the network. In this experiment, the network's output layer is composed of 4 nodes. For example, the 4–bit binary string of desired output vector 1000 is used to identify the 9th character I.

Figs. 3.5 and 3.6 show the mean-squared error versus the iteration number for both algorithms during the training of the $7 \times 7$ patterns. The learning rate was set as $lr = 0.2$, which leads to the best learning in the sense that it holds a smooth learning curve with the fastest convergence speed. The transfer functions of the hidden and output layers were set as bipolar sigmoid

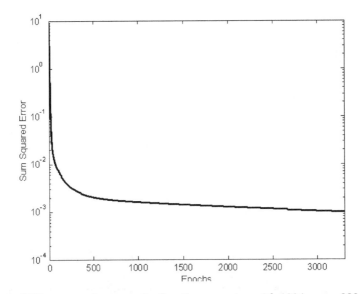

**Fig. 3.3.** SBP error performance for function mapping with 100 inputs: 3303 epochs

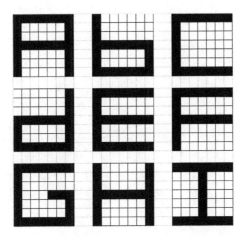

**Fig. 3.4.** The character patterns for the pattern recognition problem

function, which demonstrates a better learning performance than a unipolar sigmoid function and linear function for this problem. As can be seen from Figs. 3.5 and 3.6, the SBP algorithm tends to reach a certain mean-squared error and remains there for quite a while with little or no improvement. In contrast, the proposed DOL algorithm continues to make steady progress towards improving the $mse$ performance throughout the training process. The experiment results with different initial weights chosen randomly in the range of $[-1, 1]$ are shown in Table 3.2.

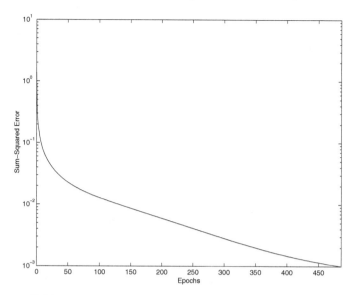

**Fig. 3.5.** DOL error performance for the characters recognition: 523 epochs

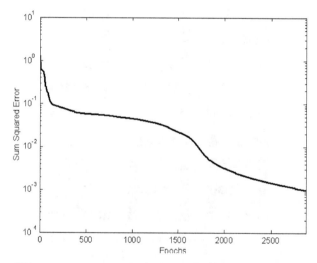

**Fig. 3.6.** SBP error performance for the characters recognition: 2875 epochs

*Remark 3.7.* From the simulation results, it is clear that the DOL algorithm has a stable and fast convergence, which is less sensitive to different initial weights and inputs as compared to the SBP algorithm. As shown in Table 3.1 and Table 3.2, although there exist complex expressions to compute the dynamical optimal learning rate in DOL, the improvement ratios of DOL over SBP with respect to both iteration numbers and *cpu* time are significant.

**Table 3.2.** Improvement ratio of DOL over SBP for 9 patterns with $mse = 0.001$

| Index | 1 | 2 | 3 | 4 | 5 | 6 | 7 | 8 | 9 |
|---|---|---|---|---|---|---|---|---|---|
| SBP($lr = 0.2$), | | | | | | | | | |
| Iterations | 1056 | 1254 | 1559 | 1784 | 1284 | 1628 | 1723 | 1068 | 2212 |
| $cpu$ time(sec) | 8.26 | 7.89 | 10.00 | 11.33 | 8.63 | 10.52 | 11.09 | 7.50 | 14.08 |
| DOL, Iterations | 599 | 523 | 371 | 599 | 599 | 563 | 599 | 599 | 599 |
| $cpu$ time(sec) | 1.13 | 1.18 | 0.78 | 1.13 | 1.13 | 1.05 | 1.11 | 1.11 | 1.13 |

Average improvement ratio: iteration ratio = 2.64, time ratio = 9.02.

## 3.5 Discussions

In the Preliminary Section, it was assumed that $u_1$ and $u_2$ are large enough in contrast to the input to the neuron such that the activation function can be represented by a linear function. Therefore the saturated regions are neglected in our derivations. This is reasonable since in most cases the inputs to the network fall within a limited bound. In fact, when $u_1$ and $u_2$ are set to be large enough, the learning performance using the ramp function will be the same as that of the linear function ($\phi(x) = x$).

Fig. 3.7 illustrates the dynamical optimal learning rate along the iterations for the function mapping problem. It can be seen that the optimal learning rate ($\beta_{opt}$) keeps varying between the range of $[0.5, 1.2]$, which is in correspondence with the fact that $lr = 0.5$ is the maximum and stable learning rate of SBP for this function mapping problem. A similar DOL learning pattern has also been observed for the pattern recognition problem.

FNN with ramp activation function in the hidden layer may have limited capacity in some applications when compared to FNN with differentiable sigmoid function. More discussions were presented in (Shah and Poon, 1999). On the other hand, FNN together with the back propagation learning is not so biologically plausible, though it has universal approximation capability. In (Poon and Shah, 1998) a tree-like perceptron with Hebbian learning was proposed and it was shown to be more computational advantageous and biologically plausible. Nevertheless, detailed analysis is beyond this scope.

## 3.6 Conclusion

In this chapter, a new dynamical optimal learning (DOL) algorithm for three-layer linear FNN is presented and its generalization ability is investigated. The proposed training algorithm has been proved to be stable and less sensitive to the setting of initial parameters or learning rates. The validation results upon the function mapping and pattern recognition problems show that the DOL

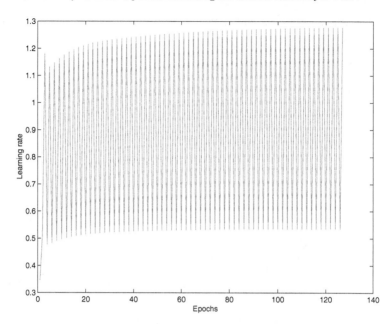

**Fig. 3.7.** The dynamical optimal learning rate for the function mapping problem

algorithm provides better generalization performance and faster convergence as compared to the SBP algorithm.

# 4

# Fundamentals of Dynamic Systems

## 4.1 Linear Systems and State Space

Consider the linear systems described by a set of ordinary differential equations:

$$\dot{\mathbf{x}} = A\mathbf{x} \tag{4.1}$$

where $\mathbf{x} \in \mathbf{R}^n$, $A$ is an $n \times n$ matrix and

$$\dot{\mathbf{x}} = \frac{d\mathbf{x}}{dt} = \begin{bmatrix} \frac{dx_1}{dt} \\ \vdots \\ \frac{dx_n}{dt} \end{bmatrix}.$$

The phase portrait of a system of differential equations such as (4.1) is the set of all solution curves in the state (or phase) space $\mathbf{R}^n$. Geometrically, the dynamical system describes the motion of the points in state space along the solution curves defined by the system of differential equations.

### 4.1.1 Linear Systems in $\mathbf{R}^2$

We consider the linear system in $\mathbf{R}^2$, i.e., $A$ is a $2 \times 2$ matrix in (4.1). There is an invertible $2 \times 2$ matrix $P$ whose columns consist of generalized eigenvectors of $A$, such that $A$ can be transformed to its similarity matrix:

$$B = P^{-1}AP.$$

Then under the linear transformation of coordinates $x = Py$, the phase portrait for the above linear system is obtained from the phase portrait for the linear system (4.2):

$$\dot{\mathbf{x}} = B\mathbf{x}. \tag{4.2}$$

In the following, we begin by describing the phase portraits for the linear system (4.2).

The matrix $B$, called the *Jordan canonical form* of $A$, can have one of the following forms:

$$B = \begin{bmatrix} \lambda & 0 \\ 0 & \mu \end{bmatrix}, \quad B = \begin{bmatrix} \lambda & 1 \\ 0 & \lambda \end{bmatrix} \quad \text{or} \quad B = \begin{bmatrix} a & -b \\ b & a \end{bmatrix}.$$

Case I. $B = \begin{bmatrix} \lambda & 0 \\ 0 & \mu \end{bmatrix}$ with $\lambda \leq \mu < 0$ or $B = \begin{bmatrix} \lambda & 1 \\ 0 & \lambda \end{bmatrix}$.

The phase portraits for the linear system (4.2) are given in Fig. 4.1, where the origin is called a *stable node* in each of these cases. If $\lambda \geq \mu > 0$ or $\lambda > 0$ in this case, the arrows in Fig. 4.1 are reversed and the origin is referred to be an *unstable node*. The stability of the node is determined by the sign of the eigenvalues: stable if $\lambda \leq \mu < 0$ and unstable if $\lambda \geq \mu > 0$. Each trajectory in Fig. 4.1 approaches the equilibrium point along a well-defined tangent line, determined by an eigenvector of $B$.

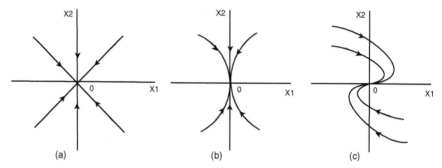

**Fig. 4.1.** The phase portraits of stable nodes at the origin. (a) $\lambda = \mu$, (b) $\lambda < \mu$, (c) $\lambda < 0$.

Case II. $B = \begin{bmatrix} \lambda & 0 \\ 0 & \mu \end{bmatrix}$ with $\lambda < 0 < \mu$.

In this case, the equilibrium point of the linear system (4.2) is referred to as a *saddle* (see Fig. 4.2). If $\mu < 0 < \lambda$, the arrows in Fig. 4.2 are reversed. The stable and unstable subspaces of (4.1) are determined by the eigenvectors of $A$.

Case III. $B = \begin{bmatrix} a & -b \\ b & a \end{bmatrix}$ with $a < 0$.

Obviously $B$ has a pair of complex eigenvalues: $\lambda = a \pm ib$. The origin is referred to as a *stable focus* (see Fig. 4.3 for the phase portrait). If $a > 0$, the trajectories spiral away from the origin as $t$ increases and the origin is called an *unstable focus*. Note that the trajectories in Fig. 4.3 do not

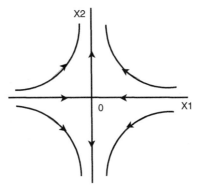

**Fig. 4.2.** A saddle at the origin.

approach the origin along any well-defined tangent lines, in contrast to Case I.

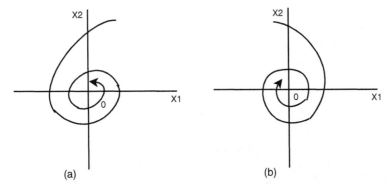

**Fig. 4.3.** The phase portraits of a stable focus at the origin. (a) $b > 0$, (b) $b < 0$.

Case IV. $B = \begin{bmatrix} 0 & -b \\ b & 0 \end{bmatrix}$.

The phase portrait for this case is shown in Fig. 4.4 and the system (4.2) is said to have a *center* at the origin.

For $\det A \neq 0$, the origin is the only equilibrium point of the linear system, i.e., $A\mathbf{x} = 0$ iff $\mathbf{x} = 0$. The following theorem tells the method to determine if the origin is a saddle, node, focus or center.

**Theorem 4.1.** *Let $\delta = \det A$, $\tau = \mathrm{trace} A$. The linear system*

$$\dot{\mathbf{x}} = A\mathbf{x}$$

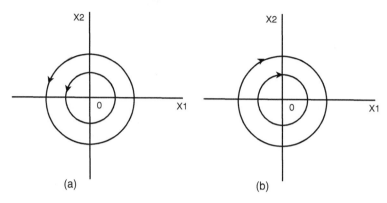

**Fig. 4.4.** The phase portraits of a center at the origin. (a) $b > 0$, (b) $b < 0$.

(i) has a saddle at the origin if $\delta < 0$.
(ii) has a node at the origin if $\delta > 0$ and $\tau^2 - 4\delta \geq 0$; it is stable if $\tau < 0$ and unstable if $\tau > 0$.
(iii) has a focus at the origin if $\delta > 0$, $\tau^2 - 4\delta < 0$ and $\tau \neq 0$; it is stable if $\tau < 0$ and unstalbe if $\tau > 0$.
(iv) has a center at the origin if $\delta > 0$ and $\tau = 0$.

This proof can be done easily by checking the eigenvalues of the matrix $A$:

$$\lambda = \frac{\tau \pm \sqrt{\tau^2 - 4\delta}}{2}.$$

If $\delta < 0$, there are two real eigenvalues of opposite sign. If $\delta > 0$ and $\tau^2 - 4\delta \geq 0$ there are two real eigenvalues of the same sign as $\tau$. If $\delta > 0$, $\tau^2 - 4\delta < 0$ and $\tau \neq 0$, there are two complex conjugate eigenvalues $\lambda = a \pm ib$. If $\delta > 0$ and $\tau = 0$ there are two purely imaginary complex conjugate eigenvalues.

### 4.1.2 Linear Systems in $\mathbf{R}^n$

Now we come to the initial value problem: for an initial point $x(0) = x_0$ to find the solution of system (4.1).

**Theorem 4.2. (The Fundamental Theorem for Linear Systems).** *Let $A$ be an $n \times n$ matrix. Then for a given $\mathbf{x}_0 \in \mathbf{R}^n$, the system given by (4.1) has a unique solution*

$$\mathbf{x}(t) = e^{At}\mathbf{x}_0. \tag{4.3}$$

To compute the matrix function $e^{At}$, one need resort to its Taylor series or the Cayley-Hamilton theorem.

For an uncoupled linear system in $\mathbf{R}^n$, $A$ is simply a diagonal matrix, $A = \text{diag}[\lambda_1, ..., \lambda_n]$, then the solution of the system is easily written as:

$$x(t) = \text{diag}[e^{\lambda_1 t}, ..., e^{\lambda_n t}]x(0).$$

Then the trend of the trajectory along each axis $x_i = e^{\lambda_i t}x_i(0)$ can be determined by the sign of the real part of the eigenvalue $\lambda_i$: it goes inward if $\mathbf{Re}\{\lambda_i\} > 0$, outward if $\mathbf{Re}\{\lambda_i\} < 0$ and neither inward nor outward if $\mathbf{Re}\{\lambda_i\} = 0$. In other words, the stability of the equilibrium is determined by the sign of the eigenvalues. If we consider the basis of eigenvectors for the state space when computing $\mathbf{x}(t)$, the vector $\mathbf{x}(t)$ can be decomposed according to individual eigenvectors. The decomposition method is widely used in engineering analysis, especially for large-scale problems.

Let $\{\mathbf{e}_i\}$ be the set of $n$ linearly independent eigenvectors for a system. This set can be used as a basis for the state space, and then $\mathbf{x}(t)$ is uniquely decomposed as

$$\mathbf{x}(t) = \sum_{i=1}^{n} \xi(t)\mathbf{e}_i \qquad (4.4)$$

where the coefficients in this expansion $\xi_i(t), i = 1, \ldots, n$ are functions of time.

Substituting the expansion into the original equation:

$$\sum_{i=1}^{n} \dot{\xi}_i(t)\mathbf{e}_i = \sum_{i=1}^{n} \xi_i(t) A \mathbf{e}_i.$$

By rearranging we can obtain

$$\sum_{i=1}^{n} \dot{\xi}_i(t)\mathbf{e}_i - \xi_i(t)A\mathbf{e}_i = \sum_{i=1}^{n} \left(\dot{\xi}_i(t) - \xi_i(t)\lambda_i\right)\mathbf{e}_i = 0,$$

where $\lambda_i$ is the eigenvalue corresponding to eigenvector $\mathbf{e}_i$. Since the set $\{\mathbf{e}_i\}$ is linear independent, to let the whole summation be zero, all the coefficients of $\mathbf{e}_i$ (the terms in the parenthesis) must be zero, i.e.,

$$\dot{\xi}_i(t) = \xi_i(t)\lambda_i, \quad i = 1, \ldots, n. \qquad (4.5)$$

As we have solved the equations of the above form, i.e., $\xi_i = e^{\lambda_i t}\xi(0)$, now we can express $\mathbf{x(t)}$ as

$$\begin{aligned}\mathbf{x}(t) &= \sum_{i=1}^{n} \xi_i(t)\mathbf{e}_i \\ &= M \begin{bmatrix} \xi_1(t) \\ \vdots \\ \xi_n(t) \end{bmatrix}\end{aligned} \qquad (4.6)$$

Given the concepts of various equilibria (node, saddle, focus, and center), we classify them according to their dynamical property, i.e., every trajectory leaving from or approaching the origin, which corresponds to the stability or otherwise of the origin. A stable node or focus is called a *sink* of the linear system, and an unstable node or focus is called a *source*.

There is a general theorem to determine the stability of the equilibrium of the linear system (simply speaking, the stability of the linear system).

**Theorem 4.3.** *The following statements are equivalent:*

(a) *For all* $\mathbf{x}_0 \in \mathbf{R}^n$, $\lim_{t \to \infty} e^{At}\mathbf{x}_0 = 0$.
(b) *All eigenvalues of $A$ have negative real parts.*

**Definition 4.4.** *A subspace $M$ of a vector space $X$ is a subset of $X$ such that if $\mathbf{x}, \mathbf{y} \in M$, $\alpha$ and $\beta$ are scalars, and $\mathbf{z} = \alpha \mathbf{x} + \beta \mathbf{y}$, then $\mathbf{z} \in M$.*

**Definition 4.5.** *The set of all eigenvectors corresponding to an eigenvalue $\lambda_i$ forms a basis for a subspace of $X$, called the eigenspace of $\lambda_i$. This eigenspace happens to be the null space of a transformation defined as $\lambda_i I - A$.*

Suppose that the $n \times n$ matrix $A$ has $k$ negative eigenvalues $\lambda_1, \cdots, \lambda_k$ and $n - k$ positive eigenvalues $\lambda_{k+1}, \cdots, \lambda_n$ and that these eigenvalues are distinct. Let $\{\mathbf{e}_1, \cdots, \mathbf{e}_n\}$ be a corresponding set of eigenvectors. Then the stable and unstable subspaces of the linear system (4.1), $E^s$ and $E^u$, are the linear subspaces spanned by $\{\mathbf{e}_1, \cdots, \mathbf{e}_k\}$ and $\{\mathbf{e}_{k+1}, \cdots, \mathbf{e}_n\}$ respectively, i.e.,

$$E^s = \text{Span}\{\mathbf{e}_1, \cdots, \mathbf{e}_k\}$$
$$E^u = \text{Span}\{\mathbf{e}_{k+1}, \cdots, \mathbf{e}_n\}.$$

If the matrix $A$ has purely imaginary eigenvalues, then there is also a center subspace $E^c$. Note that the generalized eigenvectors will be used if there is not a full set of $n$ eigenvectors (Perko, 2001).

By the fundamental theorem 4.2, the trajectory of the linear system (4.1) is given by

$$\mathbf{x}(t) = e^{At}\mathbf{x}_0$$

where $\mathbf{x}_0$ is an initial point. The mapping $e^{At} : \mathbf{R}^n \to \mathbf{R}^n$ describes the motion of points $\mathbf{x}_0 \in \mathbf{R}^n$ along the trajectories of (4.1).

A subspace $E \in \mathbf{R}^n$ is said to be *invariant* with respect to the mapping $e^{At}$ if $e^{At}\mathbf{x}_0 \subset E$ for any point $\mathbf{x}_0 \subset E$ and all $t \in \mathbf{R}$.

**Theorem 4.6.** *Let $A$ be a real $n \times n$ matrix. Then*

$$\mathbf{R}^n = E^s \oplus E^u \oplus E^c$$

*where $E^s, E^u$ and $E^c$ are the stable, unstable and center subspaces of (4.1) respectively; furthermore, $E^s, E^u$ and $E^c$ are invariant with respect to the mapping $e^{At}$ of (4.1) respectively.*

## 4.2 Nonlinear Systems

In this section we give a discussion of nonlinear systems of differential equations

$$\dot{\mathbf{x}} = f(\mathbf{x}). \tag{4.7}$$

In the previous section, we saw that any linear system $\dot{\mathbf{x}} = A\mathbf{x}$ has a unique solution in $\mathbf{R}^n$ starting from any point $\mathbf{x}_0$, which is given by $\mathbf{x}(t) = e^{At}\mathbf{x}_0$ for all $t \geq 0$. However, there such an elegant result does not exist for nonlinear systems. In general, it is not possible to solve the nonlinear system (4.7) except for some special cases. Fortunately, a large amount of qualitative properties about the local behavior of the solutions can be investigated by analyzing its equivalent linear systems as stated by the Hartman-Grobman theorem. The theorem shows that topologically the local behavior of the nonlinear system (4.7) near an equilibrium $\mathbf{x}_0$ is typically determined by the behavior of the linear system (4.1) near the origin where the matrix $A = Df(\mathbf{x}_0)$, the derivative of $f$ at $\mathbf{x}_0$.

**Definition 4.7.** *A point $\mathbf{x}_0$ is called an equilibrium point if $f(\mathbf{x}_0) = 0$. An equilibrium point $\mathbf{x}_0$ is called a hyperbolic equilibrium point of (4.7) if none of the eigenvalues of the matrix $Df(\mathbf{x}_0)$ have zero real part. The linear system (4.1) with the matrix $A = Df(\mathbf{x}_0)$ is called the linearization of (4.7) at $\mathbf{x}_0$.*

The matrix $A$ is sometimes called Jacobian (matrix) of (4.7).

Analogously as we did for linear systems, we can give the definitions of sink, source and saddle.

**Definition 4.8.** *An equilibrium point $\mathbf{x}_0$ of (4.7) is called*

1. *a sink if all of the eigenvalues of the matrix $Df(\mathbf{x}_0)$ have negative real part;*
2. *a source if all of the eigenvalues of $Df(\mathbf{x}_0)$ have positive real part;*
3. *a saddle if it is a hyperbolic equilibrium point and $Df(\mathbf{x}_0)$ has at least one eigenvalue with positive real part and at least one with negative real part.*

In the following we give a simplified clarification of the Hartman-Grobman theorem.

**Theorem 4.9.** *If $\mathbf{x}_0 \in \mathbf{R}^n$ is a hyperbolic equilibrium point of the nonlinear system (4.7), then the local behavior of (4.7) near $\mathbf{x}_0$ is topologically equivalent to the local behavior of the linear system (4.1) near the origin.*

## 4.3 Stability, Convergence and Bounded-ness

According to the Hartman-Grobman theorem, the stability of an equilibrium point $\mathbf{x}_0$ can be easily determined by examining the signs of the real parts of the eigenvalues $\lambda_i$ of the matrix $Df(\mathbf{x}_0)$ if it is hyperbolic:

1. First, a hyperbolic equilibrium $x_0$ is asymptotically stable iff $x_0$ is a sink, i.e., $\Re(\lambda_i) < 0$ for all $i = 1, \ldots, n$.
2. Second, a hyperbolic equilibrium is unstable iff. it is either a source or a saddle.

In the other words, any hyperbolic equilibrium point of (4.7) is either asymptotically stable or unstable. In contrast, the stability of nonhyperbolic equilibrium points is more difficult to determine. It is well-known whether a nonhyperbolic equilibrium point is stable, asymptotically stable or unstable is a delicate question. Quite fortunately, in analyzing nonlinear systems, the famous Lyapunov method provides a very useful way to answer this question.

To introduce the Lyapunov method, we first give a strict definition of stability in the sense of Lyapunov.

**Definition 4.10.** *Let $g_t$ denote the flow of the differential equation (4.7) defined for all $t \in \mathbf{R}$. An equilibrium point $x_0$ is stable in the sense of Lyapunov, or simply stable if for all $\epsilon > 0$ there exists a $\delta > 0$ such that for all $x \in N_\delta(x_0)$ and $t \geq 0$, then*

$$g_t(x) \in N_\epsilon(x_0).$$

*The equilibrium $x_0$ is unstable if it is not stable. And $x_0$ is asymptotically stable if it is stable and if there exists a $\delta > 0$ such that for all $x \in N_\delta(x_0)$ we have*

$$\lim_{t \to \infty} g_t(x) = x_0.$$

The following plot gives an intuitive illustration about the conception of Lyapunov stability.

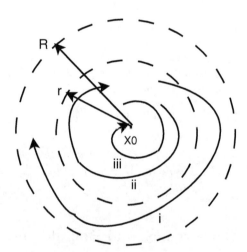

**Fig. 4.5.** Representative trajectories illustrating Lyapunov stability.

## 4.3 Stability, Convergence and Bounded-ness

In the above figure, R and r are the radius of the $\epsilon$ and $\delta$ neighborhoods of $\mathbf{x}_0$, respectively. Both trajectories i and ii indicate the stability, while the trajectory iii indicates the asymptotic stability.

**Theorem 4.11. Lyapunov Theorem** *Suppose there exists a real valued function $V(x)$ satisfying $V(\mathbf{x}_0) = 0$ and $V(\mathbf{x}) > 0$ if $\mathbf{x} \neq \mathbf{x}_0$, i.e., being positive definite. Then we have:*

(1) *If $\dot{V}(\mathbf{x}) \leq 0$ for all $\mathbf{x}$, $\mathbf{x}_0$ is stable;*
(2) *If $\dot{V}(\mathbf{x}) < 0$ for all $\mathbf{x} \neq \mathbf{x}_0$, $\mathbf{x}_0$ is asymptotically stable;*
(3) *If $\dot{V}(\mathbf{x}) > 0$ for all $\mathbf{x}$, $\mathbf{x}_0$ is unstable.*

The function $V : \mathbf{R}^n \to \mathbf{R}$ satisfying the hypotheses of the above theorem is called a *Lyapunov function*. Though the Lyapunov theorem provides a method to determine the stability of the system through the construction of a Lyapunov function $V(\mathbf{x})$, it does nothing to help us find $V(\mathbf{x})$. It does not say anything about the form of $V(\mathbf{x})$, and there is no general method for constructing Lyapunov functions. It only gives a sufficient condition for the stability, in particular, the convergence for $t \to \infty$ and is not concerned with a finite time of convergence. For nonlinear systems, finding a Lyapunov function to prove the stability is often a difficult task that requires elegant techniques and experience. The usual procedure is to firstly guess a candidate Lyapunov function and test it to see if it satisfies the hypotheses defined above.

In the following, we give some typical examples for constructing Lyapunov functions.

**Example 1:** Consider the linear system

$$\dot{\mathbf{x}} = A\mathbf{x}.$$

The candidate Lyapunov function is $V(\mathbf{x}) = \mathbf{x}^T P \mathbf{x}$, where matrix $P$ is positive definite. To verify, we compute

$$\begin{aligned} \dot{V}(\mathbf{x}) &= \dot{\mathbf{x}}^T P \mathbf{x} + \mathbf{x}^T P \dot{\mathbf{x}} \\ &= (A\mathbf{x})^T P \mathbf{x} + \mathbf{x}^T P (A\mathbf{x}) \\ &= \mathbf{x}^T (A^T P + PA) \mathbf{x}. \end{aligned}$$

Let

$$A^T P + PA = -Q. \tag{4.8}$$

If $Q$ is positive (semi)definite, then $V(\mathbf{x})$ is a Lyapunov function that demonstrates the (asymptotic) stability. Normally, we select a positive (semi)definite matrix $Q$ and find a matrix $P$ from solving equation (4.8).

In general, for linear systems, Lyapunov functions take quadratic forms.

**Example 2:** Consider a nonlinear system

$$\dot{x}_1 = -x_1 - 2x_2^2,$$
$$\dot{x}_2 = x_1 x_2 - x_2^2.$$

Obviously $\mathbf{x} = 0$ is the unique equilibrium of the system. We make a guess of the Lyapunov function:

$$V(x_1, x_2) = \frac{1}{2}x_1^2 + x_2^2.$$

Clearly $V(\mathbf{x}) > 0$ for $\mathbf{x} \neq 0$ and $V(0) = 0$. To very we compute

$$\dot{V}(\mathbf{x}) = x_1 \dot{x}_1 + 2x_2 \dot{x}_2$$
$$= x_1(-x_1 - 2x_2^2) + 2x_2(x_1 x_2 - x_2^2)$$
$$= = -x_1^2 - 2x_2^4.$$

Apparently, $\dot{V}(\mathbf{x})$ is negative definite. Thus the equilibrium is asymptotically stable.

**Example 3:** Consider a recurrent neural network with an unsaturating activation function described by:

$$\dot{z}(t) = -z(t) + \sigma\left(Wz(t) + h + (x_0 - h)e^{-t}\right) \qquad (4.9)$$

where $x_0 \in \mathbf{R}^n$, $\sigma(s) = \max(0, s)$.

Assume that the solution of (4.9) starting from the origin is bounded, i.e., $z(0) = 0$ and $|z(t)| < \infty$. If there exists a diagonal positive definite matrix $D$ such that $DW$ is symmetric, then an energy function can be constructed as follows:

$$E(t) = \frac{1}{2} z^\top(t) D(I - W) z(t) - h^\top D z(t)$$
$$- (x_0 - h)^\top D \left( z(t) e^{-t} + \int_0^t z(s) e^{-s} ds \right) \qquad (4.10)$$

for all $t \geq 0$.

Clearly, $E(0) = 0$. By the assumption that $z(t)$ is bounded, it is easy to see that $E(t)$ is also bounded. Since $z(0) = 0$, then from equation (4.9) it is obtained that

$$z_i(t) = \int_0^t e^{-(t-s)} \sigma \left( \sum_{j=1}^n w_{ij} z_j(s) + h_j + (x_{0i} - h_i) e^{-s} \right) ds$$

for $t \geq 0$ and $i = 1, \cdots, n$. Since $\sigma(\cdot) \geq 0$, obviously $z_i(t) \geq 0$ for $t \geq 0$. Thus, $\sigma(z(t)) = z(t)$ for $t \geq 0$.

Let $D = diag(d_1, \cdots, d_n)$, $d_i > 0, i = 1, \cdots, n$. Let $d = \min_{1 \leq i \leq n}(d_i)$. It follows that

$$\dot{E}(t) = \left[z(t) - ((x_0 - h)e^{-t} + Wz(t) + h)\right]^T D$$
$$\times \left[-z(t) + \sigma\left((x_0 - h)e^{-t} + Wz(t) + h\right)\right]$$
$$= [z(t) - ((x_0 - h)e^{-t} + Wz(t) + h)]^T D$$
$$\times [-\sigma(z(t)) + \sigma\left((x_0 - h)e^{-t} + Wz(t) + h\right)]$$
$$\leq -d \cdot \|\sigma(z(t)) - \sigma\left((x_0 - h)e^{-t} + Wz(t) + h\right)\|^2$$
$$= -d \cdot \|z(t) - \sigma\left((x_0 - h)e^{-t} + Wz(t) + h\right)\|^2$$
$$= -d \cdot \|\dot{z}(t)\|^2.$$

Thus, $E(t)$ is monotonically decreasing. Since $E(t)$ is bounded, there must exist a constant $E_0$ such that

$$\lim_{t \to +\infty} E(t) = E_0 < -\infty.$$

**Example 4:** The Lotka-Volterra RNNs with delays are described by the following nonlinear differential equations:

$$\dot{x}_i(t) = x_i(t) \cdot \left[h_i - x_i(t) + \sum_{j=1}^{n} \left(a_{ij} x_j(t) + b_{ij} x_j(t - \tau_{ij(t)})\right)\right] \quad (4.11)$$

for $t \geq 0, i = 1, \cdots, n$. $A = (a_{ij})_{n \times n}$ and $B = (b_{ij})_{n \times n}$ are real $n \times n$ matrices. The delays $\tau_{ij}(t)$ are non-negative continuous functions satisfying $0 \leq \tau_{ij} \leq \tau$ for all $t \geq 0$, where $\tau$ is a positive constant. If there exists a diagonal matrix $D > 0$ and a constant $k > 0$ such that

$$-D + \frac{DA + A^T D + kDBB^T D}{2} + \frac{1}{2k(1-\sigma)} I < 0,$$

then the Lyapunov candidate function of the nonlinear system (4.11) is of the form $V(t) = V_1(t) + V_2(t)$, where

$$V_1(t) = \sum_{i=1}^{n} d_i \int_{x_i^*}^{x_i(t)} \frac{\theta - x_i^*}{\theta} d\theta \quad (4.12)$$

and

$$V_2(t) = \frac{1}{2k(1-\sigma)} \sum_{i=1}^{n} \int_{t-\tau_i(t)}^{t} [x_i(s) - x_i^*]^2 ds \quad (4.13)$$

for $t \geq 0$.

Clearly, $V_1(t) \geq 0$ for all $t \geq 0$. It follows that

$$\dot{V}_1(t) = \sum_{i=1}^{n} d_i[x_i(t) - x_i^*] \cdot \frac{\dot{x}_i(t)}{x_i(t)}$$
$$= [x(t) - x^*]^\top D\left[(-I + A)(x(t) - x^*) + B(x(t - \tau(t)) - x^*)\right]$$
$$= [x(t) - x^*]^\top \left[-D + \frac{DA + A^\top D + kDBB^\top D}{2}\right](x(t) - x^*)$$
$$+ \frac{1}{2k}\|x(t - \tau(t)) - x^*\|^2 \tag{4.14}$$
$$\leq -\left(\lambda + \frac{1}{2k(1-\sigma)}\right)\|x(t) - x^*\|^2$$
$$+ \frac{1}{2k}\|x(t - \tau(t)) - x^*\|^2, \tag{4.15}$$

where $\lambda$ is the minimum eigenvalue of the matrix

$$D - \frac{DA + A^\top D - kDBB^\top D}{2} - \frac{1}{2k(1-\sigma)}I.$$

Apparently, $\lambda > 0$.

The derivative of $V_2(t)$ is computed as

$$\dot{V}_2(t) = \frac{1}{2k(1-\sigma)}\|x(t) - x^*\|^2 - \frac{1}{2k}\|x(t - \tau(t)) - x^*\|^2. \tag{4.16}$$

By applying (4.15) and (4.16), we have

$$\dot{V}(t) = \dot{V}_1(t) + \dot{V}_2(t)$$
$$\leq -\lambda\|x(t) - x^*\|^2. \tag{4.17}$$

It is shown that $V(t)$ is the Lyapunov function of system (4.11).

## 4.4 Analysis of Neuro-dynamics

In the study of neural dynamics, in addition to the Lyapunov method, there is another very useful approach that uses *energy functions*. Though it follows a similar procedure as using Lyapunov function, the energy function has some distinct features, as shown by the next definition and theorem.

Consider a neural network described by the general differential equations

$$\dot{\mathbf{x}}(t) = f(\mathbf{x}). \tag{4.18}$$

**Definition 4.12.** *The network (4.18) is called bounded, if every trajectory is bounded.*

**Definition 4.13.** *The network (4.18) is called completely stable, if every trajectory converges to an equilibrium point.*

## 4.4 Analysis of Neuro-dynamics

**Definition 4.14.** *A continuous function* $E : \mathbf{R}^n \to \mathbf{R}$ *is called an energy function of (4.18), if the derivative of E along the trajectory of satisfies* $\dot{E}(\mathbf{x}) \leq 0$ *and* $\dot{e}(\mathbf{x}) = 0$ *iff* $\dot{\mathbf{x}} = 0$.

If we make a comparison between definitions of the energy function and Lyapunov function, clearly the energy function corresponds to the counterpart of Lyapunov function for asymptotic stability, while relieving the restriction of being positive definite. There are some elaborate forms of energy functions widely used to analyze the neuro-dynamics.

1. Cohen-Grossberg network is described by

$$\frac{du_i}{dt} = a_i(u_i) \left[ b_i(u_i) - \sum_{j=1}^{N} c_{ij} g_j(u_j) \right], \quad i,j = 1, 2, ..., N, \quad (4.19)$$

where $a_i(u_i) \geq 0, c_{ij} = c_{ji}$. Its energy function is written as

$$E(u) = \frac{1}{2} \sum_{i=1}^{N} \sum_{j=1}^{N} c_{ij} g_j(u_j) - \sum_{i=1}^{N} \int_{0}^{v_i} b_i(s) \dot{g}_i(s) ds. \quad (4.20)$$

2. Hopfield network is described by

$$C_i \frac{du_i}{dt} = \frac{u_i}{R_i} + \sum_{j=1}^{N} w_{ij} g(u_j) + I_i, \quad i,j = 1, 2, ..., N, \quad (4.21)$$

where $w_{ij} = w_{ji}$. The energy function is given as

$$E(v) = -\frac{1}{2} \sum_{i=1}^{N} \sum_{j=1}^{N} w_{ij} v_i v_j + \sum_{i=1}^{N} I_i v_i - \sum_{i=1}^{N} \frac{1}{R_i} \int_{0}^{v_i} g_i^{-1}(s) ds. \quad (4.22)$$

We introduce a fundamental theorem for energy functions and omit the proof. Readers may refer to (Yi and Tan, 2004) for more details.

**Theorem 4.15.** *Suppose the network (4.18) is bounded and has an energy function* $E(\mathbf{x})$. *If the equilibrium points are isolated, then the network (4.18) is completely stable.*

Subsequently, we study how to analyze the neuro-dynamics by applying the Lyapunov function or energy function methods through some typical problems in the neural network literature.

**Example 1:** To prove that if the matrix $W$ is symmetric, the Hopfield network (4.21) is completely stable (Theorem 2.5, (Yi and Tan, 2004)).

## 4 Fundamentals of Dynamic Systems

*Proof.* From equation (4.21), it follows that

$$u_i(t) = u_i(0)e^{-\frac{1}{C_iR_i}} + \frac{1}{C_i}\int_0^t e^{-\frac{t-s}{C_iR_i}}\left(\sum_{j=1}^N w_{ij}g_j(u_j(s)) + I_i\right)ds, \quad i=1,...,N. \tag{4.23}$$

for $t \geq 0$. Then it is easy to show that

$$|u_i(t)| \leq |u_i(0)|e^{-\frac{t}{C_iR_i}} + \frac{1}{C_i}\int_0^t e^{-\frac{t-s}{C_iR_i}}\left|\sum_{j=1}^N w_{ij}g_j(u_j(s)) + I_j\right|ds$$

$$\leq |u_i(0)| + R_i\left(\sum_{j=1}^N w_{ij} + |I_i|\right). \tag{4.24}$$

Thus, the network is bounded.

Consider the candidate energy function

$$E(v) = -\frac{1}{2}\sum_{i=1}^N\sum_{j=1}^N w_{ij}v_iv_j + \sum_{i=1}^N I_iv_i - \sum_{i=1}^N \frac{1}{R_i}\int_0^{v_i} g_i^{-1}(s)ds. \tag{4.25}$$

Calculating the derivative of $E$ along the trajectories of (4.21) and using the fact $w_{ij} = w_{ji}$, it follows that

$$\frac{dE(u(t))}{dt} = -\sum_{i=1}^N \frac{du_i(t)}{dt}\left(-\frac{1}{R_i}u_i(t) + \sum_{j=1}^N w_{ij}v_j(t) + I_i\right)$$

$$= -\sum_{i=1}^N \frac{1}{C_i}\frac{dv_i(t)}{dt}\cdot\frac{du_i(t)}{dt}$$

$$= -\sum_{i=1}^N \frac{1}{C_i}\frac{dg_i(u_i(t))}{du_i(t)}\cdot\left(\frac{du_i(t)}{dt}\right)^2$$

$$\leq 0$$

since $g(\cdot)$ is a sigmoid function and $\dot{g}(\cdot) > 0$.

Clearly, $E(u)$ is an energy function of the network. It can also show that the equilibria of (4.21) are isolated, then according to Theorem 4.15, the network is completely stable. This ends the proof.

**Example 2:** Suppose $u^* = (u_1^*, \cdots, u_n^*)^T$ is the unique equilibrium of the network (4.21), show that the following function

$$V(u) = \sum_{i=1}^N C_i \int_{u_i^*}^{u_i} [g_i(s) - g_i(u_i^*)]ds \tag{4.26}$$

is a Lyapunov function.

## 4.4 Analysis of Neuro-dynamics

*Proof.* Clearly, the above function satisfies

1. $V(u) \geq 0$ for all $u \in \mathbf{R}^n$, and
2. $V(u) = 0$ if and only if $u \equiv u^*$.

Define another function

$$\Phi(u) = \sum_{i=1}^{n} [\phi_i(u_i) - \phi_i(u_i^*)] \left[ u_i - u_i^* - \frac{\phi_i(u_i) - \phi_i(u_i^*)}{\dot{\phi}_i(0)} \right].$$

By applying the properties of $\phi(\cdot)$, one can obtain that $\Phi(u) \geq 0$ for all $u \in \mathbf{R}^n$ and $\Phi(u) = 0$ if and only if $u \equiv u^*$.

By calculating the derivative of $V(u)$ along the trajectories of (4.21), it is obtained

$$\frac{dV(u)}{dt} = -\sum_{i=1}^{n} \frac{1}{R_i} [g_i(u_i(t)) - g_i(u_i^*)](u_i(t) - u_i^*)$$

$$+ \sum_{i=1}^{n} \sum_{j=1}^{n} w_{ij} [g_i(u_i(t)) - g_i(u_i^*)]$$

$$= -\sum_{i=1}^{n} [g_i(u_i(t)) - g_i(u_i^*)] \times \left[ u_i(t) - u_i^* - \frac{g_i(u_i(t)) - g_i(u_i^*)}{\dot{g}_i(0)} \right]$$

$$+ \sum_{i=1}^{n} \sum_{j=1}^{n} \left[ -\frac{1}{\dot{g}_i(0) R_i} \delta_{ij} + \frac{w_{ij} + w_{ji}}{2} \right] [g_i(u_i(t)) - g_i(u_i^*)]$$

$$\times [g_j(u_j(t)) - g_j(u_j^*)]$$

$$\leq -\sum_{i=1}^{n} [g_i(u_i(t)) - g_i(u_i^*)] \times \left[ u_i(t) - u_i^* - \frac{g_i(u_i(t)) - g_i(u_i^*)}{\dot{g}_i(0)} \right]$$

$$= -\Phi(u(t)).$$

Thus, $\frac{dV}{dt} \leq 0$ and $\frac{dV}{dt} = 0$ if and only if $u \equiv u^*$. Therefore, by applying the Lyapunov stability theorem, the network (4.21) is globally asymptotically stable.

**Example 3:** Consider the neural networks described by the following system of nonlinear differential equations

$$C_i \frac{du_i}{dt} = -\frac{u_i}{R_i} + \sum_{j=1}^{n} T_{ij} g_i(u_j) + I_i, \quad i = 1, \cdots, n. \tag{4.27}$$

Note that the coefficients $C_i$ can be absorbed on the right hand side, thus we can rewrite the equations in a more compact form

$$\frac{d\mathbf{u}}{dt} = -D\mathbf{u} + Tg(\mathbf{u}) + I \tag{4.28}$$

where $D = \text{diag}\{d_1, \cdots, d_n\}$ is a constant $n \times n$ diagonal matrix with $d_i > 0, i = 1, \cdots, n$, $T$ is a constant $n \times n$ matrix and $I \in \mathbf{R}^n$ is a constant vector. The activation function $g(\mathbf{u}) = (g_1, \cdots, g_n)^\top : \mathbf{R}^n \to \mathbf{R}^n$ is monotonically nondecreasing with $g(0) = 0$.

Before analyzing the above network, some necessary preliminaries are introduced first. We assume the activation function $g_i(\cdot)$ is nondecreasing and nonconstant, and satisfies the global Lipschitz conditions, i.e., there exists a positive constant $l_i > 0$ satisfying

$$0 \le \frac{g_i(s_1) - g_i(s_2)}{s_1 - s_2} \le l_i, \quad i = 1, \cdots, n$$

for $s_1 \ne s_2$. Let $L = \text{diag}\{l_1, \cdots, l_n\}$.

**Definition 4.16.** *A real square matrix $A$ is said to be Lyapunov diagonally stable (LDS) if there exists a diagonal matrix $\Gamma > 0$ such that the symmetric part of $\Gamma A$ is positive definite denoted by $[\Gamma A]^S = [\Gamma A + A^\top \Gamma]/2 > 0$.*

**Lemma 4.17.** *(Forti and Tesi, 1995) Let the matrix be defined by*

$$W = -T + DL^{-1}.$$

*If $W \in LDS$, i.e., $[\Gamma W]^S > 0$ for some diagonal matrix $\Gamma > 0$, then the network model (4.28) has a unique equilibrium point for each $I \in \mathbf{R}^n$.*

By applying the candidate Lyapunov function of the following form

$$V(\mathbf{x}) = \mathbf{x}^T P \mathbf{x} + \sum_{i=1}^n \beta_i \int_0^{x_i} G_i(s) ds, \qquad (4.29)$$

where $G_i(s) = g_i(s + x_i^e) - g_i(x_i^e)$, $\mathbf{x}^e$ being the unique equilibrium point of (4.28), it can be proven that the unique equilibrium is globally asymptotically stable if there exists a positive diagonal matrix $\Gamma = \text{diag}\{\gamma_1, \cdots, \gamma_n\} > 0$ such that $[\Gamma(-W + D\bar{G}^{-1})]^S > 0$. Readers are referred to (Forti and Tesi, 1995; Liang and Si, 2001) for more details.

*Proof.* Since $W \in LDS$, according to Lemma 4.17, the network (4.28) has a unique equilibrium point $\mathbf{u}^e$. By means of the coordinate translation $\mathbf{z} = \mathbf{u} - \mathbf{u}^e$, system (4.28) can be written as the following equivalent system with the unique equilibrium $\mathbf{z} = 0$:

$$\frac{d\mathbf{z}}{dt} = -D\mathbf{z} + TF(\mathbf{z}) \qquad (4.30)$$

where $F(\mathbf{z}) = (f_1(z_1), \cdots, f_n(z_n))^\top$ and $f_i(z_i) = g_i(z_i + u_i^e) - g_i(u_i^e), i = 1, \cdots, n$. It can be seen that $f_i(\cdot), i = 1, \cdots, n$ are nondecreasing and nonconstant functions satisfying global Lipschitz conditions with the Lipschitz constants $l_i > 0 (i = 1, \cdots, n)$. Thus, $0 \le (f_i(z_i))^2 \le l_i z_i f_i(z_i)$ for $i = 1, \cdots, n$.

Consider the candidate Lyapunov function of the Lur'e-Postnikov type

$$V(\mathbf{z}) = \frac{1}{2}\mathbf{z}^\top D^{-1}\mathbf{z} + \frac{k}{\epsilon}\sum_{i=1}^n \gamma_i \int_0^{z_i} f_i(s)ds, \qquad (4.31)$$

where $\epsilon \in (0,1)$ and the constant $k$ defined as

$$k = \frac{\|D^{-1}T\|_2^2}{4\lambda_{\min}([\Gamma W]^S)} \geq 0.$$

The symbol $\|\cdot\|_2$ denotes the matrix norm of the argument $A$ defined by the nonnegative square root of the maximal eigenvalue of $A^\top A$.

By calculating the time derivative of $V(\mathbf{z})$ along the solutions of system (4.28), it is obtained that

$$\begin{aligned}
\dot{V}(\mathbf{z}) &= [D^{-1}\mathbf{z} + (k/\epsilon)\Gamma F(\mathbf{z})]^\top [-D\mathbf{z} + TF(\mathbf{z})] \\
&= -\mathbf{z}^\top\mathbf{z} + \mathbf{z}^\top D^{-1}TF(\mathbf{z}) - (k/\epsilon)F(\mathbf{z})^\top [\Gamma T]^S F(\mathbf{z}) \\
&\leq -\mathbf{z}^\top\mathbf{z} + \mathbf{z}^\top D^{-1}TF(\mathbf{z}) - (k/\epsilon)F(\mathbf{z})^\top \Gamma DL^{-1}F(\mathbf{z}) \\
&\quad + (k/\epsilon)F(\mathbf{z})^\top [\Gamma T]^S F(\mathbf{z}) \\
&= -\mathbf{z}^\top\mathbf{z} + \mathbf{z}^\top D^{-1}TF(\mathbf{z}) \\
&\quad - (k/\epsilon)F(\mathbf{z})^\top [\Gamma W]^S F(\mathbf{z}) \\
&\leq -\|\mathbf{z}\|^2 + \left(\epsilon\|\mathbf{z}\|^2 + \frac{1}{4\epsilon}\|D^{-1}TF(\mathbf{z})\|^2\right) \\
&\quad - (k/\epsilon)F(\mathbf{z})^\top [\Gamma W]^S F(\mathbf{z}) \\
&\leq -(1-\epsilon)\|\mathbf{z}\|^2 + \left(\frac{1}{4\epsilon}\|D^{-1}T\|_2^2 - \frac{k}{\epsilon}\lambda_{\min}([\Gamma W])\right)\|F(\mathbf{z})\|^2 \\
&= -(1-\epsilon)\|\mathbf{z}\|^2.
\end{aligned}$$

Therefore, $V(\mathbf{z})$ is a Lyapunov function of the system (4.28). According to Lyapunov stability theory, the equilibrium point is globally asymptotically stable.

## 4.5 Limit Sets, Attractors and Limit Cycles

**Definition 4.18.** *A point $\mathbf{p} \in E$ is called an $\omega$-limit point of the trajectory $\phi(t,\mathbf{x})$ of the system (4.7), if there is a time sequence $t_n \to +\infty$ as $n \to +\infty$ such that*

$$\lim_{n\to+\infty} \phi(t_n,\mathbf{x}) = \mathbf{p}.$$

The set of all $\omega$-limit points of a trajectory $\Gamma$ is called the $\omega$-*limit set* of $\Gamma$, denoted as $\omega(\Gamma)$. An $\omega$-limit set is called an invariant set since for any point $\mathbf{p} \in \omega(\Gamma)$, all other points of the trajectory $\phi(t,\mathbf{p})$ are also the $\omega$-limit points, i.e., $\phi(t,\omega(\Gamma)) \subset \omega(\Gamma)$.

**Definition 4.19.** *A subset $A \subset E$ is called an* attractor *or* attracting set *of (4.7) if it satisfies*

1. *it is invariant, and*
2. *there is some neighborhood $U$ of $A$ such that for all $\mathbf{x} \in U$, $\phi(t, \mathbf{x}) \in U$ for all $t \geq 0$ and $\phi(t, \mathbf{x}) \to A$ as $t \to +\infty$.*

In the above definition, a *neighborhood* of a set $A$ is any open set $U$ containing $A$. It is said $\mathbf{x}(t) \to A$ as $t \to \infty$, if the distance between $\mathbf{x}(t)$ and $A$: $d(\mathbf{x}(t), A) \to 0$ as $t \to \infty$.

An attractor is a subset of the phase space of a dynamical system and it can be of different geometrical types: points, lines, surfaces and even more complicated structures. Equilibrium points and limit cycles are two simple attractors. A stable node or focus of a dynamical system, is an attractor of the system, but a saddle is not. A stable limit cycle, discussed in the next, is an attractor. The previous study of linear and nonlinear dynamical systems has enabled us to analyze complex neural networks in terms of the stability of their steady states. However, we have yet to consider another important topic in nonlinear dynamical systems theory: nonlinear oscillations, or cyclic dynamics. The analysis of cyclic dynamics is of great interest to study the nervous system and other biological systems. On one hand, nonlinear oscillation is often involved in a lot of interesting and exciting behaviors of living organisms, for example, cardiac rhythms, hormonal cycles, the rhythms of breathing and locomotion (walking, running, swimming). On the other hand, the biological rhythms are inherently nonlinear oscillations, which are largely immune to noises resulting from physiological or environmental fluctuations.

Consider a dynamic system defined by

$$\dot{\mathbf{x}} = f(\mathbf{x}) \tag{4.32}$$

with $f \in C^1(E)$ where $E$ is an open subset of $\mathbf{R}^n$. The solution $\phi(t, \mathbf{x})$ on $E$ is called a *trajectory* or *orbit* of the dynamical system. We can think of a trajectory as a motion through the point $\mathbf{x}$.

**Definition 4.20.** *A* periodic orbit *of (4.32) is a closed solution curve of (4.32).*

Suppose $\phi(t, \mathbf{x}_0)$ is a periodic orbit, then it satisfies for all $t > 0$,

$$\phi(t + T, \mathbf{x}_0) = \phi(t, \mathbf{x}_0)$$

for some $T > 0$. The minimal $T$ for which this equality holds is called the *period* of the periodic orbit (or oscillation).

**Definition 4.21.** *A* limit cycle *is a periodic orbit such that all other trajectories in its sufficiently small neighborhood spiral towards or spiral away from it. It is a* stable *limit cycle for the former case and* unstable *for the latter case.*

In exploring limit cycles or nonlinear oscillations in neuroscience, it will be necessary to develop some analytical tools to predict their existence. There is a fundamental theorem for two-dimensional systems.

**Theorem 4.22.** *(Poincaŕe-Bendixon Theorem) Suppose $f(\mathbf{x})$ is continuous differentiable. If there is an annular region satisfies two conditions: (1) the annulus contains no equilibrium points, and (2) all trajectories from its outer and inner boundaries enter it, then there exists at least one stable limit cycle in the annulus.*

The following figure provides a schematic illustration for the above theorem. The above plot shows an annular region with boundaries r1 and r2. The

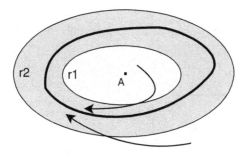

**Fig. 4.6.** A limit cycle is contained in an annulus (gray region) with boundaries r1 and r2. The inner boundary encircles an unstable equilibrium point A.

inner boundary r1 encircles an unstable equilibrium point A, such that all trajectories spiral across the r1 and enter the annulus. Also, the trajectories across r2 enter and remain in the annulus. Since there are no equilibrium in the annulus, the trajectories must approach asymptotically a closed orbit, i.e., there must exist a limit cycle.

It is important to point out that the annulus may contain more than one limit cycles. For example, in the next figure, the trajectories across the inner boundary and the outer boundary approach asymptotically limit cycles L1 and L3, respectively. Obviously, both L1 and L3 are stable, and between them there is an unstable limit cycle L2.

Now we give an example to show a neural network resulting in a limit cycle. Such kinds of neural networks are called a neural network oscillator. Let us consider the Wilson-Cowan network oscillator, as studied in (Wilson, 1999): a four-neuron network consisting of three mutually excitatory neurons E and one inhibitory neuron I. Figure 4.5 sketches the oscillator connections: the E neuron sends stimulating strength to other neurons including both E and I neurons, while the I neuron provides negative feedback to all E neurons. Neural circuits like this are typical of the cortex, where inhibitory GABA neurons comprise about 25% of the total population of cortical cells with the

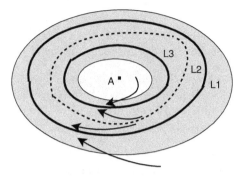

**Fig. 4.7.** A finite number of limit cycles exist in the annular region: Two stable limit cycles (L1 and L3) enclose an unstable limit cycle (L2).

rest being mainly excitatory glutamate neurons. Thus, the network depicted in Figure 4.5 may be thought as a local cortical circuit module.

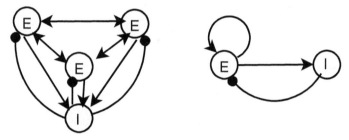

**Fig. 4.8.** Wilson-Cowan network oscillator. Excitatory connections are shown by arrows and inhibitory by solid circles. The simplified network on the right is mathematically equivalent to that on the left by symmetry if all E→E connections are identical.

The four-neuron Wilson-Cowan network (the left on Figure 4.5) can be simplified by assuming that all E neurons receive identical stimuli and have identical synaptic strengths. Under these conditions we can reduce the number of neurons in the network by virtue of symmetry, a procedure called *subsampling*. This results in a mathematically equivalent two-neuron network (the right on Figure 4.5). This kind of simplification can be generalized to any number of mutually excitatory and inhibitory neurons with identical interconnections, and the key concept is recurrent excitation coupled with recurrent inhibition. When reducing the number of E neurons, the recurrent excitations among E neurons are transformed into the self-excitation of one E neuron. The equations in terms of the spike rates are described by:

$$\frac{dE}{dt} = \frac{1}{5}(-E + S(1.6E - I + K))$$
$$\frac{dI}{dt} = \frac{1}{10}(-I + S(1.5E)) \qquad (4.33)$$

where the function $S(\cdot)$ is the so-called Naka-Rushton function

$$S(x) = \begin{cases} \frac{ax^b}{\sigma^b + x^b}, & x \geq 0 \\ 0, & x < 0. \end{cases}$$

with $b = 2, a = 100$ and $\sigma = 30$. The external input $K$ is assumed constant and the time constants for $E$ and $I$ are 5ms and 10ms respectively.

Now we prove the existence of limit cycles for $K = 20$. The isocline equations are

$$E = S(1.6E - I + K),$$
$$I = S(1.5E). \qquad (4.34)$$

From the above equations we can calculate the equilibrium point in the $E - I$ state space: $E = 12.77, I = 28.96$, which is unique. Then we can compute the Jacobian of (4.33) at the equilibrium:

$$A = \begin{pmatrix} 0.42 & -0.39 \\ 0.32 & -0.1 \end{pmatrix}.$$

Its eigenvalues are easily found to be $\lambda = 0.16 \pm 0.24i$. Hence, the only equilibrium is an unstable spiral point (focus). Given the fact that the neural response function $0 \leq S \leq 100$, it follows that trajectories entering the region bounded by $0 \leq E \leq 100, 0 \leq I \leq 100$ must stay within it. And trajectories must leave away from some small neighborhood of the unstable focus. Therefore, we can construct an annular region without containing equilibrium point. According to Poincaré-Bendixon Theorem, there must exist an asymptotically stable limit cycle.

The simulation result for $k = 20$ is plotted in Figure 4.9.

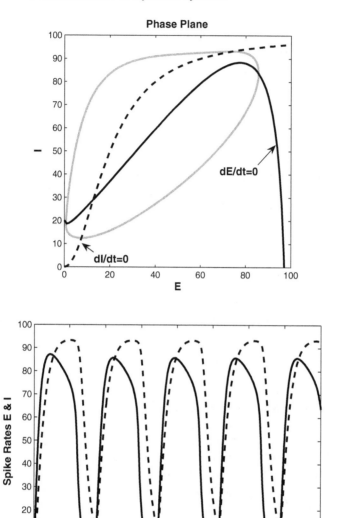

**Fig. 4.9.** Limit cycle of the Wilson-Cowan oscillator. In the above panel an asymptotic limit cycle is plotted in the phase plane along with two isoclines (upper panel). The lower panel plots $E(t)$ (solid line) and $I(t)$ (dashed line) as function of time. Adapted from Wilson (1999).

# 5
# Various Computational Models and Applications

## 5.1 RNNs as a Linear and Nonlinear Programming Solver

Mathematical models of optimization can be generally represented by a constraint set $X$ and a cost function $f$ that maps elements of $X$ into real numbers. The set $X$ consists of the available decision variables $x$ and the cost function $f(x)$ is a scalar measure of choosing $x$. In an explicit way, the optimization process is to find an optimal decision, i.e., an $x^* \in X$ such that

$$f(x^*) \leq f(x), \quad \forall x \in X, \tag{5.1}$$

where $x$ is a vector of $n$ variables $(x_1, \cdots, x_n)$ and $X$ is a subset of $R^n$, the $n$-dimensional Euclidean space.

When the constraint set $X$ is specified by linear inequalities and $f(x)$ is nonlinear, the above problem is called *nonlinear programming*. In a simpler case where $f$ is linear, it is called *linear programming*. Both linear and nonlinear programming are common in systems where a set of design parameters is optimized subject to inequality constraints.

In particular, the linear programming problem is of the form

$$\begin{aligned}
\text{minimize} \quad & c^T x \\
\text{subject to} \quad & Ax = b, \\
& Cx \leq d, \\
& l \leq x \leq u,
\end{aligned} \tag{5.2}$$

where $c, l, u \in R^n$, $A \in R^{m_E \times n}$, $C \in R^{m_I \times n}$, $b \in R^{m_E}$, $d \in R^{m_I}$.

It is considered to be a nonlinear programming if the objective function $f(x)$ is nonlinear:

$$\begin{aligned}\text{minimize} \quad & f(x)\\ \text{subject to} \quad & Ax = b,\\ & Cx \leq d,\\ & l \leq x \leq u.\end{aligned} \qquad (5.3)$$

Quadratic optimization is an important nonlinear programming problem, in which $f(x)$ is a quadratic function:

$$f(x) = \frac{1}{2}x^T Q x + c^T x, \qquad (5.4)$$

where $Q$ is a symmetric matrix.

Recurrent neural networks have been used successfully to solve constrained optimization problems (Hopfield and Tank, 1985; Kennedy and Chua, 1988; Zak et al., 1995a). Being dynamic solvers, neural networks are especially useful in real time applications with time-dependent cost functions, e.g., on-line optimization for robotics or satellite guidance. In such a dynamic solver, the characteristics of neural networks are of great concern: convergence rate, stability and complexity. So far various models of neural network solvers and the theoretical results on stability have been established, either in continuous time (Liang, 2001; Maa and Shanblatt, 1992) or discrete-time manner (Pérez-Ilzarbe, 1998).

### 5.1.1 Recurrent Neural Networks

In solving nonlinear programming problems, one has the alternative of using a dedicated electrical circuit which simulates both the objective and the constraint functions. Such an analog circuit solution is particularly attractive in real-time applications with time-dependent cost functions. Since the emergence of some successful models as a dynamical solver, e.g., Hopfields network (Hopfield and Tank, 1985) and Chuas network (Kennedy and Chua, 1988), neural networks have attracted growing interest in the past two decades (Bouzerdoum and Pattison, 1993; Sudharsanan and Sundareshan, 1991; Liang and Wang, 2000; Xia and Wang, 2000). Compared with other methods of intelligent computation, neural networks have nice properties that are amicable to dynamical analysis. On the other hand, neural networks, like those of Hopfield and Chua (Hopfield and Tank, 1985; Kennedy and Chua, 1988), are readily adapted to integrated circuits. In this chapter, the discussion is restricted to Chuas network.

The linear and nonlinear programming aforementioned, for the sake of applying dynamical analysis, can be described by:

$$\begin{aligned}\text{minimize} \quad & f(x_1, x_2, \cdots, x_n)\\ \text{subject to} \quad & g_1(x_1, x_2, \cdots, x_n) \geq 0,\\ & g_2(x_1, x_2, \cdots, x_n) \geq 0,\\ & \cdots\\ & g_q(x_1, x_2, \cdots, x_n) \geq 0,\end{aligned} \qquad (5.5)$$

## 5.1 RNNs as a Linear and Nonlinear Programming Solver

where $q$ is the number of inequality constraints. Let $g(x) = [g_1(x), \cdots, g_q(x)]$. The canonical nonlinear programming dynamic model is described by

$$C_i \frac{dx_i}{dt} = -\frac{\partial f}{\partial x_i} - \sum_{j=1}^{q} i_j \frac{\partial g_j}{\partial x_i}, \qquad (5.6)$$

for $i = 1, \cdots, n$, where $i_j = \phi_j(g_j(x))$. The nonlinearity of $\phi_j(\theta), \theta \in R$ can be chosen as

$$\phi_j(\theta) = \begin{cases} 0, & \text{if } \theta > 0, \\ \theta, & \text{else.} \end{cases}$$

Note that $\theta \cdot \phi_j(\theta) \geq 0$.

It is assumed that the nonlinear function $\phi_j(\cdot)$ and the partial derivatives of $f(\cdot)$ and $g(\cdot)$ are continuous. Thus the right-hand side of equation (5.6) is continuous. It is further assumed the solution of equation (5.6) is bounded. Consider the $C^1$ scalar function:

$$E(x) = f(x) + \sum_{j=1}^{q} \int_0^{g_j(x)} \phi_j(\theta) d\theta. \qquad (5.7)$$

By taking the time derivative, and using the fact that $i_j = g_j(f_j(v))$, it is obtained that

$$\frac{dE}{dt} = \sum_{i=1}^{n} \frac{\partial f}{\partial x_i} \cdot \frac{dx_i}{dt} + \sum_{j=1}^{q} \sum_{i=1}^{n} \phi_j(g_j(x)) \frac{\partial g_j}{\partial x_j} \cdot \frac{dx_i}{dt}$$

$$= -\sum_{i=1}^{n} (C_i \cdot \frac{dx_i}{dt}) \frac{dx_i}{dt}$$

$$= -\sum_{i=1}^{n} C_i (\frac{dx_i}{dt})^2.$$

Therefore $\frac{dE}{dt} \leq 0$. This implies that the derivative of $E(v)$ is strictly less than zero except at the equilibrium points. Since $E(x)$ is bounded from below, $E(x)$ is a Lyapunov function for the system. Hence, the trajectory $x(t)$ eventually converges to an equilibrium point $x^*$. It is noted that $x^*$ corresponds to the constrained minimum of $f(x)$.

### 5.1.2 Comparison with Genetic Algorithms

Genetic algorithms (GAs), using genetic operations (e.g., selection, crossover and mutation) that mimic the principle of natural selection, have emerged as an attractive optimization tool (Michalewicz, 1994). GA finds its advantages in solving complex problems with the objective functions without nice properties such as differentiability, where gradient search loses its capability.

Genetic operations that are common in GA are fundamentals for multiobjective evolutionary algorithms (MOEA) which has great potential in tackling real-world design problems, where multiple conflicting objectives are often confronted (Deb, 2001; Veldhuizen and Lamont, 2000; Zitzler et al., 2000; Tan et al., 2001) or even in a dynamical environment (Farina et al., n.d.).

Genetic algorithms search the decision space through the use of simulated evolution: the survival of the fittest strategy. In general, the fittest individuals of any population tend to reproduce and survive to the next generation, thus improving successive generations. However, inferior individuals can, by chance, survive and reproduce. The algorithm is stated as follows:

(1) Initialize a population $P_0$.
(2) i=1.
(3) Select $P_i$ from $P_{i-1}$ by a selection function.
(4) Perform genetic operators (crossover and mutation) on selected $P_i$.
(5) Evaluate $P_i$.
(6) Let i=i+1 and go to step (3) until termination criterion met.

The selection of individuals to produce subsequent generations plays an extremely important role in a genetic algorithm. The most widely used selection strategy is to assign a probability of selection to each individual based on its fitness value such that the better individuals an increased chance of being selected. In such a probabilistic selection, an individual can be selected more than once with all individuals in the population having a chance of being selected. A probability $P_j$ is firstly assigned to each individual. Then a sequence of $N$ random numbers $U(0,1)$ is generated and compared against the cumulative probability, $C_i = \sum_{j=1}^{i} P_j$. The appropriate individual $i$ is selected if $C_{i-1} < U(0,1) \leq C_i$. Several selection methods vary in assigning probabilities to individuals: roulette wheel, normalized geometric ranking, etc.

In the roulette wheel approach, the probability $P_j, j = 1,...,N$ is defined by

$$P_j = \frac{F_j}{\sum_{i=1}^{N} F_i}, \qquad (5.8)$$

where $F_j$ is the fitness of individual $j$. However, the method of roulette wheel is limited to the maximization of positive cost function.

Normalized geometric ranking, allowing for minimization and negativity, assigns $P_j$ based on the rank of solution $j$:

$$P_j = \frac{q}{1-(1-q)^N}(1-q)^{r-1}. \qquad (5.9)$$

where $q$ is the probability of selecting the best individual, $r$ is the rank of the individual ($r = 1$ means the best, $N$ the worst). The probabilities also satisfy the requirement $\sum_{i=1}^{N} P_i = 1$.

After selection, the population now is ready to perform genetic operations: *mutation* and *crossover*.

## 5.1 RNNs as a Linear and Nonlinear Programming Solver

Given a chromosome $s^t = (x_1, \cdots, x_j, \cdots, x_n)$, suppose the $j$-th gene is selected for mutation, then the new chromosome is $s^{t+1} = (x_1, \cdots, x'_j, \cdots, x_n)$, where the value of $x_j$ is determined by mutation functions. A quantity of chromosomes are chosen randomly with the probability $p_{um}, p_{bm}, p_{nm}$ to perform uniform mutation, boundary mutation and non-uniform mutation, respectively. Let $l_j$ and $u_j$ be the lower and upper bound, respectively, for each variable $j$. The mutation functions are next described:

(1) **Uniform mutation** simply sets $x'_j$ equal to a uniform random number $U(l_j, u_j)$.

(2) **Boundary mutation** sets $x'_j$ equal to either its lower or upper bound with equal probability: let $p = U(0, 1)$, then

$$x'_j = \begin{cases} l_j, & \text{if } p < 0.5 \\ u_j, & \text{if } p \geq 0.5. \end{cases}$$

(3) **Non-uniform mutation** sets $x'_j$ equal to a non-uniform random number:

$$x'_j = \begin{cases} x_j + (u_j - x_j)\delta(t) \\ x_j - (x_j - l_i)\delta(t). \end{cases}$$

where $\delta(t) = (r(1 - \frac{t}{T}))^b, r = U(0,1)$. $t$ and $T$ is the current and maximal generation number, respectively. $b$ is a shape parameter determining the degree of non-uniformity.

Chromosomes are randomly selected in pairs for application of crossover operators according to appropriate probabilities. Two simple but efficient methods are often used: *simple crossover* and *whole arithmetical crossover*.

(1) **Simple crossover** produces two offspring as follows: if $s^t_v = (v_1, \ldots, v_n)$ and $s^t_w = (w_1, \ldots, w_n)$ are crossed after the $k$-th position, the resulting offspring are $s^{t+1}_v = (v_1, \cdots, v_k, w_{k+1}, \ldots, w_n)$ and $s^{t+1}_w = (w_1, \cdots, w_k, \ldots, v_n)$. When the convex constraint set $X$ is considered, a slight modification is needed:

$$\begin{cases} s^{t+1}_v = (v_1, \cdots, v_k, a \cdot w_{k+1} + (1-a) \cdot v_{k+1}, \cdots, a \cdot w_n + (1-a) \cdot v_n), \\ s^{t+1}_w = (w_1, \cdots, w_k, a \cdot v_{k+1} + (1-a) \cdot w_{k+1}, \cdots, a \cdot v_n + (1-a) \cdot w_n), \end{cases}$$

where $a \in [0, 1]$.

(2) **Whole arithmetical crossover** produces the offspring by linear combination of the parents: $s^{t+1}_v = a \cdot s^t_v + (1-a) \cdot s^t_w$, and $s^{t+1}_w = a \cdot s^t_w + (1-a) \cdot s^t_v$, where $a \in [0, 1]$.

Taking a simple function optimization for an example: maximize $y = x + 10\sin(5x) + 7\cos(4x), x \in [0, 9]$. 10 random points are chosen to perform the selection, cross over and mutation, and after a few generations, the optimum is found (Figure 5.1). Figure 5.2 shows the evolution of the best solution and average solution.

Genetic algorithm and neural networks have different strategies in handling constraints. Given an optimization function, the following constraints are to be considered:

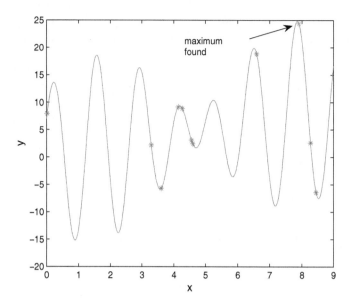

**Fig. 5.1.** Function optimization using genetic algorithm.

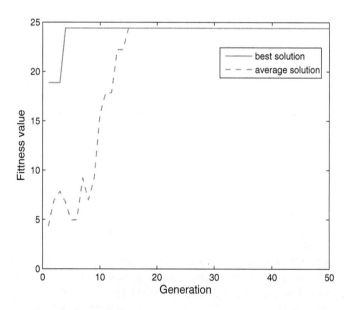

**Fig. 5.2.** Evolution of best solution and average solution.

## 5.1 RNNs as a Linear and Nonlinear Programming Solver

1. Domain constraints: $l \leq x \leq u$.
2. Equality constraints: $Ax = b$.
3. Inequality constraints: $Cx \leq d$.

In genetic algorithm, the first step is to eliminate equalities (Michalewicz, 1994). Suppose there are $p$ independent equality equations, matrix $A$ can be split vertically into two matrixes $A_1$ and $A_2$, such that $A_1^{-1}$ exists, that is:

$$A_1 x^1 + A_2 x^2 = b,$$

where $x^1 = (x_1, \cdots, x_p)$. Then it is easily seen that $x^1 = A_1^{-1} b - A^{-1} A_2 x^2$. Now the set of inequalities can be represented as $C_1 x^1 + C_2 x^2 \leq d$, which can be transformed into $C_1(A_1^{-1} b - A^{-1} A_2 x^2) + C_2 x^2 \leq d$. Consequently, after eliminating $p$ variables, the final set of constraints consists of only the following inequalities:

1. Domain constraints: $l \leq x^2 \leq u$.
2. New inequalities: $l^1 \leq A_1^{-1} b - A^{-1} A_2 x^2 \leq u^1$.
3. Original inequality constraints: $(C_2 - C_1 A_1^{-1} A_2) x^2 \leq d - C_1 A_1^{-1} b$.

A different strategy, however, is employed in neural networks implementation. The constraints should take the form of $g(x) \geq 0$, as equation (5.5) discussed in the previous section. Hence, the constraints are represented as:

$$g(x) = \begin{bmatrix} g_1(x) \\ \vdots \\ g_{2m_e + m_I}(x) \end{bmatrix} = \hat{A}x - \hat{b},$$

where

$$\hat{A} = \begin{bmatrix} A \\ -A \\ -C \end{bmatrix}, \quad \hat{b} = \begin{bmatrix} b \\ -b \\ -d \end{bmatrix}.$$

Observing the network model, equation (5.6), $i_j, \forall j$ will be zero if $g_j \geq 0$. Consequently, the constraints $g_j(x) \geq 0$ are realized by a nonlinear function $\phi(g_j(x))$. Actually, in energy function (11.3), the second term plays the role of penalty: when $g_j(x) \geq 0$, there is no penalty and the energy function is exactly $f(x)$; otherwise, the second term takes on a nonnegative value resulting in a penalty term.

**Example 1: (Linear Programming)**

Consider a linear programming problem taken from (Zak et al., 1995a):

$$\begin{aligned} \text{minimize} \quad & -8x_1 - 8x_2 - 5x_3 - 5x_4 \\ \text{subject to} \quad & x_1 + x_3 = 40, \\ & x_2 + x_4 = 60, \\ & 5x_1 - 5x_2 \geq 0, \\ & -2x_1 + 3x_2 \geq 0, \\ & x_1, x_2, x_3, x_4 \geq 0. \end{aligned} \quad (5.10)$$

To apply the network of Chua, equation (5.6), the constraints is formulated as:

$$g_1(x) = 5x_1 - 5x_2 \geq 0,$$
$$g_2(x) = -2x_1 + 3x_2 \geq 0,$$
$$g_3(x) = x_1 + x_3 - 40 \geq 0,$$
$$g_4(x) = -x_1 - x_3 + 40 \geq 0,$$
$$g_5(x) = x_2 + x_4 - 60 \geq 0,$$
$$g_6(x) = -x_2 - x_4 + 60 \geq 0,$$
$$g_7(x) = x_1 \geq 0,$$
$$g_8(x) = x_2 \geq 0,$$
$$g_9(x) = x_3 \geq 0,$$
$$g_{10}(x) = x_4 \geq 0. \tag{5.11}$$

The dynamics of Chua's network is governed by the equation

$$\dot{x} = C^{-1}\left(-c - s\nabla g(x)\phi(g(x))\right), \tag{5.12}$$

where $s$ is a positive design constant. Without loss of generality, we let $C$ be an identity matrix. Figure 5.3 depicts the convergence of each trajectory when $s = 10$. In this case, the network converges to the equilibrium $x^e = [41.72, 41.66, -0.60, 18.83]^T$. When increasing the value of $s$, e.g., $s = 100$, the network gives a more accurate solution $x^e = [40.17, 40.17, -0.06, 19.89]^T$, which is now close enough to the optimum $x^* = [40, 40, 0, 20]^T$. The evolution of $x_i, i = 1, \cdots, 4$ versus time for $s = 100$ is shown in Figure 5.4.

Using the genetic algorithm toolkit GenoCOP developed by Michalewicz (Michalewicz, n.d.), the best chromosome is found after 10 generations (with a population size of 40): $x = [40, 40, 0, 20]^T$.

**Example 2: (Nonlinear Programming)**

Consider a quadratic optimization problem which was studied in (Kennedy and Chua, 1988):

$$\text{minimize} \quad 0.4x_1 + \frac{1}{2}(5x_1^2 + 8x_2^2 + 4x_3^2) - 3x_1x_2 - 3x_2x_3$$
$$\text{subject to} \quad x_1 + x_2 + x_3 \geq 1,$$
$$x_1, x_2, x_3 \geq 0. \tag{5.13}$$

The objective function if of the form $\phi(x) = \frac{1}{2}x^T Q x + c^T x$, where

$$Q = \begin{bmatrix} 5/2 & -3/2 & 0 \\ -3/2 & 4 & -3/2 \\ 0 & -3/2 & 2 \end{bmatrix}, \quad c = [0.4, 0, 0]^T.$$

As can be seen (Figure 5.5), the network converges to the equilibrium $x^e = [0.20, 0.34, 0.45]^T$, which is close to the theoretical optimum $x^* = [0.25, 0.33, 0.42]^T$.

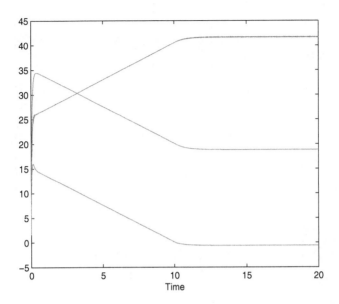

**Fig. 5.3.** State trajectories of Chua's network with $s = 10$ (Example 1).

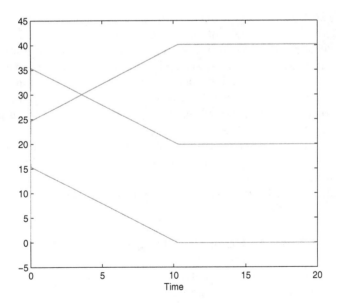

**Fig. 5.4.** State trajectories of Chua's network with $s = 100$ (Example 1).

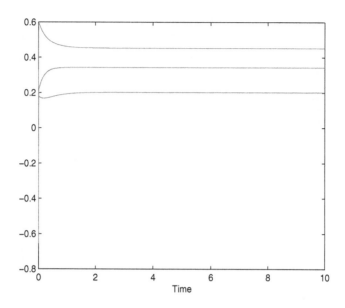

**Fig. 5.5.** State trajectories of Chua's network with $s = 100$ (Example 2).

Using the genetic algorithm toolkit GenoCOP, the best chromosome is found after 293 generations (with a population size of 70): $x = [0.25, 0.33, 0.41]^T$.

This demonstrates that neural networks can be ensured to converge to the optimum of a regular optimization problem (differentiable objective function and constraints), based on a gradient search. If the objective function is convex, the equilibrium of a neural network corresponds to the global optimum. In contrast, the performance of genetic algorithms relies more on parameter settings, such as maximal generation number and probability of applying genetic operations on chromosomes, which cannot be assured a priori. However, genetic algorithms, manipulating a large population of individuals to evolve just as natural evolution, are advantageous in solving non-regular optimization problems, which may obstruct deterministic methods.

## 5.2 RNN Models for Extracting Eigenvectors

Computing eigenvectors of a matrix is an important issue in data analysis, signal processing, pattern recognition and so on, especially for computing eigenvectors corresponding to the largest or smallest eigenvalues. There are a few existing neural network models that deals with this issue. However, they are mostly aimed to extract the eigenvectors of positive definite symmetric matrix, for example, (Luo et al., 1995; Luo et al., 1997). In this section we

## 5.2 RNN Models for Extracting Eigenvectors

introduce a recently developed RNN model (Yi et al., 2004) that applies for more general cases.

Let $A$ be an $n \times n$ real symmetric matrix. The recurrent network is described by the following system of differential equations:

$$\dot{x}(t) = x^T x A x - x^T A x x \qquad (5.14)$$

Let $\lambda_1, \cdots, \lambda_n$ be all the eigenvalues of $A$ ordered by $\lambda_1 \geq \cdots \geq \lambda_n$. Suppose that $S_i (i = 1, \cdots, n)$ is an orthonormal basis in $R^n$ such that each $S_i$ is an eigenvector of $A$ corresponding to the eigenvalue $\lambda_i$. Let $\sigma_i (i = 1, \cdots, m)$ be all the distinct eigenvalues of $A$ ordered by $\sigma_1 > \cdots > \sigma_m$.

It can be proven that each solution of (1) starting from any nonzero points in $R^n$ will converge to an eigenvector of $A$ (See the proof in (Yi and Tan, 2004)). Furthermore, given any nonzero vector $x(0) \in R^n$, if $x(0)$ is not orthogonal to $V_{\sigma_1}$, then the solution of (1) starting from $x(0)$ converges to an eigenvector corresponding to the largest eigenvalue of $A$. Replacing $A$ in network (1) with $-A$, and supposing $x(0)$ is a nonzero vector in $R^n$ which is not orthogonal to $V_{\sigma_m}$, then the solution starting from $x(0)$ will converge to an eigenvector corresponding to the smallest eigenvalue of $A$.

Symmetric matrix can be randomly generated in a simple way. Let $Q$ be any randomly generated real matrix, and $A = \frac{Q^T + Q}{2}$. Clearly, $A$ is a symmetric matrix.

A $5 \times 5$ symmetric matrix $A$ is generated as

$$A = \begin{bmatrix} 0.7663 & 0.4283 & -0.3237 & -0.4298 & -0.1438 \\ 0.4283 & 0.2862 & 0.0118 & -0.2802 & 0.1230 \\ -0.3237 & 0.0118 & -0.9093 & -0.4384 & 0.7684 \\ -0.4298 & -0.2802 & -0.4384 & -0.0386 & -0.1315 \\ -0.1438 & 0.1230 & 0.7684 & -0.1315 & -0.4480 \end{bmatrix}.$$

Using the network model, the estimated eigenvectors corresponding to $\lambda_{max}$ and $\lambda_{min}$ respectively are:

$$\xi_{max} = \begin{bmatrix} 1.0872 \\ 0.6264 \\ -0.0809 \\ -0.4736 \\ -0.0472 \end{bmatrix}, \xi_{min} = \begin{bmatrix} 0.1882 \\ 0.0600 \\ 1.3209 \\ 0.3697 \\ -0.8446 \end{bmatrix}.$$

The convergence procedure is shown in Fig. 5.6. The result is an estimation of the eigenvector, that is $\xi_{max}$ above, corresponding to the largest eigenvalue. An immediate following result is an estimation to the maximum eigenvalue of 1.2307, which is an accurate approximation to the true value with a precision of 0.0001, given the value computed by Matlab.

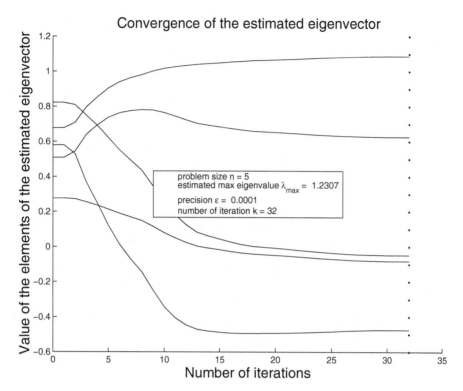

**Fig. 5.6.** Estimation of an eigenvector corresponding to the largest eigenvalue of the matrix $A$. Curtesy of Yi and Tan (2004).

## 5.3 A Discrete-Time Winner-Takes-All Network

[1]The winner-takes-all (WTA) problem has been studied extensively in neural network literature. A lot of WTA network models have been presented. Let $x_i, y_i$ denote the states and outputs of the network respectively, $i = 1, \cdots, n$, the Yis WTA network (Yi et al., 2000) is described by the system of discrete-time equations:

$$x_i(k+1) = \frac{(c+b)(x_i(k)+\gamma)^\alpha}{\sum_{j=1}^n (x_j(k)+\gamma)^\alpha} - \gamma \quad (5.15a)$$

$$y_i(k+1) = h(x_i(k)) \quad (5.15b)$$

where $c \geq 1, \alpha > 1$ and $\beta > \gamma > 0$ are constants. $y_i, i = 1, \cdots, n$ are the output values of the states $x_i$ through the nonlinearity $h(x)$. The diagram of the network is shown in Fig. 5.7.

---

[1] Reuse of the materials of "Winner-take-all discrete recurrent neural networks", 47(12), 2000, 1584–1589, IEEE Trans. on Circuits and Systems-II, with permission from IEEE.

## 5.3 A Discrete-Time Winner-Takes-All Network

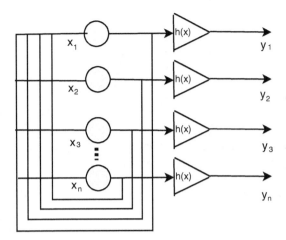

**Fig. 5.7.** The discrete-time WTA network (5.15).

The activation function $h(x)$ is a piecewise linear function defined as

$$h(x) = \begin{cases} c, & \text{if } x \geq c \\ x, & \text{if } x \in (0, c) \\ 0, & \text{if } x \leq 0 \end{cases}$$

Let $\Omega_0$ denote the set of all initial values $\mathbf{x} = (x_1, \cdots, x_n)^\top$ satisfying

$$0 < x_{k_1}(0) \leq x_{k_2}(0) \leq \cdots \leq x_{k_{n-1}} < x_{k_n} < c.$$

The initial condition confines that each component $x_i$ satisfies $0 < x_i < c$ and the largest component is unique.

Suppose $\mathbf{x}(0) \in \Omega_0$. Then the network possesses the following dynamical properties:

i. If $x_i(0) < x_j(0)$, then $x_i(k) < x_j(k)$ for all $k \geq 0$;
ii. If $x_i(0) = x_j(0)$, then $x_i(k) = x_j(k)$ for all $k \geq 0$.

Furthermore, in the steady state, for any $\mathbf{x}(0) \in \Omega_0$, if $x_m(0) > x_j(0)$, $m \neq j, j = 1, \cdots, n$, then

$$x_i(k) \to \begin{cases} c + \beta - \gamma, & \text{if } i = m \\ -\gamma, & \text{otherwise} \end{cases}$$

as $k \to +\infty$.

From this, it is revealed that the initial order of the neuron states remains unchanged in evolving the network. After convergence, one neuron reaches the maximal value as the winner, while the rest are all suppressed. For more details, the readers are referred to (Yi et al., 2000).

Based on the above analysis, we obtain the following conditions for the parameters of the WTA network:

$$\begin{cases} \alpha > 1 \\ \beta > \gamma > 0 \\ \frac{\beta-c}{2} < \gamma < \min\{\beta, (c+\beta)/n\}. \end{cases}$$

In the above conditions, $\alpha > 1$ is used to control the convergence rate. The larger the value of $\alpha$ is, the less iteration numbers will be. The simulation results are shown in Figs. 5.8 and 5.9 for the set of parameters: $n = 10, c = 1, \alpha = 2, \beta = 0.2, \gamma = 0.11$. The initial values of $x_i$ are randomly generated from the range $(0, 1)$. It is shown that there is only one winner after the competition which converges to the value of $c + \beta - \gamma = 1.09$ and all the losers converge to the value of $-\gamma = -0.11$ (see Fig. 5.8). It is shown that the output value of the winner neuron is 1 while those of all the losers are 0 (see Fig. 5.9).

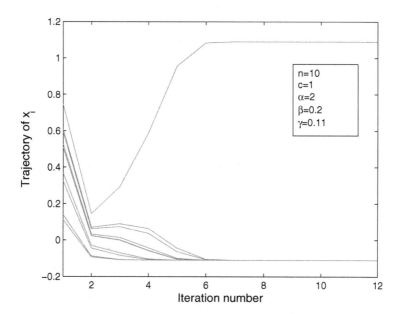

**Fig. 5.8.** The convergence of the network states $x_i$.

## 5.4 A Winner-Takes-All Network with LT Neurons

Hahnloser (Hahnloser, 1998) proposed a winner-takes-all network with linear threshold (LT) neurons. A generalization to this model is described by the system of differential equations:

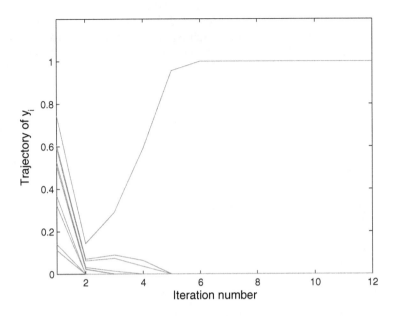

**Fig. 5.9.** The convergence of the outputs of the network.

$$\dot{x}_i(t) = -x_i(t) + w_i\sigma(x_i(t)) - L + h_i, \tag{5.16a}$$

$$\tau \dot{L}(t) = -L(t) + \sum_{j=1}^{n} w_j \sigma(x_j(t)), \tag{5.16b}$$

where $w_i > 0, i = 1, \cdots, n$ denote the local excitation strengths, $h_i > 0$ the external input currents, $L$ the global inhibitory neuron, and $\tau > 0$ a time constant. The function $\sigma(x)$ is a linear threshold nonlinearity $\sigma(x) = \max(0, x)$.

It has been shown that the network (5.16) performs the winner-takes-all computation if $w_i > 1$ and $\tau < \frac{1}{w_i - 1}$, i.e., in steady states the network is stabilized at

$$x_k(t \to \infty) = h_k, \tag{5.17a}$$
$$x_i(t \to \infty) = h_i - w_k h_k \le 0, \ i \ne k, \tag{5.17b}$$
$$L(t \to \infty) = w_k h_k, \tag{5.17c}$$

where $k$ is labeling the winning neuron. It should be pointed out that the winner $x_k$ is not always the neuron which receives the largest external input current. Furthermore, the steady states are not unique. However, the potential winners can be inferred from the above formulation of steady states: these are neurons satisfying the inequality (5.17b). It implies that the product of the winner's input $h_k$ with $w_k$ must be larger than any other input.

72     5 Various Computational Models and Applications

Given the parameters $w_i = 2$ for all $i$, $\tau = 0.5$, $h = (1,2,3)^\top$, according to the inequality (5.17b), both the neurons 2 and 3 can be the winners. We ran the simulation for 20 random initial points. It is shown that either $x_3$ or $x_2$ wins the competition (for example, see Fig. 5.10 and Fig. 5.11 respectively). The phase portrait for $x_1 - L$ is plotted in Fig. 5.12.

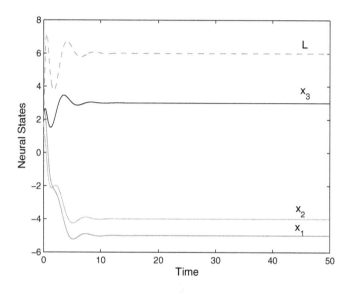

**Fig. 5.10.** The neural states in 50 seconds. Initial point is randomly generated and the neuron $x_3$ becomes the winner. $w_i = 2, \tau = 0.5, h = (1,2,3)^\top$

If we look the competition as selecting a winner from a set of inputs, there exist different selections, which depend on the starting points $\mathbf{x}(0)$. This fact is known as the *multistability of selection* (Hahnloser et al., 2000). In (Hahnloser et al., 2000), silicon circuits were built to emulate the behaviors of the network 5.18, which added local neighboring interactions to the above dynamics:

$$\frac{C}{E_k}\frac{dE_k}{dt} + E_k = \sigma\left(e_k + \sum_{l=-2}^{2} \alpha_l E_{k+l} - \beta I\right), \quad (5.18a)$$

$$I = i + \sum_{k=1}^{n} E_k, \quad (5.18b)$$

where $E_k, k = 1, \cdots, n$ and $I$ are output currents of excitatory neurons and the inhibitory neuron, $C > 0$ the capacitance constant. The excitatory neurons interact through self ($\alpha_0$), nearest neighbor ($\alpha_1$ and $\alpha_{-1}$), and next nearest

5.4 A Winner-Takes-All Network with LT Neurons    73

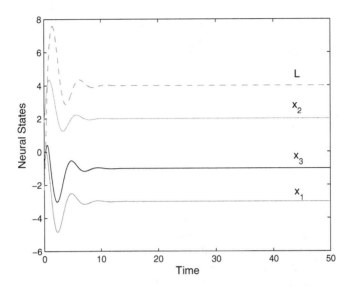

**Fig. 5.11.** The neural states in 50 seconds. Initial point is randomly generated. Differing from the previous figure, the neuron $x_2$ becomes the winner for the same set of parameters.

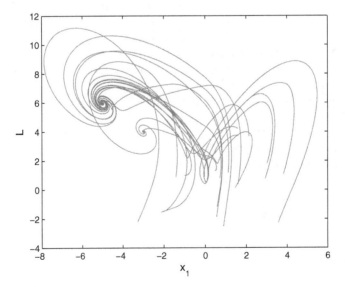

**Fig. 5.12.** The phase portrait in the $x_1 - L$ plane. 20 initial points are randomly generated. It is shown that there exist two steady states.

neighbor ($\alpha_2$ and $\alpha_{-2}$) connections. The inhibitory neuron $I$ sums the currents from excitatory neurons, and this in turn inhibits them with a strength $\beta$. In the excitatory-inhibitory network, it is known that slow inhibition can lead to oscillatory or explosive instability. Detailed analysis for the oscillation behaviors of LT networks is given in a subsequent chapter.

## 5.5 Competitive-Layer Model for Feature Binding and Segmentation

In (Wersing et al., 2001a), a competitive-layer model (CLM) was proposed for feature binding and sensory segmentation. The CLM consists of a set of $L$ layers of feature-selective neurons. The activity of a neuron at position $r$ in layer $\alpha$ is denoted by $x_{r\alpha}$, and the set of the neuron activities $x_{r\alpha}$ share a common position $r$ in each layer is denoted as a feature column $r$. Each column is associated with a particular feature described by a parameter vector $m_r$.

In the CLM architecture, for each input feature, there is a corresponding neuron in each layer. A vertical WTA circuit implements a topographic competition between layers. Lateral interactions characterize compatibility between features and guide the binding process. A binding between two features, represented by columns $r$ and $r$, is expressed by simultaneous activities $x_{r\hat{\alpha}} > 0$ and $x_{r\hat{\alpha}}$ that share a common layer $\hat{\alpha}$. Therefore, binding is achieved by having each activated column $r$ assign its feature to one (or several) of the layers, interpreting the activity $x_{r\alpha}$ as a measure of for the certainty of that assignment. All the neurons in a column $r$ are equally driven by an external input $h_r$ which is interpreted as the significance of the detection of feature $r$ by a preprocessing step. The afferent input $h_r$ is fed to the activities $x_{r\alpha}$ with a connection strength $J_r > 0$.

While in each layer $\alpha$, the activities are coupled by the lateral interaction $f_{rr}^{\alpha}$ which corresponds to the degree of compatibility between features $r$ and $r$ and is symmetric. In other words, the contextual information is stored in the lateral interactions. The vertical WTA circuit, which uses mutual inhibitory interactions with symmetric connection strengths $I_r^{\alpha\beta} = I_r^{\beta\alpha} > 0$, is to enforce dynamical assignment of the input features to the layers. Based on these configurations, the dynamics of the network can be described by the following system of differential equations:

$$\dot{x}_{r\alpha} = -x_{r\alpha} + \sigma \left( x_{r\alpha} - \sum_{\beta} I_r^{\alpha\beta} x_{r\beta} + \sum_{r'} f_{rr'}^{\alpha} x_{r'\alpha} + J_r h_r \right), \quad (5.19)$$

where $\sigma(x) = \max(0, x)$ is a nonsaturating linear threshold transfer function.

A typical feature example is local edge element characterized by position and orientation, $m_r = (x_r, y_r, \theta_r)$. More complex features were used by (Ontrup and Ritter, 1998) for texture segmentation.

## 5.5 Competitive-Layer Model for Feature Binding and Segmentation

The work (Wersing et al., 2001a) proposed some theoretical analysis for the convergence and assignment of the CLM network dynamics.

**Theorem 5.1.** Let $k_{r\alpha} = I_r^{\alpha\alpha} - f_{rr}^\alpha - \sum_{r' \neq r} \max(0, f_{rr'}^\alpha)$.

i. If $k_{r\alpha} > 0$, then the CLM dynamics is bounded.
ii. If $k_{r\alpha} > 0$ and $0 \leq x_{r\alpha}(0) \leq M$ for all $r, \alpha$, $M = \max_r(J_r h_r / k_{r\alpha})$, then $0 \leq x_{r\alpha}(t) \leq M$ for all $r, \alpha$ and $t > 0$.

**Theorem 5.2.** Let $F_{r\alpha} = \sum_r f_{rr}^\alpha x_{r\alpha}$ denote the lateral feedback of neuron $(r, \alpha)$. If the lateral interaction is self-excitatory, i.e., $f_{rr}^\alpha > 0$ for all $r, \alpha$, and the vertical interactions satisfy: $I_r^{\alpha\alpha} I_r^{\beta\beta} \leq (I_r^{\alpha\beta})^2$ for all $\alpha, \beta$, then an attractor of the CLM has in each column $r$ either

i. At most one positive activity $x_{r\hat{\alpha}}$ with

$$x_{r\hat{\alpha}} = \frac{J_r h_r}{I_r^{\hat{\alpha}\hat{\alpha}}} + \frac{F_{r\hat{\alpha}}}{I_r^{\hat{\alpha}\hat{\alpha}}},$$

$$x_{r\beta} = 0, \quad \text{for all } \beta \neq \hat{\alpha},$$

where $\hat{\alpha} = \hat{\alpha}(r)$ is the index of the maximally supporting layer characterized by $F_{r\hat{\alpha}} > F_{r\beta}$ for all $\beta \neq \hat{\alpha}$; or
ii. All activities $x_{r\alpha}, \alpha = 1, \cdots, L$ in a column $r$ vanish and $F_{r\alpha} \leq -J_r h_r$ for all $\alpha = 1, \cdots, L$.

Theorem 5.1 asserts that factors $k_{r\alpha}$ control the stability margin. Theorem 5.2 states that as long as there are self-excitatory interactions within each layer and the vertical cross-inhibition is sufficiently strong, the CLM converges to a unique assignment of features to the layer with maximal lateral feedback $F_{r\alpha}$. It is worthy to note that due to the layered topology, this does not require arbitrarily large vertical couplings. Also note that the lack of an upper saturation is essential for the WTA behavior because it allows the exclusion of spurious ambiguous states. Nonzero activities are stable only due to a dynamical equilibrium and not due to saturation. The proofs of the theorems can be found in (Wersing et al., 2001a). The CLM dynamics can be simulated in parallel by any differential equation integrator like the Euler or Runge-Kutta method. However, if simulated in a serial manner, an alternative approach that replaces the explicit trajectory integration by a rapid search for fixed-point attractors can be used (Wersing et al., 2001a). This is done by iteratively solving the fixed-point equations, known as Guass-Seidel method. The implementation of the algorithm is described as follows:

**Algorithm 5.1** *The CLM for Feature Binding*

1. Initialize all $x_{r\alpha}$ with small random values such that

$$x_{r\alpha} \in [h_r/L - \epsilon, h_r/L + \epsilon].$$

Initialize $T = T_c$.

2. Do $N \cdot L$ times: choose $(r, \alpha)$ randomly and update $x_{r\alpha} = \max(0, \xi)$, where

$$\xi = \frac{J_r h_r - \sum_{\beta \neq \alpha} I_r^{\alpha\beta} + \sum_{r' \neq r} f_{rr'}^{\alpha} x_{r'\alpha}}{I_r^{\alpha\alpha} - f_{rr}^{\alpha} + T}.$$

3. Decrease $T$ by $T := \eta T$ with $0 < \eta < 1$. Go to step 2 until convergence.

Before applying the CLM to feature binding and segmentation, a preprocessing step is required to acquire the interaction values. The edge features are firstly generated by applying $3 \times 3$ pixel Sobel $x$ and $y$ operators. The co-circular interaction is employed to produce the interaction values, which is defined as: given two edges at positions $\mathbf{r}_1$ and $\mathbf{r}_2$ with a difference vector $\mathbf{d} = \mathbf{r}_1 - \mathbf{r}_2$, $d = \|\mathbf{d}\|$ and $\hat{\mathbf{d}} = \mathbf{d}/d$, and unit orientation vector $\hat{\mathbf{n}}_1 = (n_1^x, n_1^y), \hat{\mathbf{n}}_2 = (n_2^x, n_2^y)$, the co-circular interaction is:

$$f^{cocirc}((\mathbf{r}_1, \hat{\mathbf{n}}_1), (\mathbf{r}_2, \hat{\mathbf{n}}_2)) = \theta(a_1 a_2 q)\left(e^{-d^2/R^2 - C^2 S}\right) - I e^{-2d^2/R^2},$$

where $a_1 = n_1^x \hat{d}_y - n_1^y \hat{d}_x$, $a_2 = n_2^x \hat{d}_y - n_2^y \hat{d}_x$, $q = \hat{\mathbf{n}}_1 \cdot \hat{\mathbf{n}}_2$, and the function $\theta(x) = 1$ for $x \geq 0$ and $\theta(x) = 0$ otherwise. The degree of co-circularity is given by $C = |\hat{\mathbf{n}}_1 \cdot \hat{\mathbf{d}}| - |\hat{\mathbf{n}}_2 \cdot \mathbf{d}|$, the parameter $S > 0$ controls the sharpness of the co-circularity and $I > 0$ controls the inhibition strength.

## 5.6 A Neural Model of Contour Integration

In (Li, 1998), a neural model that performs contour integration was proposed to emulate what is said to take place in the primary visual cortex. The model is composed of recurrently connected excitatory neurons and inhibitory interneurons, receiving visual input via oriented receptive fields (RF) resembling those found in the primary visual cortex. The network is able to form smooth contours of an image, through intracortical interactions modifying initial activity patterns from input, selectively amplifying the activities of edges. The model is outlined as: Visual inputs are modeled as arriving at discrete spatial locations. At each location $i$ there is a hypercolumn composed of $K$ neuron pairs. Each pair $(i, \theta)$ has RF center $i$ and preferred orientation $\theta = k\pi/K$ for $k = 1, 2, \cdots, K$, and is called (a neural representation of) an edge segment.

The excitatory cells receives the visual input; where its output quantifies the response or salience of the edge segment and projects to higher visual areas. The inhibitory cells are treated as interneurons. When an input image contains an edge at $i$ with orientation $\beta$ and input strength $\hat{I}_{i\beta}$, then the input that edge segment $i\theta$ receives is $I_{i\theta} = \hat{I}_{i\beta}\phi(\theta - \beta)$, where $\phi(\theta - \beta) = e^{-|\theta - \beta|/(\pi/8)}$ is the orientation tuning curve. The neural dynamics is described as:

$$\dot{x}_{i\theta} = -\alpha_x x_{i\theta} - \sum_{\Delta\theta} \psi(\Delta\theta) g_y(y_{i,\theta+\Delta\theta}) + J_0 g_x(x_{i\theta})$$
$$+ \sum_{j\neq i,\theta'} J_{i\theta,j\theta'} g_x(x_{j\theta'}) + I_{i\theta} + I_0, \qquad (5.20a)$$

$$\dot{y}_{i\theta} = -\alpha_y y_{i\theta} + g_x(x_{i\theta}) + \sum_{j\neq i,\theta'} W_{i\theta,j\theta'} g_x(x_{j\theta'}) + I_c, \qquad (5.20b)$$

where $\alpha_x, \alpha_y$ are the membrane time constants, $I_o, I_c$ are the background inputs to excitatory cells and inhibitory cells, respectively. $\psi(\Delta\theta)$ is an even function modeling inhibition within a hypercolumn and decrease with $|\Delta\theta|$, $g_x(), g_y$ are sigmoid-like nonlinear and nondecreasing functions.

The self-excitatory connection is $J_0$ is set equal to 0.8. The long-range synaptic connections $J_{i\theta,j\theta}$ and $W_{i\theta,j\theta}$ are determined as follows. Assume the distance between two edge elements $i\theta$ and $j\theta$ is $d$, the angles between the edge elements and the line connecting them are $\theta_1$ and $\theta_2$, $|\theta_1| \leq |\theta_2| \leq \pi/2$, and $\theta_{1,2}$ are positive or negative depending on whether the edges rotate clockwise or counterclockwise toward the connecting line in no more than a $\pi/2$ angle. Denote $\beta = 2|\theta_1| + 2\sin(|\theta_1 + \theta_2|)$, $\Delta\theta = \theta - \theta$ with $|\Delta\theta| \leq \pi/2$, then the synaptic connections are calculated by:

$$J_{i\theta,j\theta'} = \begin{cases} 0.126 e^{(-\beta/d)^2 - 2(\beta/d)^7 - d^2/90} & \text{if } 0 < d \leq 10 \text{ and } \beta < \pi/2.69 \\ & \text{or } 0 < d \leq 10.0 \text{ and } \beta < \pi/1.1 \\ & \text{and } |\theta_{1,2}| < \pi/5.9, \\ 0 & \text{otherwise.} \end{cases}$$

$$W_{i\theta,j\theta'} = \begin{cases} 0 & \text{if } d = 0 \text{ or } d \geq 10 \text{ or } \beta < \frac{\pi}{1.1} \\ & \text{or } |\Delta\theta| \geq \frac{\pi}{3} \text{ or } |\theta_1| < \frac{\pi}{11.999}, \\ 0.14(1 - e^{-0.4(\beta/d)^{1.5}}) e^{-(\Delta\theta/(\pi/4))^{1.5}} & \text{otherwise.} \end{cases}$$

## 5.7 Scene Segmentation Based on Temporal Correlation

Segmentation of nontrivial images is one of the most difficult tasks in image processing, which has a lot of applications, such as industrial inspection, autonomous target acquisition and so on. Segmentation subdivides an image into its constituent objects (or regions), and the level to which the subdivision is carried depends on the problem being solved. There has been numerous image segmentation algorithms in image processing discipline, generally based on one of two basic properties of intensity values: discontinuity and similarity (Gonzalez and Woods, 2002). However, it is not a trivial task to characterize the outcome of segmentation using a neural network.

Neural networks have proven to be a successful approach to pattern recognition. Unfortunately, little work has been devoted to scene segmentation which is generally regarded as a part of the preprocessing (often meaning manual segmentation) stage. Scene segmentation is a particularly challenging task for neural networks, partly because traditional neural networks lack the representational power for encoding multiple objects simultaneously. Apart from the spatial coding proposals like the CLM (Wersing et al., 2001a), another proposed approach relies on temporal correlation to encode the feature binding and segmentation (von der Malsburg, 1973; von der Malsburg, 1981; von der Malsburg and Buhmann, 1992). The correlation theory of von der Malsburg (1981) asserts that an object is represented by the temporal correlation of the firing activities of the scattered cells that encode different features of the object. Multiple objects are represented by different correlated firing patterns that alternate in time, each corresponding to a single object. Temporal correlation offers an elegant way of representing multiple objects in neural networks. On the other hand, the discovery of synchronous oscillations in the visual cortex has triggered much interest in exploring oscillatory correlation to solve the problems of segmentation and figure-ground segregation.

A special form of temporal correlation is oscillatory correlation. In (Wang and Terman, 1997), the segmentation is implemented on the basis of locally excitatory globally inhibitory oscillator networks (LEGION). The basic unit of LEGION is a neural oscillator $i$ which consists of an excitatory cell $x_i$ and an inhibitory cell $y_i$, and the dynamics is described as:

$$\dot{x}_i = 3x_i - x_i^3 + 2 - y_i + I_i H\left(p + \exp(-\alpha t) - \theta\right) + S_i + \rho \quad (5.21a)$$

$$\dot{y}_i = \epsilon \cdot \left(\gamma(1 + \tanh(\frac{x_i}{\beta})) - y_i\right) \quad (5.21b)$$

Here $H$ stands for the Heaviside step function, which is defined as $H(v) = 1$ if $v \geq 0$ and $H(v) = 0$ if $v < 0$. $I_i$ represents external stimulation which is assumed to be applied from time 0 on, and $S_i$ denotes the coupling from other oscillators in the network. $\rho$ denotes the amplitude of Gaussian noise. $\epsilon$ is a small positive constant, $\gamma, \beta$ also constants. The coupling term $S_i$ is given by:

$$S_i = \sum_{k \in N(i)} W_{ik} H(x_k - \theta_x) - W_z H(z - \theta_{xz}) \quad (5.22)$$

where $W_{ik}$ is the connection from $k$ to $i$, $N(i)$ is the neighborhood of $i$.

The algorithm to implement the LEGION is summarized as follows:

**Algorithm 5.2** *The LEGION Algorithm for Sensory Segmentation*

1. *Initialize*
   *1.1 Set $z(0) = 0$;*
   *1.2 Form effective connections*

   $$W_{ij} = I_M/(1 + |I_i - I_k|), k \in N(i)$$

   *1.3 Find leaders*

   $$P_i = H(\sum_{k \in N(i)} W_{ik} - \theta_p)$$

## 5.7 Scene Segmentation Based on Temporal Correlation

*1.4 Place all the oscillators randomly on the left branch, i.e., $x_i(0)$ takes a random value between $LC_x$ and $LK_x$.*

2. *Find one oscillator $j$ satisfying (1) $x_j(t) > x_k(t)$, where $k$ is currently on the left branch, and (2) $P_j = 1$. Then*

$$x_j(t+1) = RK_x; z(t+1) = 1;$$
$$x_k(t+1) = x_k(t) + (LK_x - x_j(t)), \text{ for } k \neq j.$$

3. *Iterate until stop:*
   **if** $(x_i(t) = RK_x \text{ and } z(t) > z(t-1))$

$$x_i(t+1) = x_i(t)$$

   **else if** $(x_i(t) = RK_x \text{ and } z(t) \leq z(t-1))$

$$x_i(t+1) = LC_x; z(t+1) = z(t) - 1$$

   **if** $x_i(t+1) = 0$ *go to step 2.*
   **else**

$$S_i(t+1) = \sum_{k \in N(i)} W_{ik} H(x_k(t) - LK_x) - W_z H(z(t) - 0.5)$$

   **if** $S_i(t+1) > 0$

$$x_i(t+1) = RK_x; z(t+1) = z(t) + 1$$

   **else**

$$x_i(t+1) = x_i(t).$$

# 6

# Convergence Analysis of Discrete Time RNNs for Linear Variational Inequality Problem

[1]In this chapter, we study the convergence of a class of discrete recurrent neural networks to solve Linear Variational Inequality Problems (LVIPs). LVIPs have important applications in engineering and economics. Not only the networks exponential convergence for the case of positive definite matrix is established, but its global convergence for positive semidefinite matrice is also proved. Conditions are derived to guarantee the convergences of the network. Comprehensive examples are discussed and simulated to illustrate the results.

## 6.1 Introduction

Let $\Omega$ be a nonempty subset of $R^n$, $A$ be an $n \times n$ matrix and $b \in R^n$. The linear variational inequality problem (LVIP) is to find a vector $x^* \in D$ such that
$$(x - x^*)^T (Ax^* + b) \geq 0, \quad \forall x \in D. \tag{6.1}$$

LVIP is a very general problem. In fact, many problems such as linear equation, linear complementarity problem, linear and quadratic optimization problems, etc. can be described by LVIP. Thus, finding solutions for LVIP has importance not only in theory but also from a application-wise perspective.

If $A$ is a symmetric matrix, then the VIP$(D, A, b)$ is equivalent to the following quadratic optimization problem
$$\min \left\{ \frac{1}{2} x^T A x + x^T b \, \middle| \, x \in D \right\}.$$

In this chapter, we assume that the convex subset $D$ is defined by
$$D = \{x \in R^n | x_i \in [c_i, d_i], i = 1, \cdots, n\}$$

---

[1] Reuse of the materials of "Global exponential stability of discrete-time neural networks for constrained quadratic optimization", 56(1), 2004, 399–406, Neurocomputing, with permission from Elsevier.

where $c_i$ and $d_i$ are constants such that $c_i \leq d_i (i = 1, \cdots, n)$.

It has been proven that $x^* \in R^n$ is a solution of LVIP if and only if it satisfies the following equation

$$x^* = f\Big(x^* - \alpha(Ax^* + b)\Big). \tag{6.2}$$

Thus, solving the LVIP is equivalent to solve the above equation. In the following, we employ discrete recurrent neural networks, described as

$$x(k+1) = f\Big(x(k) - \Pi\left(Ax(k) + b\right)\Big) \tag{6.3}$$

for all $k \geq 0$, where $\Pi$ is a diagonal matrix.

The function $f(\theta) = (f_1(\theta), \cdots, f_n(\theta))^T$ is defined by

$$f_i(\theta) = \begin{cases} c_i, & \text{if } \theta < c_i \\ d_i, & \text{if } \theta > d_i \\ \theta, & \text{otherwise} \end{cases}$$

for $i = 1, \cdots, n$.

This model of discrete time recurrent neural networks has its advantages in digital computer simulations over continuous time models of neural networks. Moreover, it can be easily implemented in digital hardware.

## 6.2 Preliminaries

**Definition 6.1.** *For each $\alpha$, a vector $x^* \in R^n$ is called a equilibrium point of (1) if and only if*

$$x^* = f\Big(x^* - \alpha\left(Ax^* + b\right)\Big).$$

In (Pérez-Ilzarbe, 1998), it has been shown that if $A$ is a positive definite matrix, then (6.3) has a unique equilibrium point.

For each $x = (x_1, \cdots, x_n)^T \in R^n$, we denote $\|x\| = \sqrt{\sum_{i=1}^{n} x_i^2} = \sqrt{x^T x}$.

**Definition 6.2.** *The neural network (6.3) is said to be globally exponentially convergent, if the network possesses a unique equilibrium $x^*$ and there exist constants $\eta > 0$ and $M \geq 1$ such that*

$$\|x(k+1) - x^*\| \leq M \|x(0) - x^*\| \exp(-\eta k)$$

*for all $k \geq 0$. The constant $\eta$ is called a lower bound of the convergence rate of the network 6.3).*

**Lemma 6.3.** *We have*

$$[f(x) - f(y)]^T [f(x) - f(y)] \leq [x - y]^T [x - y]$$

*for all $x, y \in R^n$.*

The proof of this lemma can be derived directly from the definition of the function $f_c^d$. Details are hence omitted.

## 6.3 Convergence Analysis: $A$ is a Positive Semidefinite Matrix

In this section, we will consider the LVIP when $A$ is a positive semidefinite matrix. We will use the following neural network to solve this problem. The neural network is as follows

$$x(k+1) = f\Big(x(k) - \alpha(Ax(k) + b)\Big) \tag{6.4}$$

where $\alpha > 0$ is some constant.

Let

$$y(k) = x(k) - \alpha(Ax(k) + b)$$

then, we have $x(k+1) = f(y(k))$, that is

$$x_i(k+1) = f_i\Big(y_i(k)\Big), (i = 1, \cdots, n) \tag{6.5}$$

for all $k \geq 0$.

For any $x_0 \in R^n$, we will use $x(k, x_0)$ to denote the solution of (6.4) starting from $x_0$.

Constructing a function as follows

$$\begin{aligned}E(x(k)) = &-\frac{1}{2}x^T(k+1)(I - \alpha A)x(k+1) \\ &- \alpha x^T(k+1)b + x^T(k+1)y(k) \\ &- \sum_{j=1}^{n} \int_0^{y_i(k)} f_i(s)ds\end{aligned}$$

for all $k \geq 0$. This function will play an important role for analyzing the network's convergence.

Since $f$ is bounded, it is easy to see that $E(x(k))$ is bounded for all $k \geq 0$.

**Lemma 6.4.** *If* $0 < \alpha \leq \frac{1}{\lambda_{max}(A)}$, *then we have*

$$\begin{aligned}&E(x(k+1)) - E(x(k)) \\ &\leq \sum_{j=1}^{n} \int_{y_i(k)}^{y_i(k+1)} \Big[f_i(y(k)) - f_i(s)\Big]ds \\ &\leq 0\end{aligned}$$

*for all* $k \geq 0$.

*Proof.* Denote

$$z(k+1) = x(k+2) - x(k+1)$$

for all $k \geq 0$.

Since $A$ is positive semidefinite and $0 < \alpha \leq \frac{1}{\lambda_{max}(A)}$, then $(I - \alpha A)$ is also a positive semidefinite matrix.

It follows that

$$E(x(k+1)) - E(x(k))$$
$$= -\frac{1}{2} z^T(k+1)(I - \alpha A)z(k+1)$$
$$+ x^T(k+2)y(k+1) - x^T(k+1)y(k)$$
$$- z^T(k+1)y(k+1) - \sum_{j=1}^{n} \int_{y_i(k)}^{y_i(k+1)} f_i(s) ds$$
$$\leq x^T(k+1)[y(k+1) - y(k)]$$
$$- \sum_{j=1}^{n} \int_{y_i(k)}^{y_i(k+1)} f_i(s) ds$$
$$= \sum_{j=1}^{n} \int_{y_i(k)}^{y_i(k+1)} \left[ x_i(k+1) - f_i(s) \right] ds$$
$$= \sum_{j=1}^{n} \int_{y_i(k)}^{y_i(k+1)} \left[ f_i(y(k)) - f_i(s) \right] ds$$
$$\leq 0$$

for all $k \geq 0$. This completes the proof.

Denote

$$\Omega^e = \left\{ x^e \in R^n | x^e = f\left(x^e - \alpha(Ax^e + b)\right) \right\}.$$

**Theorem 6.5.** *If* $0 < \alpha \leq \frac{1}{\lambda_{max}(A)}$, *then for any* $x_0 \in R^n$, *we have*

$$\lim_{k \to +\infty} x(k, x_0) = x^e \in \Omega^e.$$

*Proof.* For any $x_0 \in R^n$, let $x(k, x_0)$ be the solution of (6.4) starting from $x(0)$. Obviously, $x(k, x_0) \in D$ for all $k \geq 1$. Let $\Omega(x_0)$ be the $\Omega - limit$ of $x(k, x_0)$, it is easy to see that $\Omega(x_0) \neq \phi$ and $\Omega(x_0) \subseteq D$. Since

$$E(x(k+1, x_0)) \leq E(x(k, x_0))$$

for all $k \geq 0$, together with $x(k, x_0) \in D$ for all $k \geq 0$, it follows that

$$\lim_{k \to +\infty} E(x(k, x_0)) = E_0 < +\infty.$$

For any $x^* \in \Omega(x_0)$, let $x(k, x^*)$ be the solution of (6.4) starting form $x^*$. Obviously, $x(k, x^*) \in \Omega(x_0)$ for all $k \geq 0$. By the continuity of the function of $E$, we have $E(x(k, x^*)) = E_0$ for all $k \geq 0$. Then, it follows that

$$\sum_{i=1}^{n} \int_{y_i(k)}^{y_i(k+1)} \Big[f_i(y(k)) - f_i(s)\Big] ds = 0$$

for all $k \geq 0$. Since each $f_i (i = 1, \cdots, n)$ is an nondecreasing function, then it must have $y(k+1) = y(k)$ for all $k \geq 0$, and so

$$x(k+1, x^*) = x(k, x^*) = x^*$$

for all $k \geq 0$. This shows that $x^*$ is actually an equilibrium point of (6.4). Then, $\Omega(x_0) \subseteq \Omega^e$. It is easy to see that $\Omega(x_0) = \{x^*\}$, then it follows that

$$\lim_{k \to +\infty} x(k, x_0) = x^*.$$

This completes the proof.

## 6.4 Convergence Analysis: $A$ is a Positive Definite Matrix

In this section, we consider the convergence of (6.3) under the assumption that $A$ is a positive definite matrix.

Since $A$ is a positive definite matrix, then all of its eigenvalues are positive. Let $\lambda_i > 0 (i = 1, \cdots, n)$ be all the eigenvalues of $A$. Denote $\lambda_{min}$ and $\lambda_{max}$ the smallest and largest eigenvalues of $A$, respectively.

Define a continuous function

$$r(\alpha) = \max_{1 \leq i \leq n} |1 - \alpha \lambda_i|$$

for $\alpha > 0$. Obviously, $r(\alpha) > 0$ for all $\alpha > 0$. This function will play an important role in the estimation of the convergence rate of the network. Let us first attain some insight into this function.

Denote

$$\alpha^* = \frac{2}{\lambda_{max} + \lambda_{min}}$$

obviously, $\alpha^* > 0$.

**Theorem 6.6.** *We have*

(1).
$$r(\alpha) = \begin{cases} 1 - \lambda_{min}\alpha, & 0 < \alpha \leq \alpha^* \\ \lambda_{max}\alpha - 1, & \alpha^* \leq \alpha < +\infty. \end{cases}$$

(2). $r(\alpha) < 1$ if and only if

$$\alpha \in \left(0, \frac{2}{\lambda_{max}}\right).$$

(3).
$$\frac{\lambda_{max} - \lambda_{min}}{\lambda_{max} + \lambda_{min}} = r(\alpha^*) \leq r(\alpha)$$

for all $\alpha > 0$.

*Proof.* By the definition of $r(\alpha)$, it is easy to see that

$$r(\alpha) = \max\left\{|1-\alpha\lambda_{\min}|, |1-\alpha\lambda_{\max}|\right\}.$$

Since

$$|1-\alpha^*\lambda_{min}| = \frac{\lambda_{max}-\lambda_{min}}{\lambda_{max}+\lambda_{min}} = |1-\alpha^*\lambda_{max}|$$

then by a simple calculation it follows that

$$r(\alpha) = \begin{cases} 1-\lambda_{min}\alpha, & 0 < \alpha \leq \alpha^* \\ \lambda_{max}\alpha - 1, & \alpha^* \leq \alpha < +\infty. \end{cases}$$

Using this expression, it is easy to derive the results of the theorem. Details are omitted and the proof is completed.

Fig. 6.1 shows an intuitive explanation about the above theorem. It is easy to see that $\alpha^*$ is the minimum point of $r(\alpha)$, and $r(\alpha) < 1$ if and only if $0 < \alpha < 2/\lambda_{max}$.

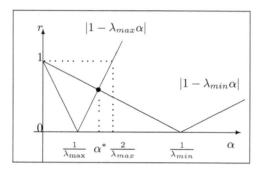

**Fig. 6.1.** The function $r(\alpha)$.

**Theorem 6.7.** *For each $\alpha$, if*

$$0 < \alpha < \frac{2}{\lambda_{max}}$$

*then the network (6.3) is globally exponentially converge with a lower bound of convergence rate*

$$\eta(\alpha) = -\ln r(\alpha) > 0.$$

*Proof.* Since $\lambda_i (i=1,\cdots,n)$ are all the eigenvalues of $A$. Then, it is easy to see that $(1-\alpha\lambda_i)^2 (i=1,\cdots,n)$ are all the eigenvalues of the matrix $(I-\alpha A)^2$.

For any $x(0) \in R^n$, let $x(k)$ be the solution (6.3) starting from $x(0)$. Since $\alpha > 0$, then the network has a unique equilibrium $x^*$. Using the previous given lemmas, it follows that

$$\begin{aligned}
\|x(k+1) - x^*\|^2 &= \left\| f\Big(x(k) - \alpha\,(Ax(k) + b)\Big) \right. \\
&\quad \left. - f\Big(x^* - \alpha\,(Ax^* + b)\Big) \right\|^2 \\
&\leq \|(I - \alpha A)(x(k) - x^*)\|^2 \\
&= [x(k) - x^*]^T (I - \alpha A)^2 [x(k) - x^*] \\
&\leq \max_{1 \leq i \leq n} \left[(1 - \alpha \lambda_i)^2\right] \|x(k) - x^*\|^2 \\
&= r(\alpha)^2 \|x(k) - x^*\|^2
\end{aligned}$$

for all $k \geq 0$. That is

$$\begin{aligned}
\|x(k+1) - x^*\| &\leq r(\alpha) \|x(k) - x^*\| \\
&= r(\alpha)^k \|x(0) - x^*\|
\end{aligned}$$

for all $k \geq 0$, and so

$$\|x(k+1) - x^*\| \leq \exp(-\eta(\alpha)k)\|x(0) - x^*\|$$

for all $k \geq 0$, where

$$\eta(\alpha) = -\ln r(\alpha).$$

Since $0 < \alpha < 2/\lambda_{max}$, $\eta(\alpha) > 0$ for each $\alpha$. The above shows that the network is globally exponentially convergent and $\eta(\alpha) > 0$ is a lower bound of global exponential convergence rate. This completes the proof.

## 6.5 Discussions and Simulations

In order to guarantee that the network has a large convergence rate, one needs to choose the parameter $\alpha$ in the network (6.3) so that $r(\alpha)$ to be as small as possible. From the theorems in last section, we know that $\alpha^*$ is the minimum point of the function $r(\alpha)$. Hence, we can choose $\alpha^*$ as the parameter $\alpha$ in the network (6.3). In this way, the network will be guaranteed to have a large lower bound of global exponential convergence rate. However, if we choose $\alpha^*$ as the parameter, we need to calculate the largest and smallest eigenvalues of the matrix of $A$ in advance. This will introduce additional computations especially when the dimension of the matrix $A$ is very high. On the other hand, from the implementation point of view, the network would be difficult to be implemented if it contains some eigenvalues in the network. However, since

$$tra(A) = \sum_{i=1}^{n} a_{ii} > \lambda_{max}$$

when $n \geq 2$, we have

$$\frac{2}{tra(A)} \in \left(0, \frac{2}{\lambda_{max}}\right)$$

for $n \geq 2$. As $n = 1$ is a trivial and very special case, we do only consider the situation of $n \geq 2$. Then, we can choose the parameter $\alpha$ in the network (3) by

$$\alpha = tra(A) = \sum_{i=1}^{n} a_{ii}.$$

Hence, the network (6.3) becomes

$$x(k+1) = f\left(x(k) - \frac{2}{tra(A)}(Ax(k) + b)\right) \qquad (6.6)$$

for all $k \geq 0$. Obviously, the network has a lower bound convergence rate

$$\eta = -\ln r\left(\frac{2}{tra(A)}\right) > 0.$$

It is quite easy to be implemented.

In order to further improve the numerical stability and convergence speed of the network, we can use the preconditioning technique in (Pérez-Ilzarbe, 1998). The aim of the preconditioning technique is to make the difference between the largest and smallest eigenvalues of the matrix $A$ as small as possible. This method is as follows. First, to define a diagonal matrix $P$ with the diagonal elements $p_{ii} = 1/\sqrt{a_{ii}}(i = 1, \cdots, n)$. Then, do some transformations such that

$$\tilde{A} = PAP, \quad \tilde{b} = Pb, \quad \tilde{c} = P^{-1}c, \quad \tilde{d} = P^{-1}d.$$

Since $tra(\tilde{A}) = n$, the network (4) becomes

$$\begin{cases} \tilde{x}(k+1) = f\left(\tilde{x}(k) - \frac{2}{n}(\tilde{A}\tilde{x}(k) + \tilde{b})\right) \\ x(k+1) = P\tilde{x}(k+1) \end{cases} \qquad (6.7)$$

for $k \geq 0$.

Next, we describe the algorithm of using the network to find the minimizer of the quadratic optimization problem. Firstly, we apply the preconditioning technique to the quadratic optimization problem. Secondly, we randomly choose an initial value. Since the network is globally exponentially convergent, any trajectory of the network starting from any point will converge to the minimizer of (6.1). Usually, choosing the origin as an initial value is recommended. Finally, set a error estimation parameter $\epsilon > 0$ and then ran the network until the distance of $x(k+1)$ and $x(k)$ is less than $\epsilon$ for some $k$.

*Example 1*: Consider the quadratic optimization problem of (6.1) with

## 6.5 Discussions and Simulations

$$A = \begin{bmatrix} 0.180 & 0.648 & 0.288 \\ 0.648 & 2.880 & 0.720 \\ 0.288 & 0.720 & 0.720 \end{bmatrix}$$

and

$$b = \begin{bmatrix} 0.4 \\ 0.2 \\ 0.3 \end{bmatrix}, \quad c = -d = \begin{bmatrix} -20 \\ -20 \\ -20 \end{bmatrix}$$

This example was studied in (Pérez-Ilzarbe, 1998) by using the neural network form (6.3) and it was shown that the network is globally convergent for this example but it was no known whether it is globally exponentially converge or not. We will use our network in this example to show it is globally exponentially convergent.

By applying the preconditioning technique, we have

$$\tilde{A} = \begin{bmatrix} 1.0000 & 0.9000 & 0.8000 \\ 0.9000 & 1.0000 & 0.5000 \\ 0.8000 & 0.5000 & 1.0000 \end{bmatrix}$$

and

$$\tilde{b} = \begin{pmatrix} 0.9428 \\ 0.1179 \\ 0.3536 \end{pmatrix}, \tilde{c} = \begin{pmatrix} -8.4853 \\ -33.9411 \\ -16.9706 \end{pmatrix}, \tilde{d} = \begin{pmatrix} 8.4853 \\ 33.9411 \\ 16.9706 \end{pmatrix}$$

We set a error estimation $\epsilon = 0.0001$ and randomly select a point $(14.6673, 7.5177, 0.1975)^T$ as an initial value and then ran the network. The network takes 46 steps to converge exponentially to the minimizer point $x^* = (-20.0000, 3.3796, 4.2038)^T$ with the above error estimation. See Fig. 6.2 for the convergence illustration.

Next, we randomly select 50 points in $R^3$ as the initial values of the network. The network globally exponentially converges to the minimizer point $x^* = (20.0000, -3.3790, -4.2049)^T$ with a error estimation of 0.0001. Fig. 6.3 shows the spatial global exponential convergence of the network.

*Example 2*: Consider the quadratic optimization problem of (1) with bound constraints:
$$\min\{E(x) = (x_1 - 2x_2)^2 : -1 \leq x_1, x_2 \leq 1\} \tag{6.8}$$

it's obvious that
$$A = \begin{bmatrix} 2 & -4 \\ -4 & 8 \end{bmatrix}$$

and
$$b = \begin{bmatrix} 0 \\ 0 \end{bmatrix}, \quad c = -d = \begin{bmatrix} -1 \\ -1 \end{bmatrix}$$

This example was studied in (Bouzerdoum and Pattison, 1993) by using the continuous neural network. We will use our network to this example to show it is globally exponentially converge.

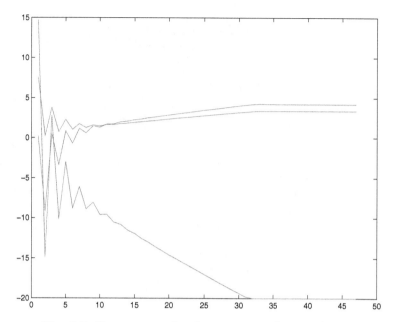

**Fig. 6.2.** Convergence of each component of one trajectory.

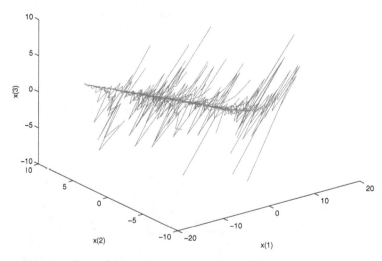

**Fig. 6.3.** Spatial convergence of the network with 50 trajectories.

By applying the preconditioning technique, we have

$$\tilde{A} = \begin{bmatrix} 1.0000 & -1 \\ -1 & 1.0000 \end{bmatrix}$$

and

$$\tilde{b} = \begin{pmatrix} 0 \\ 0 \end{pmatrix}, \tilde{c} = \begin{pmatrix} -1.4142 \\ -2.8284 \end{pmatrix}, \tilde{d} = \begin{pmatrix} 1.4142 \\ 2.8284 \end{pmatrix}$$

We set an error estimation $\epsilon = 0.0001$ and randomly select 30 points as initial values and then ran the network (5). It can be seen that all the trajectories from the initial point converge exponentially to the minimal $E = 0$. See Fig. 6.4 for the convergence illustration.

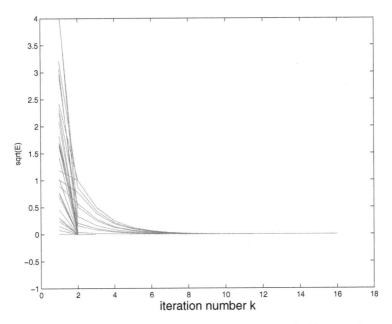

**Fig. 6.4.** Convergence of random initial 30 points in Example 2.

Next, to observe the global convergence behavior of the network, we randomly select 300 points generated uniformly in $R^2$ as the initial states of the network. The spatial representation of these trajectories is shown in Fig. 6.5. Its obvious that all the trajectories starting from the inside and the outside of $\Omega$ converge to the attractive set $\Omega^*$ of the network for this optimization problem, i.e., $\Omega^* = \{x \in \Omega \mid x_1 = 2x_2\}$, is an attractive set of the network. Fig. 6.5 shows the spatial global exponential convergence of the network.

*Example 3*: Consider the following bound-constrained quadratic minimization problem:

$$\min \{E(x) : -1 \leq x_i \leq 1 \forall i\} \tag{6.9}$$

with the following positive semidefinite $A$ matrix

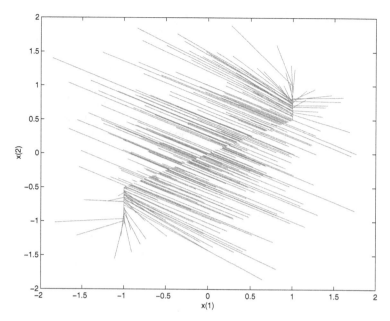

**Fig. 6.5.** Spatial convergence of the network with 300 trajectories.

$$A = \begin{bmatrix} 2 & 2 & 1 \\ 2 & 2 & 1 \\ 1 & 1 & 2 \end{bmatrix}$$

which is associated with the eigenvalues of $0, 1.2679, 4.7321$. When the vector $b$ is give by

$$b = \begin{bmatrix} -0.5 \\ 0.5 \\ -0.5 \end{bmatrix}$$

For a point to correspond to this constrained minimum it is a necessary and sufficient condition that it satisfies the Kuhn-Tucker optimality conditions given by

$$\frac{\partial E}{\partial x_i}\Big|_{x_i = x_i^*} \begin{cases} \geq 0 \text{ if } x_i^* = c_i \\ = 0 \text{ if } x_i^* \in (c_i, d_i) \\ \leq 0 \text{ if } x_i^* = d_i \end{cases} \quad (6.10)$$

for all $i$. Because $Ax + b = 0$ has no solutions in global area, from the Kuhn-Tucker optimality conditions it can be concluded that the minimum must exist in one of the edges of the cuboid determined by the bounds $c_i \leq x_i \leq d_i$. we have a single constrained minimum $x^* = [1 \ -1 \ 0.25]^T$. Fig. 6.6 illustrating the neural network solution to this problem with randomly selecting 100 points.

Now if the vector $b$ is given by

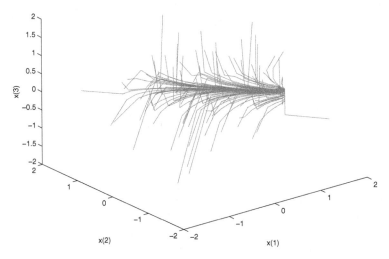

**Fig. 6.6.** Convergence of the network (5) in Example 3.

$$b = \begin{bmatrix} -0.5 \\ -0.5 \\ 0.5 \end{bmatrix}$$

We note that $Ax+b = 0$ has infinite solutions, so there exist an infinite number of minima within the bounds, i.e., there are an infinite number of constrained minima. The neural network solution yields a different equilibrium point, and the different constrained minimum can be obtained depending on the initial states. We run the network with randomly selecting 200 initial points to show the results. Figs. 6.7-6.9 give the spatial representation of neural network solution trajectories in the plane $(x_1, x_2)$, $(x_1, x_3)$ and $(x_2, x_3)$, respectively with randomly selecting 200 starting points. Fig. 6.10 shows the trajectories convergence in 3-D space.

*Example 4*: To illustrate our network's global convergence, we randomly generate a $7 \times 7$ positive semidefinite matrix $A$ with two zero eigenvalues. Then consider the following bound-constrained quadratic minimization problem of (1) with

$$\begin{bmatrix} 1.1328 & 0.6474 & 0.7573 & 0.8870 & 1.1265 & 1.2224 & 0.7975 \\ 0.6474 & 1.5935 & 0.8961 & 1.2513 & 1.2763 & 1.6427 & 1.3190 \\ 0.7573 & 0.8961 & 1.4415 & 1.4391 & 1.2856 & 1.7246 & 0.6725 \\ 0.8870 & 1.2513 & 1.4391 & 2.0326 & 1.6573 & 2.4805 & 1.1611 \\ 1.1265 & 1.2763 & 1.2856 & 1.6573 & 1.8291 & 2.2310 & 1.3451 \\ 1.2224 & 1.6427 & 1.7246 & 2.4805 & 2.2310 & 3.1369 & 1.6343 \\ 0.7975 & 1.3190 & 0.6725 & 1.1611 & 1.3451 & 1.6343 & 1.3384 \end{bmatrix}$$

94    6 Convergence Analysis of Discrete Time RNNs for LVIP

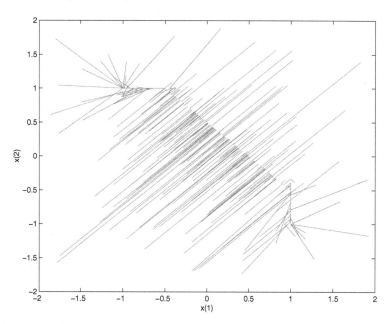

**Fig. 6.7.** Spatial representation of the trajectories in the plane $(x_1, x_2)$ in Example 3.

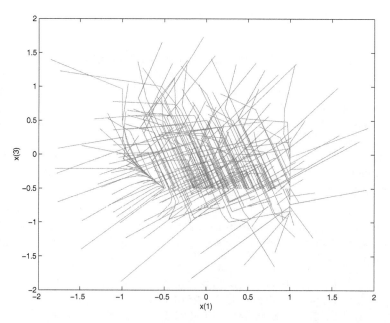

**Fig. 6.8.** Spatial representation of the trajectories in the plane $(x_1, x_3)$ in Example 3.

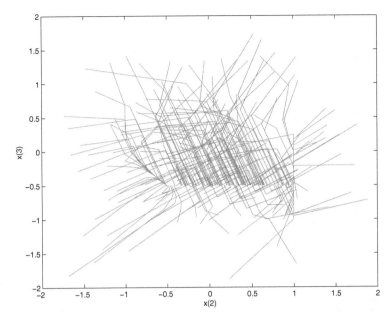

**Fig. 6.9.** Spatial representation of the trajectories in the plane $(x_2, x_3)$ in Example 3.

and

$$b = \begin{bmatrix} -0.2828 \\ -0.0594 \\ -0.2695 \\ -0.3892 \\ -0.3259 \\ 0.2476 \\ -0.3554 \end{bmatrix}, c = -d = 4 \begin{bmatrix} -1 \\ -1 \\ -1 \\ -1 \\ -1 \\ -1 \\ -1 \end{bmatrix}$$

The positive semidefinite matrix $A$ has two zero eigenvalues:

$$0, 0, 0.1395, 0.3964, 0.7147, 1.0481, 10.2060.$$

We choose the zero point as an initial state of the network, and then ran the network, where it converges to the minimizer

$$x^* = (-0.4727, -2.5063, 1.3459, 3.7015, -0.1440, -4.0000, 4.0000)^T$$

with 492 iteration steps with an error estimation $\epsilon = 0.0001$. Fig. 6.11 shows the convergence behavior for the components of the trajectories of the network starting from zero.

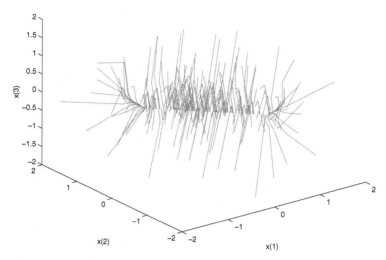

**Fig. 6.10.** Spatial representation of solution trajectories in 3-D space in Example 3.

## 6.6 Conclusions

In this chapter we have studied a discrete time recurrent neural network for the problem of linear variational inequality problem. We have considered two situations: 1). $A$ is positive semidefinite, 2). $A$ is positive definite. Convergence of the networks are analyzed and conditions are given for guaranteeing convergence. Simulation results illustrated the applicability of the proposed theory.

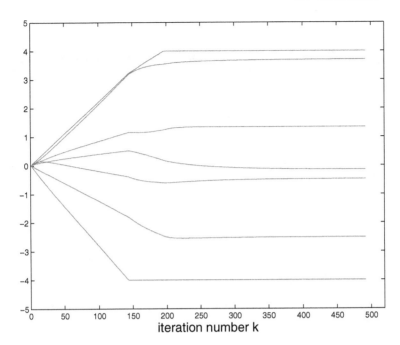

**Fig. 6.11.** Solution trajectories of each components in Example 4.

# 7

# Parameter Settings of Hopfield Networks Applied to Traveling Salesman Problems

## 7.1 Introduction

[1]Since the seminal work of Hopfield in using neural networks to solve the traveling salesman problem (TSP) (Hopfield and Tank, 1985), there has been an increasing interest in applying Hopfield-like neural networks to solve classical combinatorial problems (Aiyer et al., 1990; Peng et al., 1993). However, the difficulty in setting the network parameters often leads to the convergence of the network dynamics to invalid solutions, let alone optimal ones, which makes it a nontrivial work to find the optimal regions for the network parameters. Several methods for analyzing the convergence of the network to valid states have been proposed. Aiyer, et al. (Aiyer et al., 1990) analyzed the behavior of continuous Hopfield networks (CHNs) based on the eigenvalues of the connection matrix and derived the parameter settings for TSP. Abe (Abe, 1993) obtained the convergence and suppression conditions assuming a piecewise-linear activation function by comparing the values of the energy at the vertices of a unit hypercube. Peng et al (Peng et al., 1993) suggested the local minimum escape (LME) algorithm, which intends to improve the local minima problem of CHNs by combining a network disturbing technique with the Hopfield networks local minima searching property. Chaotic neural network provides another promising approach to solve TSP due to its global search ability and remarkable improvements with less local minima have been reported (Chen and Aihara, 1995; Hasegawa et al., 2002a; Nozawa, 1992; Wang and Smith, 1998; Wang and Tian, 2000). Cooper (Cooper, 2002) developed the higher-order neural networks (HONN) for mapping TSP and studied the stability conditions of valid solutions. It has been widely recognized that the H-T formulation (Hopfield and Tank, 1985) of the energy function is not ideal for the TSP, since the nature of the formulation causes infeasible solutions to

---

[1] Reuse of the materials of "On parameter settings of Hopfield networks applied to traveling salesman problems", 52(5), 2005, 994–1002, IEEE Trans. on Circuits and Systems-I, with permission from IEEE.

occur most of the time (Wilson and Pawley, 1988b). The inter-relationships among the parameters indicate that the H-T formulation for TSP does not have good scaling properties and only a very narrow range of parameter combinations that result in valid and stable solutions to TSP, which explains the unsatisfactory percentage of valid tours generated by approaches using the H-T formulation of the TSP (Kamgar-Parsi and Kamgar-Parsi, 1992), (Brandt et al., 1988). Therefore, it is believed that finding a better representation in mapping TSP onto CHN and searching for effective ways to optimally select the penalty parameters for the network are two important approaches, which have attracted many efforts in both areas with some useful results being reported (see (Smith, 1999) and the references therein). However, the valid solutions of the network obtained by these improvements often cannot be guaranteed. More recently, Talaván et al (Talaván and Yáñez, 2002a) presented a parameter setting procedure based on the stability conditions of CHN energy function and hence the analytical conditions which guaranteed the equilibrium points to be valid solutions, and the parameter settings developed illustrated a more promising performance. In order to avoid the poor scaling properties of the H-T formulation and to find an efficient method for selecting the optimal parameters, it is ideal to combine the two approaches and obtain a systematic way to set the parameters optimally. Inspired by the idea, this work employs an enhanced formulation of mapping TSP onto CHN. A systematic method is then developed to ensure the convergence of valid solutions and the suppression of spurious steady states of the network by investigating the dynamical stability of the network on a unit hypercube. Consequently, a theoretical method for selecting the penalty parameters is obtained. A brief description of TSP and CHN is given in Section 7.2. The enhanced formulation of mapping TSP is described in Section 7.3. The dynamical stability analysis of the network to represent valid solutions is analyzed in Section 7.4 and the suppression of spurious states is discussed in Section 7.5. A set of parameter setting criterion for various size of TSP is presented in Section 7.6. The simulations are given in Section 7.7 to illustrate the theoretical findings. Finally, conclusions are drawn in Section 7.8.

## 7.2 TSP Mapping and CHN Model

The traveling salesman problem, a classical combinatorial optimization problem widely studied by mathematicians and artificial intelligence specialists, can be formulated as follows: Given a set of $N$ cities, find the shortest path linking all the cities such that all cities are visited exactly once. This problem is known as NP (Nondeterministic Polynomial, a terminology of computational complexity) hard problem (Zhang, 2000), which is difficult to deal with or even intractable when the size of the problem, as measured by the number of cities, becomes large.

For $N$ cities, TSP can be mapped into a fully connected neural network consisting of $N \times N$ neurons, which are grouped into $N$ groups of $N$ neurons, where each group of the $N$ neurons represents the position in the tour of a particular city. The network's output is a vector of the form, for example, $\mathbf{v} = (0\ 1\ 0\ 1\ 0\ 0\ 0\ 0\ 1)^t$ or in a 2-D form of

$$\mathbf{v} = \begin{pmatrix} 0 & 1 & 0 \\ 1 & 0 & 0 \\ 0 & 0 & 1 \end{pmatrix}$$

which represents a tour of three cities. Let $u_{x,i}$ and $v_{x,i}$ denotes the current state and the activity (output) of the neuron $(x, i)$ respectively, where $x, i \in 1, \ldots, N$ is the city index and the visit order, respectively. Each neuron is also subjected to a bias of $i^b_{x,i}$. The strength of connection between neuron $(x, i)$ and neuron $(y, j)$ is denoted by $T_{xy,ij}$. Let $\mathbf{u}, \mathbf{v}, \mathbf{i^b}$ be the vectors of neuron states, outputs and biases, the dynamics of CHN can be described by a system of differential equations,

$$\frac{d\mathbf{u}}{dt} = -\frac{\mathbf{u}}{\tau} + T v + \mathbf{i^b}, \tag{7.1}$$

where $T = T^t \in \Re^{N^2 \times N^2}$ is a constant matrix and $\tau$ is a time constant. The activation function is a hyperbolic tangent

$$v_{x,i} = \phi(u_{x,i}) = \frac{1}{2}(1 + \tanh(\frac{u_{x,i}}{u_0})), u_0 > 0. \tag{7.2}$$

An energy function exists such that

$$\mathcal{E} = -\frac{1}{2}\mathbf{v}^t T v - (\mathbf{i^b})^t \mathbf{v} \tag{7.3}$$

and the existence of equilibrium states for the CHN can be ensured.

The Hopfield-Tank (H-T) formulation of the energy function mapping the TSP is described by

$$E = \frac{A_0}{2} \sum_x \sum_i \sum_{j \neq i} v_{x,i} v_{x,j} + \frac{B_0}{2} \sum_i \sum_x \sum_{y \neq x} v_{x,i} v_{y,i}$$
$$+ \frac{C_0}{2} (\sum_x \sum_i v_{x,i} - N)^2 + \frac{D_0}{2} \sum_x \sum_{y \neq x} \sum_i d_{xy} v_{x,i}(v_{y,i+1} + v_{y,i-1}), \tag{7.4}$$

where $d_{xy}$ is the distance from City $x$ to City $y$, and the scaling parameters $a, b, c, d$ are positive constants. The first and the second terms represent the constraint that at most one neuron of the array $\mathbf{v}$ activated at each row and column, respectively. The third term represents the constraint that the total number of neurons activated is exactly $N$. The fourth term measures the tour length corresponding to a given tour $\mathbf{v}$, where the two terms inside the parenthesis stand for two neighboring visiting cities of $v_{x,i}$ implying the tour length is calculated twice (it is for the convenience of mapping between (7.3) and (7.4), since a scaling constant $\frac{1}{2}$ appears in the energy functions). The energy function reaches a local minimum when the network is at a valid tour state.

With this formulation, the Hopfield network has the connection strengths and the external input given by

$$T_{xi,yj} = -\{A_0 \delta_{x,y}(1-\delta_{i,j}) + B_0(1-\delta_{x,y})\delta_{i,j} + C_0$$
$$+ D_0(\delta_{i,j-1} + \delta_{i,j+1})d_{x,y}\}, \tag{7.5}$$
$$i^b_{x,i} = C_0 N, \tag{7.6}$$

where $\delta_{i,j}$ is Kronecker's delta.

However, the H-T formulation does not work well and has produced some disappointing results when it is applied to TSP, since most of the time the network converges to infeasible solutions. To stabilize valid solutions, the formulation (7.4) was modified slightly in (Hopfield and Tank, 1985) and the works thereafter by replacing the bias $i^b = C_0 N$ with an effective bias $\tilde{i}^b = C_0 \tilde{N}$, where $\tilde{N}$ is larger enough than $N$. However, replacing $N$ with $\tilde{N}$ forces the network to violate the constraints, which implies that the potential of this modification is very limited. In fact, only a few percentage of trials yield valid tours and the percentage of valid solutions reaches a maximum for some value of $\tilde{N}$ and drops to zero when $\tilde{N}$ is further increased (Kamgar-Parsi and Kamgar-Parsi, 1992).

## 7.3 The Enhanced Lyapunov Function for Mapping TSP

From the mathematical programming view, the TSP can be described as a quadratic 0-1 programming problem with linear constraints,

$$\text{minimize } \mathcal{E}^{obj}(\mathbf{v}) \tag{7.7}$$

$$\text{subject to } S_i = \sum_{x=1}^{N} v_{xi} = 1 \quad \forall i \in \{1, \cdots, N\}, \tag{7.8}$$

$$S_x = \sum_{i=1}^{N} v_{xi} = 1 \quad \forall x \in \{1, \cdots, N\}, \tag{7.9}$$

and a redundant constraint $S = \sum_x \sum_i v_{x,i} = N$; $v_{xi} \in \{0,1\}$ and $\mathcal{E}^{obj}$ is the tour length described by a valid 0-1 solution $\mathbf{v}$.

Bearing this in mind, to map the TSP onto the Hopfield network, the Lyapunov function is set such that

$$\mathcal{E}^{lyap}(\mathbf{v}) = \mathcal{E}^{obj}(\mathbf{v}) + \mathcal{E}^{cns}(\mathbf{v}). \tag{7.10}$$

## 7.3 The Enhanced Lyapunov Function for Mapping TSP

The modified penalty function is given as

$$\mathcal{E}^{cns}(\mathbf{v}) = \frac{A}{2}\sum_x(\sum_i v_{x,i} - 1)^2 + \frac{B}{2}\sum_i(\sum_x v_{x,i} - 1)^2. \quad (7.11)$$

where the first term represents the constraint that each row has exactly one neuron activated, and the second term represents the constraint that each column has exactly one neuron on fire. The scaling parameters $A$ and $B$ play the role of balancing the constraints.

In order to force trajectories toward the vertices of the hypercube, it is necessary to add to $\mathcal{E}^{lyap}$ an additional penalty function of the form

$$\mathcal{E}^{drv}(\mathbf{v}) = \frac{C}{2}\sum_{x=1}^{N}\sum_{i=1}^{N}(1 - v_{x,i})v_{x,i} = \frac{C}{2}(\frac{N^2}{4} - \sum_{x=1}^{N}\sum_{i=1}^{N}(v_{x,i} - \frac{1}{2})^2), \quad (7.12)$$

which introduces a gradient component to drive the output states of the network outward from $\frac{1}{2}$, thus the trajectories will tend toward the vertices of the hypercube. Finally, the enhanced Lyapunov function for TSP is formulated as

$$\mathcal{E}^{lyap}(\mathbf{v}) = \frac{A}{2}\sum_x(\sum_i v_{x,i} - 1)^2 + \frac{B}{2}\sum_i(\sum_x v_{x,i} - 1)^2$$
$$+ \frac{C}{2}\sum_x\sum_i(1 - v_{x,i})v_{x,i}$$
$$+ \frac{D}{2}\sum_x\sum_{y\neq x}\sum_i d_{xy}v_{x,i}(v_{y,i+1} + v_{y,i-1}). \quad (7.13)$$

With this new formulation, the connection matrix and the external input of the Hopfield network are computed as follows,

$$T_{xi,yj} = -\{A\delta_{x,y} + B\delta_{i,j} - C\delta_{x,y}\delta_{i,j} + D(\delta_{i,j-1} + \delta_{i,j+1})d_{x,y}\}, \quad (7.14)$$
$$i^b_{x,i} = A + B - \frac{C}{2}. \quad (7.15)$$

Though a similar energy formulation was proposed in (Brandt et al., 1988) and (Tachibana et al., 1995), where $A$ and $B$ were simply set to 1 and only two parameters ($C$ and $D$) were allowed to change, there was a lack of a systematic way on setting the parameters. Some advantages of the modified Lyapunov function have been confirmed by Brandt et al (Brandt et al., 1988). As stated in their work, the new formulation has relatively sparse connections as contrast to the H-T formulation with $(4N - 3)N^2$ and $N^4$ connections, respectively. It is clear that the reduction of the connection weights will become more significant as the size of the TSP is increased. By a simple analysis method of dynamical stability, B. Kamgar-Parsi et al (Kamgar-Parsi and Kamgar-Parsi, 1992) showed that the H-T formulation is not stable, which explained the reason that Hopfield and Tank had to choose a larger value $\tilde{N}$ in their formulation instead of the city number $N$ itself. Following the same analysis approach, it is easy to show that the modified formulation is stable. Given the above preliminary investigation, it is preferable to choose the modified Lyapunov function for solving the TSP. In the following sections, by investigating the dynamical stability of the network, the theoretical results which provide an analytic method to optimally set the scaling parameters $A, B, C$ and $D$ are presented. As a consequence, the convergence to valid solutions can be assured at high percentages. Additionally, the new parameter setting scheme avoids the drawback that the connection weights depend on the distances between inter cities (Kamgar-Parsi and Kamgar-Parsi, 1992; Brandt et al., 1988), hence it is easy to apply the obtained results to various sizes of TSPs.

## 7.4 Stability Based Analysis for Network's Activities

In this section, conditions are derived so that all neurons activities are either firing at state 1 or off at state 0, i.e., the network output vector $\mathbf{v} = (v_{x,i})$ converges to the vertices satisfying the constraints. There are three main approaches on how to determine the convergence states of the Hopfield network, i.e., relying on the analysis of the eigenvalue of the connection matrix (Aiyer et al., 1990), comparing the values of the energy at vertices of the unit hypercube (Abe, 1993) and the first derivatives of the Lyapunov function for the stability analysis (Talaván and Yáñez, 2002a; Matsuda, 1998; Cooper, 2002).

The Lyapunov function of CHN mapping of TSP is a quadratic function of the neuron's activity $v_{x,i}$ defined on the vertices of a unit hypercube in $\Re^{N^2}$, therefore the stability of the equilibria of the network is determined by their derivatives. As the continuous variables of the Lyapunov function, the activity of the neuron $\mathbf{v} = (v_{x,i})$ is an equilibrium of the network if and only if $\frac{\partial \mathcal{E}^{lyap}}{\partial v_{x,i}} \geq 0 (\forall v_{x,i} = 0)$, $\frac{\partial \mathcal{E}^{lyap}}{\partial v_{x,i}} = 0 (\forall v_{x,i} \in (0,1))$ and $\frac{\partial \mathcal{E}^{lyap}}{\partial v_{x,i}} \leq 0 (\forall v_{x,i} = 1)$.

**Lemma 7.1.** *(Matsuda, 1998) The activity which converges to a vertex of the hypercube is asymptotically stable if it is verified*

$$\frac{\partial \mathcal{E}^{lyap}}{\partial v_{x,i}} = \begin{cases} > 0 (v_{x,i} = 0) \\ < 0 (v_{x,i} = 1) \end{cases} \quad (7.16)$$

*for all the components $(x, i)$. Conversely, it is unstable if*

$$\frac{\partial \mathcal{E}^{lyap}}{\partial v_{x,i}} = \begin{cases} < 0 (v_{x,i} = 0) \\ > 0 (v_{x,i} = 1) \end{cases} \quad (7.17)$$

*holds for all the components $(x, i)$.*

Hence, the properties of the equilibria of the network can be investigated by the first partial derivative of the Lyapunov function with respect to $v_{x,i}$. Let $\mathcal{E}_{x,i}(\mathbf{v}) = \frac{\partial \mathcal{E}^{lyap}}{\partial v_{x,i}}$ and from (7.13), it is obtained,

$$\mathcal{E}_{x,i}(\mathbf{v}) = A(S_x - 1) + B(S_i - 1) + \frac{C}{2}(1 - 2v_{x,i}) + D \sum_{y \neq x} d_{x,y}(v_{y,i-1} + v_{y,i+1}). \tag{7.18}$$

Define the lower bound and the upper bound of the derivative components of the neurons at a firing state and an off state, $\underline{\mathcal{E}}^0(\mathbf{v}) = \min_{v_{x,i}=0} \mathcal{E}_{x,i}(\mathbf{v})$ and $\bar{\mathcal{E}}^1(\mathbf{v}) = \max_{v_{x,i}=1} \mathcal{E}_{x,i}(\mathbf{v})$, respectively. Consequently, the inequalities of (7.16) is reformulated with a more compact form which gives stronger conditions for the stability,

$$\begin{cases} \underline{\mathcal{E}}^0(\mathbf{v}) > 0, \\ \bar{\mathcal{E}}^1(\mathbf{v}) < 0. \end{cases} \tag{7.19}$$

Let $d_L = \min_{x,y} d_{x,y}$ and $d_U = \max_{x,y} d_{x,y}$ be the lower bound and the upper bound of the Euclidean distance between any two cities.

## 7.5 Suppression of Spurious States

To represent a valid tour, there must be one and only one neuron firing in each row $x$ and each column $i$ of $\mathbf{v}$. In other words, the network activities representing a valid tour exist on the vertices of a unit hypercube. Define the hypercube by $H = \{v_{x,i} \in [0,1]^{N \times N}\}$, its vertex set $H_C = \{v_{x,i} \in \{0,1\}^{N \times N}\}$, then the valid tour set is $H_T = \{v_{x,i} \in H_C | S_x = 1, S_i = 1\}$. Obviously, $H_T \subset H_C \subset H$.

To ensure the network is not converging to invalid solutions (i.e., *spurious steady states*), all the spurious states have to be suppressed for both cases of $\mathbf{v} \in H_C - H_T$ and $\mathbf{v} \in H - H_C$.

**Definition 7.2.** *Given a vertex point* $\mathbf{v} \in H_C$, *its adjacent vertex point* $\mathcal{A}_{x,i}$ *is defined by changing only one element* $v_{x,i}$ *by its complementary* $1 - v_{x,i}$.

Hence, the adjacent vertices set of $\mathbf{v}$ can be defined by $\mathcal{A} = \mathcal{A}^+ \cup \mathcal{A}^-$, where

$$\mathcal{A}^+ = \{\mathcal{A}_{x,i} \in H_C | v_{x,i} = 0, \forall x, i \in \{1, \ldots, N\}\},$$

$$\mathcal{A}^- = \{\mathcal{A}_{x,i} \in H_C | v_{x,i} = 1, \forall x, i \in \{1, \ldots, N\}\}.$$

**Definition 7.3.** *Any point* $\mathbf{v} \in H$ *of the hypercube* $\Re^N$ *is an interior point* $\mathbf{v} \in H - H_C$ *if and only if a component* $v_{x,i} \in (0,1)$ *exists. An interior point* $\mathbf{v} \in H - H_C$ *becomes an edge point if*

$$|\{(x,i) \in \{1, 2, \cdots, N\}^2 | v_{x,i} \in (0,1)\}| = 1,$$

where $|\cdot|$ denotes the cardinal operation.

**Definition 7.4.** *Let* $\mathbf{v} \in H - H_C$ *be an edge point with* $v_{x^0,i^0} \in (0,1)$, *its adjacent vertex point* $\mathbf{v}' = (v'_{x,i}) \in H_C$ *is defined by increasing or decreasing the nonintegral component* $v_{x^0,i^0}$, *i.e.,*

$$v'_{x,i} = \begin{cases} 1 \text{ or } 0, & \text{if } (x,i) = (x^0, i^0), \\ v_{x,i}, & \text{otherwise.} \end{cases}$$

Fig. 7.1 gives an illustration of the vertex point, edge point and interior point.

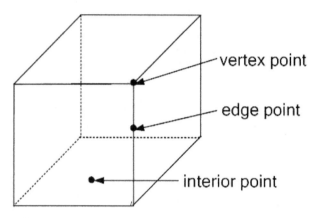

**Fig. 7.1.** Vertex point, edge point and interior point.

As pointed out by (Park, 1989), an interior point that is not an edge point has a null probability of being the convergence point of the continuous Hopfield neural networks. Therefore it needs to suppress the states of the network output being an edge point.

**Lemma 7.5.** *The vertex* $\mathbf{v} \in H_C - H_T$ *is unstable if the derivatives of the Lyapunov function satisfy*

$$\max_{S=N-1} \underline{\mathcal{E}}^0(\mathbf{v}) < 0 < \min_{S=N} \bar{\mathcal{E}}^1(\mathbf{v}) \tag{7.20}$$

*or*

$$\max_{S=N} \underline{\mathcal{E}}^0(\mathbf{v}) < 0 < \min_{S=N+1} \bar{\mathcal{E}}^1(\mathbf{v}) \tag{7.21}$$

*Proof.* Since the energy always increases if more than $N$ neurons fire, the spurious states (invalid tours) only occur when $N$ neurons or less fire. Let $k \in \{0, 1, 2, \cdots, N\}$, given an invalid tour $\mathbf{v}^k \in H_C - H_T$ such that $S = \sum_x \sum_i v^k_{x,i} = k$ and suppose

$$\underline{\mathcal{E}}^0(\mathbf{v}^k) = \max_{S=k} \underline{\mathcal{E}}^0(\mathbf{v}), \tag{7.22}$$

there exists a row $x'$ and a column $i'$ such that $v_{x',i'}^k = 0$, thus another invalid tour can be chosen from its neighborhood set $\mathcal{A}^+$, i.e., $\mathbf{v}^+ = \mathcal{A}_{x',i'}(\mathbf{v}^k)$. Obviously, $S = \sum_x \sum_i v_{x,i}^+ = k+1$ and

$$\max_{S=k+1} \underline{\mathcal{E}}^0(\mathbf{v}) \geq \underline{\mathcal{E}}^0(\mathbf{v}^+). \tag{7.23}$$

To show the derivation, 3 exhaustive cases are considered in the following.

Case 1 : $v_{x,i} = 1$, and $v_{x',i'}$ changes from 0 to 1, for $x' \neq x, i' \neq i$.
For example,

$$\begin{pmatrix} v_{x,i} & 0 & 0 & 0 \\ 0 & 0 & 1 & 0 \\ 0 & 0 & 0 & 1 \\ 0 & 0 & 0 & 0 \end{pmatrix} \longrightarrow \begin{pmatrix} v_{x,i} & * & a & * \\ * & \circ & \Diamond & \circ \\ b & d & c & d \\ * & \circ & \Diamond & \circ \end{pmatrix}.$$

The above matrix can be partitioned into 4 non-overlapped regions $a$, $b$, $c$ and $d$ according to their relationships with $v_{x,i}$. For example, region $a$ represents the row $(*, a, *)$. Region $d$ indicate either of its two consecutive columns of $v_{x,i}$, while region $c$ denotes a column that is not next to $v_{x,i}$. Observed from equation (7.18), the value of $\mathcal{E}_{x,i}$ can only be changed by the elements in regions $a$, $b$ and $d$. Thus, by applying equation (7.18), it can be determined how the energy derivative $\mathcal{E}_{x,i}$ changes. If $v_{x',i'}$ is in region $a$, then the first term in equation (7.18) will increase, due to the increment of $S_x$. If $v_{x',i'}$ is in region $b$, then the second term in equation (7.18) will increase, due to the increment of $S_i$. If $v_{x',i'}$ is in region $c$, then $\mathcal{E}_{x,i}$ remains the same. If $v_{x',i'}$ is in region $d$, then the fourth term in equation (7.18) will increase, due to the increment of either $v_{y,i-1}$ or $v_{y,i+1}$. As such,

$$\mathcal{E}_{x,i}(\mathbf{v}^+) \geq \mathcal{E}_{x,i}(\mathbf{v}^k), \quad \forall (x,i) \in \{1, 2, \cdots, N\}^2.$$

Case 2 : $v_{x,i} = 0$, and $v_{x',i'}$ changes from 0 to 1, for $x' \neq x, i' \neq i$.
Again, the energy derivative will increase due to parameter $A, B, D$, if $v_{x',i'}$ is in Region $a, b$ and $d$ respectively. It stays the same if $v_{x',i'}$ is in Region $c$. Thus,

$$\mathcal{E}_{x,i}(\mathbf{v}^+) \geq \mathcal{E}_{x,i}(\mathbf{v}^k), \quad \forall (x,i) \in \{1, 2, \cdots, N\}^2.$$

Case 3 : $v_{x,i}$ itself changes from 0 to 1.
In this case, by applying the energy derivative formulation (7.18), it can be shown that

$$\mathcal{E}_{x,i}(\mathbf{v}^k) = -A - B + \frac{C}{2} + D(d_{xy} + d_{yz}),$$

$$\mathcal{E}_{x,i}(\mathbf{v}^+) = -\frac{C}{2} + D(d_{xy} + d_{yz}).$$

To satisfy $\mathcal{E}_{x,i}(\mathbf{v}^+) \geq \mathcal{E}_{x,i}(\mathbf{v}^k)$, it holds that $A + B \geq C$. This criterion will be necessary for the non-convergence of valid solutions and must be considered when setting the parameters $A, B, C$.

Therefore, from the above three cases, it is observed that

$$\underline{\mathcal{E}}^0(\mathbf{v}^+) \geq \underline{\mathcal{E}}^0(\mathbf{v}^k). \tag{7.24}$$

Applying (7.22) and (7.23), for large enough $A, B$, it is obtained that

$$\max_{S=k+1} \underline{\mathcal{E}}^0(\mathbf{v}) \geq \underline{\mathcal{E}}^0(v^k) = \max_{S=k} \underline{\mathcal{E}}^0(\mathbf{v}), \quad \forall \mathbf{v} \in H_C - H_T. \tag{7.25}$$

It can be proven analogously that,

$$\min_{S=k+1} \bar{\mathcal{E}}^1(\mathbf{v}) \leq \min_{S=k} \bar{\mathcal{E}}^1(\mathbf{v}), \quad \forall \mathbf{v} \in H_C - H_T. \tag{7.26}$$

From the conditions (7.17) for instability, any spurious steady state, $\mathbf{v} \in H_C - H_T$ is unstable if the conditions (7.20) or (7.21) are satisfied. It completes the proof.

The next two propositions relate the derivative bounds to the network parameters.

**Proposition 7.6.** *Any invalid tour* $\mathbf{v} \in H_C - H_T$, $S = \sum_x \sum_i v_{x,i} = N - 1$, *the following bound is obtained*

$$\max_{S=N-1} \underline{\mathcal{E}}^0(\mathbf{v}) \leq -A - B + \frac{C}{2} + 3Dd_U.$$

*Proof.* For any point $\mathbf{v} \in H_C$, let $I_0$ denote the index set of its columns whose elements are equal to 0, i.e., $I_0 = \{i \in \{1, 2, \cdots, N\} : S_i = 0\}$ and $N_0 = |I_0|$ denotes the cardinal. Now consider $S = N - 1$, $I_0 \neq \emptyset$ and $N_0 \geq 1$, then $x'$ and $i'$ exist such that $v_{x',i'} = 0$ and $S_{x'} = 0$, e.g.,

$$\mathbf{v} = \begin{pmatrix} 0 & 0 & 0 & 0 \\ 1 & 0 & 0 & 1 \\ 0 & v_{x',i'} & 0 & 0 \\ 0 & 0 & 0 & 1 \end{pmatrix}.$$

Therefore, from (7.18),

$$\mathcal{E}_{x',i'}(\mathbf{v}) = -A - B + \frac{C}{2} + D \sum_{y \neq x} d_{x,y}(v_{y,i'-1} + v_{y,i'+1})$$

$$\leq -A - B + \frac{C}{2} + Dd_U(S_{i'-1} + S_{i'+1}).$$

Taking the minimum over all $i \in I_0$, we obtain

$$\mathcal{E}^0(\mathbf{v}) \leq -A - B + \frac{C}{2} + Dd_U \min_{i \in I_0}(S_{i-1} + S_{i+1}).$$

Applying the technical result (*Lemma A. 2*, (Talaván and Yáñez, 2002a)),

$$\min_{i \in I_0}(S_{i-1} + S_{i+1}) \leq 4 - \frac{2(N-S)}{N_0}. \tag{7.27}$$

and taking into account that $S_{i-1} + S_{i+1} \in \mathcal{N}$, the following is obtained,

$$\min_{i \in I_0}(S_{i-1} + S_{i+1}) \leq [4 - \frac{2(N-S)}{N_0}] \leq 3.$$

Hence it proves that

$$\mathcal{E}^0(\mathbf{v}) \leq -A - B + \frac{C}{2} + 3Dd_U.$$

**Proposition 7.7.** *Any invalid tour* $\mathbf{v} \in H_C - H_T$, $S = \sum_x \sum_i v_{x,i} = N$, *the following bound is obtained*

$$\min_{S=N} \bar{\mathcal{E}}^1(\mathbf{v}) \geq min\{B, A + Dd_L, (N-1)A\} - \frac{C}{2}.$$

*Proof.* To prove $\min_{S=N} \mathcal{E}^1(\mathbf{v}) \geq min\{B, A + Dd_L, (N-1)A\} - \frac{C}{2}$, the following 3 exhaustive cases for $v_{x,i} = 1$ are considered.

Case 1 : $\exists v_{x',i'} = 1 : S_{x'} \geq 1, S_{i'} \geq 2$, e.g.,

$$\mathbf{v} = \begin{pmatrix} 0 & 0 & 0 & 0 \\ 0 & 1 & 0 & 0 \\ 0 & 0 & 0 & 1 \\ 1 & 0 & 0 & 1 \end{pmatrix}.$$

From equation (7.18), it is derived

$$\mathcal{E}_{x',i'} = A(S_{x'} - 1) + B(S_{i'} - 1) - \frac{C}{2} + D \sum_{y \neq x} d_{x,y}(v_{y,i'-1} + v_{y,i'+1})$$

$$\geq B - \frac{C}{2}. \tag{7.28}$$

Case 2 : $\exists v_{x',i'} = 1 : S_{x'} \geq 2, S_{i'} = 1$, e.g.,

$$\mathbf{v} = \begin{pmatrix} 0 & 0 & 0 & 0 \\ 0 & 0 & 1 & 0 \\ 1 & 1 & 0 & 0 \\ 0 & 0 & 0 & 1 \end{pmatrix}.$$

Then

$$\mathcal{E}_{x',i'} = A(S_{x'} - 1) - \frac{C}{2} + Dd_{x,y}$$

$$\geq A - \frac{C}{2} + Dd_L. \qquad (7.29)$$

Case 3 : $\exists v_{x',i'} = 1 : S_{x'} = N, S_{i'} = 1$, e.g.,

$$\mathbf{v} = \begin{pmatrix} 0 & 0 & 0 & 0 \\ 0 & 0 & 0 & 0 \\ 1 & 1 & 1 & 1 \\ 0 & 0 & 0 & 0 \end{pmatrix}.$$

It follows that

$$\mathcal{E}_{x',i'}(\mathbf{v}) \geq A(N-1) - \frac{C}{2}. \qquad (7.30)$$

Considering (7.28)-(7.30), it holds

$$\min_{S=N} \mathcal{E}^1(\mathbf{v}) \geq \min\{B, A + Dd_L, A(N-1)\} - \frac{C}{2}.$$

This completes the proof.

**Theorem 7.8.** *Any invalid tour* $\mathbf{v} \in H_C - H_T$ *is not a stable equilibrium point for CHN if the following criteria are satisfied,*

$$\min\{B, A + Dd_L, (N-1)A\} - \frac{C}{2} > 0, \qquad (7.31)$$

$$-A - B + \frac{C}{2} + 3Dd_U < 0, \qquad (7.32)$$

$$A + B \geq C \qquad (7.33)$$

*Proof.* It is already obtained that $A + B \geq C$ from the proof of Lemma 7.5. Then by applying Lemma 7.5 and the above two propositions, the rest of the conditions can be obtained.

**Lemma 7.9.** *Suppose the edge point* $\mathbf{v} \in H - H_C$ *with* $v_{x^0,i^0} \in (0,1)$ *an equilibrium of the CHN with* $T_{xi,xi} < 0 \ \forall (x,i) \in \{1,\ldots,N\}^2$, *then it is unstable if the following condition is satisfied,*

$$\mathcal{E}_{x,i} \notin (0, A + B - C), \quad \forall (x,i)|v_{x,i} = 1 \qquad (7.34)$$

or

$$\mathcal{E}_{x,i} \notin (-A - B + C, 0), \quad \forall (x,i)|v_{x,i} = 0. \qquad (7.35)$$

## 7.5 Suppression of Spurious States

*Proof.* According to the result in (Talaván and Yáñez, 2002a), the stability of any edge point will be avoided if

$$\mathcal{E}_{x,i}(\mathbf{v}') \notin (0, -T_{xi,xi}), \quad \forall (x,i) | v'_{x,i} = 1, \forall \mathbf{v}' \in H_C \tag{7.36}$$

or

$$\mathcal{E}_{x,i}(\mathbf{v}') \notin (T_{xi,xi}, 0), \quad \forall (x,i) | v'_{x,i} = 0, \forall \mathbf{v}' \in H_C. \tag{7.37}$$

Derived from (7.14), we obtain

$$T_{xi,xi} = -A - B + C. \tag{7.38}$$

Hence, for $T_{xi,xi} < 0$, the following condition should be met

$$A + B > C. \tag{7.39}$$

The proof is straightforward.

Therefore, the conditions for the nonconvergence of an edge point $\mathbf{v} \in H - H_C$ are easily obtained.

**Theorem 7.10.** *Any invalid solution $\mathbf{v} \in H - H_C$ is not a stable equilibrium point if*

$$\min\{B, A + Dd_L, (N-1)A\} - \frac{C}{2} > A + B - C, \tag{7.40}$$

$$3Dd_U - \frac{C}{2} < 0, \tag{7.41}$$

$$A + B > C. \tag{7.42}$$

*Proof.* According to Lemma 7.9 and the previous two propositions, it is easy to derive the conditions.

*Remark 7.11.* By making all spurious states unstable, the valid states would necessarily be stable since the Hopfield network is convergent by design. This explains why the convergence condition of valid solutions can be omitted. Considering the valid state, it gives that

$$\mathcal{E}_{x,i}(\mathbf{v}) = \frac{C}{2} + Dd_{x,y} \geq \frac{C}{2} + Dd_L,$$

for $v_{x,i} = 0$ and

$$\mathcal{E}_{x,i}(\mathbf{v}) = -\frac{C}{2} + D(d_{x,y} + d_{yz}) \leq -\frac{C}{2} + 2Dd_U,$$

for $v_{x,i} = 1$. Recall the condition (7.19), any valid state $\mathbf{v} \in H_T$ is stable if $-\frac{C}{2} + 2Dd_U < 0$. Not surprisingly, it can be omitted since it is implied by the condition (7.41) in the above theorem.

## 7.6 Setting of Parameters

As observed from Theorem 7.8 and Theorem 7.10, the conditions in Theorem 7.8 can be omitted since they are implied by those in the latter one. Thus, the following conditions ensure nonconvergence of all spurious states $v \in H - H_T$.

$$3Dd_U - \frac{C}{2} < 0, \tag{7.43}$$

$$A + B > C, \tag{7.44}$$

$$\min\{B, A + Dd_L, (N-1)A\} - \frac{C}{2} > A + B - C. \tag{7.45}$$

The following approach is proposed for setting the numerical values of the parameters, based on the above 3 criteria. However, there is no need to adhere to this requirement, as any other values satisfying the above criteria will be equally valid. Note that $A$ cannot be set to be equal to $B$, as this will violate either condition (7.44) or (7.45). This can be shown by a simple calculation: It holds that $A = B > C/2$ from equation (7.44), while it contradicts that $B - C/2 > 2B - C$ given by equation (7.45). Another contradiction also occurs when derived from equation (7.45). Consequently, the parameter setting is suggested as the following.

$$D = C/(10d_U),$$
$$A = C/2 - Dd_L/10,$$
$$B = A + Dd_L,$$
$$C = \text{any desired value.}$$

This allows $C$ to be set to any arbitrary value while ensuring the criteria (7.43)-(7.45) are satisfied.

## 7.7 Simulation Results and Discussions

In this section, the criteria (7.43)-(7.45) will be validated by simulations and the performance compared with that obtained by the parameter settings of Talaván et al (Talaván and Yáñez, 2002a). Here the algorithm introduced by them is adopted. The continuous Hopfield network may encounter stability problems when it is implemented by discretization. Interested readers are suggested to refer to the work of Wang (Wang, 1998; Wang, 1997b) for detailed analysis.

*Ten-city TSP:* A 10-city example is designed to illustrate the performance of both the energy functions. In each case, the parameter $C$ was varied and 1000 trials were carried out. The initial states were randomly generated by $v_{x,i} = 0.5 + \alpha u, \forall (x, i) \in \{1, 2, \cdots, N\}^2$ where $\alpha$ is a small value and $u$ is a uniform random value in $[-0.5, 0.5]$.

## 7.7 Simulation Results and Discussions

The simulation results are shown in Tables 7.1 and 7.2 in terms of the number of good and invalid solutions as well as the minimum and average tour length. The cases of optimum state and near-optimum state including the 10-city configurations are shown in Figure 8.1 and Figure 8.2, respectively. The number of good solutions includes the optimum and near-optimum states (within 25% of the optimum length).

**Table 7.1.** Performance of the parameter settings obtained from Talaván

| $C$ | Good | Invalid | Min length | Ave length |
|---|---|---|---|---|
| 100K | 16 | 411 | 18.1421 | 25.2547 |
| 10K | 44 | 405 | 19.0618 | 25.6066 |
| 1K | 13 | 394 | 19.0618 | 25.5345 |
| 100 | 16 | 351 | 17.3137 | 25.3033 |
| 10 | 52 | 55 | 18.8089 | 24.1888 |
| 1 | 135 | 14 | 17.3137 | 22.8208 |
| 0.1 | 195 | 0 | 22.0837 | 23.4533 |
| 0.01 | 192 | 14 | 16.4853 | 22.1017 |
| 0.001 | 267 | 29 | 16.4853 | 21.6479 |

**Table 7.2.** Performance of the new parameter settings

| $C$ | Good | Invalid | Min length | Ave length |
|---|---|---|---|---|
| 100K | 220 | 22 | 16.4853 | 22.1737 |
| 10K | 226 | 27 | 16.4853 | 22.1040 |
| 1K | 227 | 3 | 15.3137 | 21.9492 |
| 100 | 233 | 2 | 15.3137 | 21.7767 |
| 10 | 232 | 2 | 15.3137 | 21.93333 |
| 1 | 215 | 1 | 15.3137 | 22.0452 |
| 0.1 | 223 | 0 | 16.4853 | 21.7484 |
| 0.01 | 208 | 5 | 16.4853 | 22.0157 |
| 0.001 | 204 | 11 | 16.4853 | 22.1432 |

*Remark 7.12.* Ideally, it is fair to have all neurons starting from the same initial state. However, this cannot be implemented in practice. All neurons with the same initial state will lead to non-convergence of the network, as all neurons on each row will always have the same potential. Therefore, a small random value $\alpha \in [-10^{-3}, 10^{-3}]$ is added to break the symmetry. The value of $\alpha$ is chosen to be small so that the randomness introduced does not play a significant role in determining the final state.

114    7 Parameter Settings of Hopfield Networks Applied to TSP

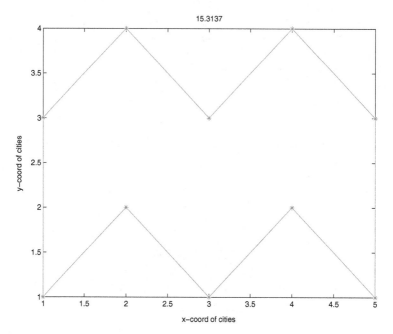

**Fig. 7.2.** Optimum tour state

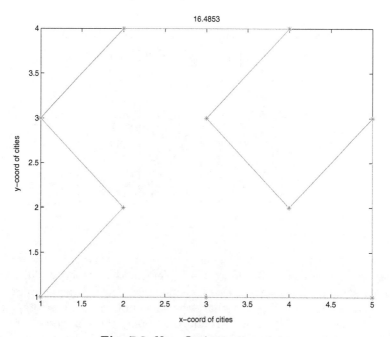

**Fig. 7.3.** Near-Optimum tour state

*Remark 7.13.* The performance is compared with that obtained from the work (Talaván and Yáñez, 2002a) where the H-T formulation of the Lyapunov function was employed. The new parameter settings of the network can successfully suppress the number of invalid states generated by the parameter settings of (Talaván and Yáñez, 2002a) to a minimal level. Beyond that, the performance of the network with H-T formulation deteriorates greatly. As shown in Table I, the results of the parameter settings with H-T formulation depend greatly on the value of $C$. As a compromise, the network associated with H-T formulation can be said to perform well when $C$ ranges from 0.1 to 1. On the contrary, as can be seen from Table II, the new parameter settings give consistently good results for a wide range of values of $C$, ranging from 0.1 to $10^5$. It can be observed that the average tour length is always approximately 22, the minimum tour length is approximately 16, and the percentage of good solutions is about 22%, regardless of the values of $C$.

*Thirty-city TSP:* The 30-city TSP was used in the work (Hopfield and Tank, 1985; Abe, 1993), where the cities were generated randomly from [0, 1]. The new formulation and H-T formulation were validated on different values of parameter $C$ (0.1, 1, 10, 100, 1000). For every value 50 valid tours were found and each time the 30-city was generated. Figure 7.4 and Figure 7.5 show the comparison of average tour length and minimal tour length, respectively. The results again confirm the advantages of the modified energy function employed. It is shown that good results can only be obtained using parameters limited to a small range for the H-T formulation. The modified energy function with the new parameter setting, however, is able to produce good results from a wide range of parameter values.

## 7.8 Conclusion

In order to guarantee the feasibility of CHN for solving TSP, the dynamical stability conditions of CHN defined on a unit hypercube have been proven in this work. The features of an enhanced Lyapunov function mapping TSP onto CHN have been investigated, which substantiates its advantages over the H-T formulation. The results revealed the relationships between the network parameters and solution feasibility, under which a systematic method for parameter settings in solving the TSP has been obtained. With this new parameter settings, the quality of the solutions has been improved in the sense that the network gives consistently good performance for a wide range of parameter values as compared to the most recent work of (Talaván and Yáñez, 2002a). The proposed criteria also ensure the convergence of the network to valid solutions as well as the suppression of spurious states, which facilitates the procedure of parameter settings and can be easily applied to various sizes of TSPs.

116    7 Parameter Settings of Hopfield Networks Applied to TSP

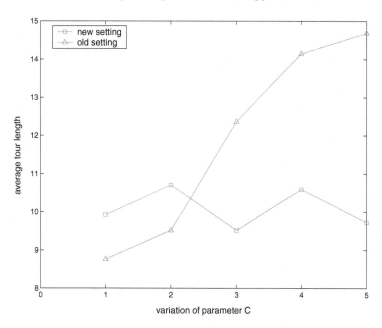

**Fig. 7.4.** Comparison of average tour length between the modified formulation (new setting) and H-T formulation (old setting).

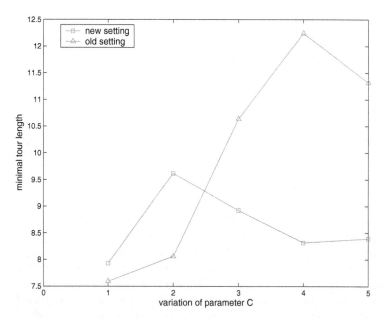

**Fig. 7.5.** Comparison of minimal tour length between the modified formulation (new setting) and H-T formulation (old setting).

# 8

# Competitive Model for Combinatorial Optimization Problems

## 8.1 Introduction

[1]In the last chapter, the method of applying Hopfield networks in solving the traveling salesman problem (TSP) has been analyzed. From the view of mathematical programming, the TSP can be described as a quadratic 0-1 programming problem with linear constraints,

$$\text{minimize } E^{obj}(\mathbf{v}) \tag{8.1}$$

$$\text{subject to } S_i = \sum_{x=1}^{n} v_{xi} = 1 \quad \forall i \in \{1, \cdots, n\}, \tag{8.2}$$

$$S_x = \sum_{i=1}^{n} v_{xi} = 1 \quad \forall x \in \{1, \cdots, n\}, \tag{8.3}$$

and a redundant constraint $S = \sum_x \sum_i v_{x,i} = n$, where $v_{xi} \in \{0,1\}$ and $E^{obj}$ is the total tour length described by a valid 0-1 solution matrix $\mathbf{v}$. The energy function to be minimized in the network is

$$E(\mathbf{v}) = E^{obj}(\mathbf{v}) + E^{cns}(\mathbf{v}), \tag{8.4}$$

where $E^{cns}$ is the constraints described by (8.2) and (8.3).

In the seminal work of (Hopfield and Tank, 1985), the optimization problem was solved using a highly-interconnected network of nonlinear analog neurons. However, the convergence of network to valid states and preferably quality ones depended heavily on the settings of the penalty terms in the energy function. It required careful setting of these parameters in order to obtain valid quality solutions, which is often a difficult based on a trial-and-error approach.

---

[1] Reuse of the materials of "A columnar competitive model for solving combinatorial optimization problems", 15(6), 2004, 1568–1573, IEEE Trans. on Neural Networks, with permission from IEEE.

It has been a continuing research effort to improve the performance of the Hopfield network since it was proposed (see the review (Smith, 1999)). The behavior of the Hopfield network was analyzed based on the eigenvalues of the connection matrix (Aiyer et al., 1990) and the parameter settings for TSP was derived. The local minimum escape (LME) algorithm (Peng et al., 1993) was proposed to improve the local minimum by combining the network disturbing technique with the Hopfield network's local minima search property. Most recently, a parameter setting procedure based on the stability conditions of the energy function was presented (Talaván and Yáñez, 2002a). Although these methods have been successful to some extent for improving the quality and validity of solutions, spurious states were often obtained. Moreover, the existing methods require a large volume of computational resources, which restricts their practical applications.

## 8.2 Columnar Competitive Model

The dynamics of Hopfield networks can be described by a system of differential equations with the activation function as a hyperbolic tangent. Let $\mathbf{v}, \mathbf{i^b}$ be the vectors of neuron activities and biases, and $\mathbf{W}$ be the connection matrix, then the energy function of the Hopfield network for the high-gain limit expression exists such that

$$E = -\frac{1}{2}\mathbf{v}^T \mathbf{W} \mathbf{v} - (\mathbf{i^b})^T \mathbf{v}. \tag{8.5}$$

Hopfield has shown that the network will converge to local minima of energy function (8.5) if $\mathbf{W}$ is symmetric (Hopfield and Tank, 1985).

In this section, the columnar competitive model (CCM) which is constructed by incorporating winner-takes-all (WTA) into the network in a column-wise form is introduced. Competitive learning by WTA has been recognized to play an important role in many areas of computational neural networks, such as feature discovery and pattern classification (Rumelhart and Zipser, 1985; Wang, 1997c; Sum et al., 1999; Yi et al., 2000). Nevertheless, the potential of WTA as a means of eliminating all spurious states is seen due to its intrinsic competitive nature that can elegantly reduce the number of penalty terms, and hence the constraints of the network for optimization. The WTA mechanism can be described as: given a set of $n$ neurons, the input to each neuron is calculated and the neuron with the maximum input value is declared the winner. The winner's output is set to '1' while the remaining neurons will have their values set to '0'.

The neurons of the CCM are evolved based on this WTA learning rule. The competitive model handles the columnar constraints (8.2) elegantly, due to its competitive property of ensuring one '1' per column. As in two dimensional forms, $\mathbf{v} = \{v_{x,i}\}, \mathbf{i^b} = \{i^b_{x,i}\}$, where the subscript $x, i \in \{1, \ldots, n\}$ denotes the city index and the visit order, respectively. The strength of connection between neuron $(x, i)$ and neuron $(y, j)$ is denoted by $W_{xy,ij}$. Hence, the associated energy function can be written as

$$E(\mathbf{v}) = \frac{K}{2}\sum_x \sum_i (v_{x,i} \sum_{j \neq i} v_{x,j}) + \frac{1}{2}\sum_x \sum_{y \neq x} \sum_i d_{xy} v_{x,i}(v_{y,i+1} + v_{y,i-1}), \quad (8.6)$$

where $K > 0$ is a scaling parameter and $d_{xy}$ is the distance between cities $x$ and $y$. Comparing (8.5) and (8.6), the connection matrix and the external input of the network are computed as follows,

$$W_{xi,yj} = -\{K\delta_{xy}(1 - \delta_{ij}) + d_{xy}(\delta_{i,j+1} + \delta_{i,j-1})\}, \quad (8.7)$$
$$\mathbf{i}^\mathbf{b} = 0, \quad (8.8)$$

where $\delta_{i,j}$ is the Kronecker's delta.

The input to a neuron $(x, i)$ is calculated as

$$\begin{aligned} Net_{x,i} &= \sum_y \sum_j (W_{xi,yj} v_{yj}) + \mathbf{i}^\mathbf{b} \\ &= -\sum_y d_{xy}(v_{y,i-1} + v_{y,i+1}) - K\sum_{j \neq i} v_{x,j}. \end{aligned} \quad (8.9)$$

The WTA is applied based on a column-by-column basis, with the winner being the neuron with the largest input. The WTA updating rule is thus defined as

$$v_{x,i} = \begin{cases} 1, & \text{if } Net_{x,i} = \max\{Net_{1,i}, Net_{2,i}, \cdots, Net_{n,i}\} \\ 0, & \text{otherwise} \end{cases} \quad (8.10)$$

Hereafter, $v_{x,i}$ is evaluated by the above WTA rule. The algorithm of implementing the competitive model is summarized as follows:

### Algorithm 8.1 Competitive Model Algorithm

1. *Initialize the network, with each neuron having a small initial value $v_{x,i}$. A small random noise is added to break the initial network symmetry. Compute the W matrix using (8.7).*
2. *Select a column (e.g., the first column). Compute the input $Net_{x,i}$ of each neuron in that column.*
3. *Apply WTA using (8.10), and update the output state of the neurons in that column.*
4. *Go to the next column, preferably the one immediately on the right for convenience of computation. Repeat step 3 until the last column in the network is done. This constitutes the first epoch.*
5. *Go to step 2 until the network converges (i.e., the states of the network do not change).*

## 8.3 Convergence of Competitive Model and Full Valid Solutions

For the energy function (8.6) of CCM, the critical value of the penalty-term scaling parameter $K$ plays a predominant role in ensuring its convergence and driving the network to converge to valid states. Meanwhile, it is known that the stability of the original Hopfield networks is guaranteed by the well-known Lyapunov energy function. However, the dynamics of the CCM is so different from the Hopfield network, thus the stability of the CCM needs to be investigated.

In this section, our efforts are devoted to such two objectives, i.e., determining the critical value of $K$ that ensures the convergence of full valid solutions and proving the convergence of the competitive networks under the WTA updating rule. The following theorem is drawn.

**Theorem 8.1.** *Let $K > 2d_{max} - d_{min}$, where $d_{max}$ and $d_{min}$ is the maximum and the minimum distance, respectively. Then the competitive model defined by (8.6)–(8.10) is always convergent to valid states.*

*Proof.* The proof is composed of two parts based on the two objectives respectively.

(i) While the WTA ensures that there can only be one '1' per column, it does not guarantee the same for each row. This responsibility lies with the parameter $K$ of the penalty term. Without any loss of generality, it is assumed that after some updating, the network reaches the following state, which is given by

$$V = \begin{pmatrix} 0 & v_{1,i} & 0 & 0 & 0 \\ K & v_{2,i} & 0 & \cdots & 0 \\ 0 & v_{3,i} & 1 & \cdots & 0 \\ \vdots & \vdots & \vdots & \ddots & \vdots \\ 0 & v_{n,i} & 0 & \cdots & 1 \end{pmatrix}$$

Let the first row be an all-zero row. The input to each neuron in the $i$-th column is computed by

$$Net_{1,i} = -(d_{12} + d_{13}),$$
$$Net_{2,i} = -(K + d_{23}),$$
$$Net_{3,i} = -(K + d_{23}),$$
$$Net_{n,i} = -(K + d_{2n} + d_{3n}).$$

## 8.3 Convergence of Competitive Model and Full Valid Solutions

For valid state, $v_{1,i}$ occupying the all-zero row will have to be the winner, i.e., $Net_{1,i} = \max_{x=1,\ldots,n} \{Net_{x,i}\}$. Therefore, it is verified that

$$d_{12} + d_{13} < K + d_{23},$$
$$d_{12} + d_{13} < K + d_{2n} + d_{3n}.$$

To satisfy both conditions, it is sufficient for $K$ to satisfy

$$K > d_{12} + d_{13} - d_{23}, \tag{8.11}$$

since $d_{2n} + d_{3n} > d_{23}$ straightforwardly.

Let $d_{max} = \max\{d_{xy}\}$, $d_{min} = \min\{d_{xy}\}$. Firstly, assume the 'worst' scenario for deriving (8.11), i.e., $d_{12} = d_{13} = d_{max}$, $d_{23} = d_{min}$, it holds

$$K > 2d_{max} - d_{min}. \tag{8.12}$$

Secondly, assume the 'best' scenario, i.e., $d_{12} = d_{13} = d_{min}$, $d_{23} = d_{max}$, the following is obtained

$$K > 2d_{min} - d_{max}. \tag{8.13}$$

Obviously, condition (8.12) is the sufficient condition for guaranteed convergence to fully valid solutions, while (8.13) is the necessary condition for convergence to some valid solutions. Although a specific case has been assumed, the results obtained are regardless of the specific case.

**(ii)** To investigate the dynamical stability of the CCM, a $n \times n$ network (for $n$ cities) is considered. After the $n$-th WTA updating, the network would have reached the state with only one '1' per column, but may have more than one '1' per row.

Suppose $\mathbf{v}^t$ and $\mathbf{v}^{t+1}$ are two states before and after WTA updating respectively. Consider $p$-th column, and let neuron $(a, p)$ be the only active neuron before updating, i.e., $v_{a,p}^t = 1$ and $v_{i,p}^t = 0, \forall i \neq a$. After updating, let neuron $(b, p)$ be the winning neuron, i.e., $v_{b,p}^{t+1} = 1$, $v_{i,p}^{t+1} = 0, \forall i \neq b$.

The energy function (8.6) can be further broken into two terms $E_p$ and $E_o$, i.e., $E = E_p + E_o$, where $E_p$ stands for the energy of the consecutive columns $p-1, p$ and $p+1$ and of the rows $a$ and $b$. $E_o$ stands for the energy of the rest columns and rows.

$E_p$ is calculated by

$$E_p = \frac{K}{2}\left(\sum_i v_{a,i} \sum_{j \neq i} v_{a,j} + \sum_i v_{b,i} \sum_{j \neq i} v_{b,j}\right) + \sum_x \sum_y d_{xy} v_{x,p}(v_{y,p+1} + v_{y,p-1}).$$

Accordingly, $E_o$ is computed by

$$E_o = \frac{K}{2} \sum_{x \neq a,b} \sum_i (v_{x,i} \sum_{j \neq i} v_{x,j}) + \frac{1}{2} \sum_x \sum_y \sum_{i \neq p-1, p, p+1} d_{xy} v_{x,i}(v_{y,i+1} + v_{y,i-1})$$
$$+ \frac{1}{2}\sum_x \sum_y d_{xy} v_{x,p-1} v_{x,p-2} + \frac{1}{2}\sum_x \sum_y d_{xy} v_{x,p+1} v_{x,p+2}.$$

Thus, it can be seen that only $E_p$ will be affected by the states of the column $p$. It is assumed that the neuron $(a, p)$ is the only active neuron in the $p$-th column before updating, i.e., $v_{a,p}^t = 1$ and $v_{i,p}^t = 0$ for all $i \neq a$, and the neuron $(b, p)$ wins the competition after updating, i.e., $v_{b,p}^{t+1} = 1$ and $v_{i,p}^{t+1} = 0$ for all $i \neq b$. Let $E^t$ and $E^{t+1}$ be the network energy before updating and after updating, respectively. To investigate how $E$ changes under the WTA learning rule, the following two cases are considered.

Case 1 : Row $a$ contains $m$ '1' $(m > 1)$, row $b$ is an all-zero row.
The input to neuron $(a, p)$ and $(b, p)$ is computed by (8.14) and (8.15), respectively.

$$Net_{a,p} = -K \sum_{j \neq p} v_{a,j} - \sum_y d_{ay}(v_{y,p-1} + v_{y,p+1}). \tag{8.14}$$

$$Net_{b,p} = -K \sum_{j \neq p} v_{b,j} - \sum_y d_{by}(v_{y,p-1} + v_{y,p+1}). \tag{8.15}$$

Considering the WTA learning rule (8.10), it leads to $Net_{b,p} = \max_i \{Net_{i,p}\}$. It implies that

$$K \sum_{j \neq p} v_{b,j} + \sum_y d_{by}(v_{y,p-1}+v_{y,p+1}) < K \sum_{j \neq p} v_{a,j} + \sum_y d_{ay}(v_{y,p-1}+v_{y,p+1}). \tag{8.16}$$

Obviously, $\sum_{j \neq p} v_{b,j} = 0$, $\sum_{j \neq p} v_{a,j} = m-1$. Therefore, it is derived from (8.16) that

$$\sum_y d_{by}(v_{y,p-1} + v_{y,p+1}) - \sum_y d_{ay}(v_{y,p-1} + v_{y,p+1}) < K(m-1). \tag{8.17}$$

Now considering the energy function (8.6), $E_p^t$ and $E_p^{t+1}$ are computed by (8.18) and (8.19), respectively.

$$E_p^t = \frac{K}{2} m(m-1) + \sum_y d_{ay} v_{a,p}(v_{y,p+1} + v_{y,p-1}). \tag{8.18}$$

$$E_p^{t+1} = \frac{K}{2}(m-1)(m-2) + \sum_y d_{by} v_{b,p}(v_{y,p+1} + v_{y,p-1}). \tag{8.19}$$

Thus, it is clear that

$$E_p^{t+1} - E_p^t = -K(m-1) + \sum_y d_{by} v_{b,p}(v_{y,p+1} + v_{y,p-1}) - \sum_y d_{ay} v_{a,p}(v_{y,p+1} + v_{y,p-1}). \tag{8.20}$$

From equation (8.17), it is obtained $E_p^{t+1} - E_p^t < 0$. It implies that

$$E^{t+1} - E^t < 0. \tag{8.21}$$

Case 2 : Row $a$ contains only one '1', row $b$ is an all-zero row.
According to the WTA updating rule, again it holds that $Net_{b,p} = \max_{i}\{Net_{i,p}\}$. Similar to case 1, it is obtained

$$K\sum_{j\neq p} v_{b,j} + \sum_{y} d_{by}(v_{y,p-1}+v_{y,p+1}) < K\sum_{j\neq p} v_{a,j} + \sum_{y} d_{ay}(v_{y,p-1}+v_{y,p+1}). \quad (8.22)$$

Obviously, $\sum_{j\neq p} v_{b,j} = 0$, $\sum_{j\neq p} v_{a,j} = 0$. Therefore, we obtain

$$\sum_{y} d_{by}(v_{y,p-1}+v_{y,p+1}) - \sum_{y} d_{ay}(v_{y,p-1}+v_{y,p+1}) < 0. \quad (8.23)$$

On the other hand, $E_p^t$ and $E_p^{t+1}$ are now computed as follows,

$$E_p^t = \sum_{y} d_{ay} v_{a,p}(v_{y,p+1}+v_{y,p-1}). \quad (8.24)$$

$$E_p^{t+1} = \sum_{y} d_{by} v_{b,p}(v_{y,p+1}+v_{y,p-1}). \quad (8.25)$$

Again, it is obtained
$$E^{t+1} - E^t < 0. \quad (8.26)$$

Based on the above two cases, the CCM is always convergent under the WTA updating rule. This completes the proof.

It is noted that the WTA tends to drive the rows with more than one active neurons to reduce the number of active neurons, and in turn drives one of the neurons in all-zero rows to become active. Once there is no row that contains more than one active neurons (equivalently there is no all-zero row), then a valid state is reached and the state of the network stops the transition (in this case $E^{t+1} - E^t = 0$).

## 8.4 Simulated Annealing Applied to Competitive Model

Simulated Annealing (SA), a stochastic process, is known to improve the quality of solutions when solving combinatorial optimization problems (Papageorgiou et al., 1998). It derives its name from the analogy of thermodynamics and metallurgy. Metal is first heated to a high temperature $T_0$, thus causing the metal atoms to vibrate vigorously, resulting in a highly disordered structure. The metal is then cooled slowly, allowing the atoms to rearrange themselves in an orderly fashion, which corresponds to a structure of minimum energy.

Optimum or near optimum results can usually be obtained, but at the cost of long computational time, due to SA's slow convergence. WTA, on the other hand, offers faster convergence, but produces solutions that are of lower quality. It is thus ideal to combine these 2 techniques, so that a fully valid solution set, preferably one with mostly optimum or near optimum tour distances, can be obtained in a reasonably short time.

When $K$ is at least $2d_{max} - d_{min}$, all states are valid, but the solutions are of average quality (fewer states with optimum tour distance). As $K$ decreases, it becomes more probable for the network to converge to states of optimum tour distance, but there is also a corresponding increase in the number of spurious states. When $K$ is less than $2d_{min} - d_{max}$, the network never converges to valid states, regardless of the initial network states. Therefore, by setting a small $K$ and slowly increasing it, the quality of solutions can be increased, while preserving the validity of all states at the same time. The algorithm in conjunction with the simulated annealing schedule for the parameter $K$ can be implemented in the following way:

**Algorithm 8.2 Competitive Model with SA**

1. *Initialize the network, with $v_{x,i}$ having a small initial value, added with small random noise. Initialize $K = d_{max}$ and $\epsilon > 0$ that determines how fast $K$ reaches $2d_{max} - d_{min}$.*
2. *Do N times: Select a column i to compute the input $Net_{x,i}$ for each neuron in that column, then apply WTA and update the output of the neurons. This constitutes a whole epoch.*
3. *Increase K by*

$$K = d_{max} + (0.5 \cdot \tanh(t - \epsilon) + 0.5)(d_{max} - d_{min}),$$

*where t is the current epoch number. Go to step 2 until convergence.*

## 8.5 Simulation Results

It is known that the solutions of the double-circle problems are very hard to obtain by the original Hopfield network (Tachibana et al., 1995). In this section, the 24-city example that was analyzed in (Tachibana et al., 1995) is employed to verify our theoretical results. There are 12 cities uniformly distributed in the outer circle, while the rest are allocated in the inner circle, with the radius equal to 2 and 1.9, respectively. In this work, 500 simulations have been performed for each case of the Hopfield network and CCM. The simulation results are given in Tables 8.1 and 8.2. The term 'good' indicates the number of tours with distance within 150% of the optimum distance. Figures 8.1 and 8.2 depict the optimum tour and the near-optimum tour generated by the CCM.

As can be seen from the simulation results, the original Hopfield network generated a large number of invalid solutions w.r.t various weighting factor $C$. Its minimum and average distances are also far away from the optimum solution of 13.322. Comparatively, the new competitive model generated less

Table 8.1. The performance of original Hopfield model for the 24-city example

| $C$ | Valid | Invalid | Good | Minimum length | Average length |
|---|---|---|---|---|---|
| 10 | 302 | 198 | 111 | 35.2769 | 50.9725 |
| 1 | 285 | 215 | 108 | 26.2441 | 37.3273 |
| 0.1 | 366 | 134 | 168 | 25.2030 | 36.3042 |
| 0.01 | 360 | 140 | 159 | 27.5906 | 36.9118 |
| 0.001 | 372 | 128 | 153 | 21.6254 | 32.5432 |

Table 8.2. The performance of CCM for the 24-city example

| $K$ | Valid | Invalid | Good | Minimum length | Average length |
|---|---|---|---|---|---|
| $d_{max}$ | 438 | 62 | 248 | 13.3220 | 18.1172 |
| $d_{max} + d_{min}$ | 453 | 47 | 256 | 13.3220 | 17.9778 |
| $2d_{max} - d_{min}$ | 500 | 0 | 18 | 15.1856 | 24.1177 |

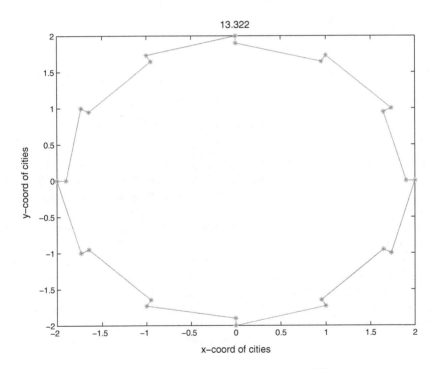

Fig. 8.1. Optimum tour generated by CCM

spurious states and produced better quality solutions in terms of the minimum and average tour lengths. In addition, the CCM has a rapid convergence rate of less than 5 epochs on average. It can be seen that when $K$ is set to a small value (e.g., $d_{max}$), with a small portion of invalid solutions, the CCM

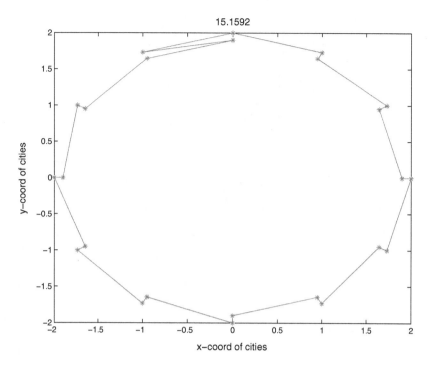

**Fig. 8.2.** Near-optimum tour generated by CCM

can easily reach the optimum solution. When $K$ is increased to $2d_{max} - d_{min}$, all the spurious states were eliminated, which was achieved at the minor expense of the solution quality. Clearly, these findings are in agreement with the theoretical results obtained.

When SA is applied to the new WTA model, its performance increases significantly (Table 8.3). All the states generated are valid, and has approximately 65% good solutions (optimum or near optimum states) for all cases. The average tour distance is about 18, which is much better than the Hopfield model and of the WTA model (without SA). Its convergence rate is slightly slower than the WTA model (without SA) and requires 9 epochs on average. It is nevertheless much faster than that of the Hopfield model.

The theoretical results are also validated upon the 48-city example which has a similar coordinates configuration as the 24-city example. 100 experiments are performed for each case and the results are given in Table 8.4. It can be seen that when $K$ was increased from $d_{max}$ to $2d_{max} - d_{min}$, all invalid solutions were suppressed with the minimum and average lengths increased, which is consistent with the findings observed in Table 8.4.

*Remark 8.2.* Since for non-proper values of $K$ (e.g., $d_{max}$ and $d_{max} + d_{min}$), the convergence of the CCM is not guaranteed, thus for the sake of comparing the performance of the CCM with different parameter settings, only the results

**Table 8.3.** The performance of CCM with SA for the 24-city example

| $\epsilon$ | Valid | Invalid | Good | Minimum length | Average length |
|---|---|---|---|---|---|
| 0 | 500 | 0 | 18 | 15.1856 | 24.1177 |
| 1 | 500 | 0 | 203 | 13.3404 | 22.2688 |
| 2 | 500 | 0 | 218 | 13.3220 | 19.9771 |
| 3 | 500 | 0 | 259 | 13.3220 | 18.7153 |
| 5 | 500 | 0 | 256 | 13.3220 | 18.6785 |
| 10 | 500 | 0 | 301 | 13.3220 | 18.4080 |
| 15 | 500 | 0 | 307 | 13.3220 | 18.3352 |
| 20 | 500 | 0 | 289 | 13.3220 | 18.4643 |

**Table 8.4.** The performance of CCM for the 48-city example

| | $K$ | Valid | Invalid | Good | Minimum length | Average length |
|---|---|---|---|---|---|---|
| $d_{max}$ | 90 | 10 | 0 | | 24.6860 | 32.7950 |
| $d_{max} + d_{min}$ | 92 | 8 | 0 | | 24.9202 | 33.8892 |
| $2d_{max} - d_{min}$ | 100 | 0 | 0 | | 35.9438 | 42.4797 |

of the convergent cases in the experiments recorded are recorded. It is also noted that when the coordinates configuration becomes more complex, good solutions become more difficult to be achieved as well.

The WTA with SA model is now applied to TSP of various city sizes, with their city topology similar to that of 24 cities, i.e., their coordinates evenly distributed around 2 circles of radius 1.9 and 2.0. The $\epsilon$ is set to 5 for the simulations. The results are shown in Table 8.5.

**Table 8.5.** The performance of CCM with SA for various city sizes

| City size | Simulations | Valid | Invalid | Minimum length | Average length | Perf. ratio |
|---|---|---|---|---|---|---|
| 30 | 250 | 250 | 0 | 13.6922 | 22.3159 | 0.6136 |
| 100 | 250 | 250 | 0 | 28.2552 | 35.6920 | 0.7916 |
| 500 | 100 | 100 | 0 | 257.1380 | 283.2810 | 0.9060 |
| 1000 | 50 | 50 | 0 | 454.3827 | 490.6963 | 0.9260 |
| 3000 | 5 | 5 | 0 | 1395.5 | 1454.2 | 0.9455 |

Perf. ratio=Minimum length/Average length

The WTA with SA model is seen to perform satisfactory, with an average performance ratio of 0.85. Less computing resources are used, thereby allowing city sizes up to 3000 be successfully solved. This is quite remarkable, as none of the existing algorithms proposed in literature has been able to solve such a large city size in practice, perhaps due to the extensive amount of time and resources required.

## 8.6 Conclusion

A new columnar competitive model (CCM) incorporating a WTA learning rule has been proposed for solving the combinatorial optimization problems, which has guaranteed convergence to valid states and total suppression of spurious states in the network. Consistently good results, superior to that of the Hopfield network in terms of both the number and quality of valid solutions were obtained using computer simulations. It also has the additional advantage of fast convergence, and utilizing comparatively lower computing resources, thereby allowing it to solve large scale combinatorial problems. The dynamics analysis of the CCM implies that the CCM is incapable of hill-climbing transitions and thus being trapped at local minima is still a problem. With the full validity of solutions ensured by competitive learning, probabilistic optimization techniques such as LME (Peng et al., 1993) and simulated annealing (Papageorgiou et al., 1998) can be incorporated into the competitive model to further improve the solution quality.

In addition, with some modification to the energy function, the competitive model can also be extended to solve other combinatorial optimization problems, such as the 4-color map.

# 9

# Competitive Neural Networks for Image Segmentation

## 9.1 Introduction

Image segmentation is an important task in image processing. It involves partitioning the image into meaningful sub-regions or grouping objects with the same attribution. This has important applications in a wide variety of areas, such as the biomedical, metrological and even the military, as distinctive features in the raw images may appear unclear to the human eyes.

Image segmentation is however, a difficult process, as it depends on a wide variety of factors such as the complexity of the image, the objects of interest or the number of classes required. While one mode of grouping may successfully segment the image into meaningful sub-regions, the same mode of grouping may not work for another image. Even if the image remains the same, there are many different ways of segmenting the image, all of which can be considered good segmentations, depending on the image's objects of interest, and to a large extent, the user's own subjectivity. The image segmentation process is made worse when noise is present, which unfortunately, is often the case.

A number of algorithms based on histogram analysis, region growing, edge detection or pixel classification, have been proposed (Pavlidis and Liow, 1990; Gonzalez and Woods, 2002). Neural networks with applications to various stages of image processing was addressed in a review (Egmont-Pertersen et al., 2002). In particular, neural networks have been used to solve the image segmentation problem (Cheng et al., 1996; Wu et al., 2001; Gopal et al., 1997; Phoha and Oldham, 1996). Generally, the method involves mapping the problem into a neural network by means of an energy function, and allowing the network to converge so as to minimize the energy function. The final state should ideally correspond to the optimum solution, with the (local) minimum energy. More often than not, the network converges to invalid states, despite the presence of penalty terms. Researchers (Cheng et al., 1996; Wersing et al., 2001a) have incorporated the Winner-Takes-All (WTA) algorithm to avoid the need for penalty terms, while ensuring the validity of solutions.

This chapter introduces a new design method of competitive neural networks for performing image segmentation. The previous design method (Cheng et al., 1996) is also studied for comparative purposes. The comparison study shows that the new model of network has high computational efficiency and better performance of noise suppression and image segmentation. Two stochastic optimization techniques to improve the quality of solutions, simulated annealing (SA) and local minima escape (LME) (Peng et al., 1993), are introduced to the competitive network.

This chapter is organized as follows: A brief introduction of image segmentation and its mapping to neural network is given in Section 9.2. The competitive model is addressed in Section 9.3 and its proof for convergence is proposed in Section 9.4. SA algorithm and LME algorithm are described in detail in Section 9.5 and 9.6, respectively. Computer simulation results are presented in Section 9.7, and conclusions drawn in Section 9.8.

## 9.2 Neural Networks Based Image Segmentation

The image segmentation problem can be mapped onto a neural network, with the cost function as the Lyapunov energy function of the network. The iterative updating of the neuron states will eventually force the network to converge to a stable state, preferable a valid one with the lowest energy. Hence, this can be seen as an optimization problem, and the best segmentation will be one with the lowest energy.

The neural network model (Lin et al., 1992) for the image segmentation problem consists of $L \times L \times C$ neurons, with $L \times L$ being the image size, and $C$ the number of classes. The network uses spatial information and is thus image size dependent. This would not be desirable in terms of the computational time and resources when the Image size or the number of classes required is large. A similar constraint satisfaction network was addressed for segmentation based on perceptual organization method (Mohan and Nevatia, 1992).

Another segmentation method employs the gray level distribution (Cheng et al., 1996) instead of spatial information, and thus has the advantages of being image size independent, using fewer neurons and requiring lesser computational time and resources. In addition, the number of segmentation classes is stated in advance.

For an image with $N$ gray levels and $C$ classes of sub-objects, it can be mapped to a network consisting of $N \times C$ neurons, with each row representing one gray level of the image and each column a particular class, subjected to the following constraints:

$$\sum_{i}^{c} v_{x,i} = 1, x = 1, \ldots, N, \tag{9.1}$$

$$\sum_{i}^{c} \sum_{x}^{N} v_{x,i} = N. \tag{9.2}$$

Each row should consist of only one '1', and the column which this '1' falls under will indicate this particular gray level's class. The strength of connection between neuron $(x, i)$ and neuron $(y, j)$ is denoted as $W_{xi,yj}$. A neuron in this network would receive inputs from all other neurons weighted by $W_{xi,yj}$ and also an additional bias, $i^b$. Mathematically, the total input to the neuron $(x, i)$ is given as

$$Net_{xi} = \sum_{y}^{N} \sum_{j}^{c} W_{xi,yj} v_{yj} + i^b. \tag{9.3}$$

With a symmetrical $W$, the energy like function exists, thereby implying the existence of equilibrium points. The image segmentation can hence be expressed as a minimization of an energy function of the form:

$$E = -\frac{1}{2} v^t W v - (i^b)^t v. \tag{9.4}$$

The energy function proposed in (Cheng et al., 1996) is as

$$E = \sum_{x=1}^{n} \sum_{y=1}^{n} \sum_{i=1}^{c} \frac{1}{\sum_{y=1}^{n} h_y v_{y,i}} v_{x,i} d_{x,y} h_y v_{y,i}, \tag{9.5}$$

and the connection weights and external input are given by

$$w_{xi,yj} = -\frac{1}{\sum_{y=1}^{n} h_y v_{y,i}} d_{x,y} h_y, \tag{9.6}$$

$$i^b = 0. \tag{9.7}$$

## 9.3 Competitive Model of Neural Networks

The competitive network is now minimizing the following energy function

$$E = \sum_{x}^{n} \sum_{i}^{c} \sum_{y}^{n} v_{xi} v_{yj} h_x h_y d_{xy}. \tag{9.8}$$

The network incorporates the competitive learning rule, winner-takes-all (WTA), into its dynamics. The neuron with the largest net-input in the same row is declared as the winner, and the updating of all neurons in the same row is as follows:

$$v_{x,i} = \begin{cases} 1, & \text{if } Net_{x,i} = \max\{Net_x\}, \\ 0, & \text{otherwise.} \end{cases} \tag{9.9}$$

WTA automatically satisfies the two constraints, (9.1)-(9.2), specified earlier. This relieves the need for penalty terms in the energy function, and hence avoids the hassle of setting their parametric values. All network states upon convergence will be valid states, i.e., no gray levels will be classified into two or more different classes, and each gray level will be assigned a class.

The architecture of the competitive model is similar to that suggested by (Cheng et al., 1996). The difference lies in the treatment of connection weights and the energy function. The connection weights are formulated by

$$W_{xi,yj} = \delta_{ij}(1 - \delta_{xy})h_x h_y d_{xy}. \tag{9.10}$$

The external input $i^b$ is set to zero. Recall equation (10.8), the net-input to neuron $(x, i)$ is computed by

$$Net_{xi} = -\sum_{y}^{n} h_x h_y d_{xy} v_{yi}. \tag{9.11}$$

The implementation of the competitive network is as follows:

**Algorithm 9.1 Competitive Model Based Image Segmentation**

1. *Input $N$ gray levels of image and its associated histogram, as well as number of classes $C$. Compute distance matrix, according to the following equation, for all gray levels*

$$d_{xy} = (g_x - g_y)^2.$$

2. *Initialize the neural network by randomly assigning one '1' per row, and setting the rest of neurons in the same row to '0'. Do the same for all other rows, while ensuring that there is at least one '1' per column.*
3. *Do $N$ times: Choose one row randomly and calculate $Net_{x,i}$ to all neurons in the same row, according to equation (9.11). Then apply WTA mechanism (9.9) to update all the neuron states within the same row. This constitutes 1 epoch*
4. *Go to step 2 until convergence, i.e. the network state, $V = (v_{x,i})$, for previous epoch is the same as current epoch.*

## 9.4 Dynamical Stability Analysis

The proposed energy function of the CHN, is always convergent during the network evolutions. The proof is given as follows.

Let the proposed energy function be re-written as

$$E = \sum_{i}^{c} E_i, \tag{9.12}$$

where $E_i = \sum_{x}^{n} \sum_{y}^{n} v_{xi} v_{yi} h_x h_y d_{xy}$, which represents the energy of all neurons in each column.

Consider row $r$ and let neuron $(r, i)$ be the winning neuron before WTA updating. Upon updating, let neuron $(r, j)$ be the winning neuron. Without loss of generality, let $r$ be the first row. Suppose $V(k)$ and $V(k+1)$ be the network states before WTA updating and after WTA updating, respectively:

$$V(k) = \begin{pmatrix} 0 & 1 & \cdots & 0 & 0 \\ v_{21} & v_{2i} & \cdots & v_{2j} & v_{2c} \\ v_{31} & v_{3i} & \cdots & v_{3j} & v_{3c} \\ \vdots & \vdots & \vdots & \vdots & \vdots \\ v_{n1} & v_{ni} & \cdots & v_{nj} & v_{nc} \end{pmatrix},$$

$$V(k+1) = \begin{pmatrix} 0 & 0 & \cdots & 1 & 0 \\ v_{21} & v_{2i} & \cdots & v_{2j} & v_{2c} \\ v_{31} & v_{3i} & \cdots & v_{3j} & v_{3c} \\ \vdots & \vdots & \vdots & \vdots & \vdots \\ v_{n1} & v_{ni} & \cdots & v_{nj} & v_{nc} \end{pmatrix}.$$

For any time $k$, the energy of the network can be broken into three terms as below,

$$E(k) = E_i(k) + E_j(k) + E_o$$
$$= \sum_x^n \sum_y^n v_{xi}(k) v_{yi}(k) h_x h_y d_{xy}$$
$$+ \sum_x^n \sum_y^n v_{xj}(k) v_{yj}(k) h_x h_y d_{xy}$$
$$+ E_o, \tag{9.13}$$

where $E_o$ is the energy of all other neurons, not in column $i$ or $j$. Note that $E_o$ remains unchanged after updating. It is derived that

$$E(k+1) - E(k)$$
$$= 2 \sum_{x=2}^n h_1 h_x d_{1x} v_{xj}(k+1) - 2 \sum_{x=2}^n h_1 h_x d_{1x} v_{xi}(k)$$
$$= 2 \sum_{x=2}^n h_1 h_x d_{1x} v_{xj}(k) - 2 \sum_{x=2}^n h_1 h_x d_{1x} v_{xi}(k).$$

Since neuron $(r, j)$ is the winning neuron, $Net_{rj} > Net_{ri}$. It implies that

$$-\sum_x^n h_r h_x d_{rx} v_{xj}(k) > -\sum_x^n h_r h_x d_{rx} v_{xi}(k), \tag{9.14}$$

that is

$$\sum_{x}^{n} h_r h_x d_{rx} v_{xj}(k) < \sum_{x}^{n} h_r h_x d_{rx} v_{xi}(k). \tag{9.15}$$

Hence it is derived that

$$E(k+1) - E(k) < 0. \tag{9.16}$$

This shows that the energy of the network is decreasing. The convergence is thus proven.

## 9.5 Simulated Annealing Applied to Competitive Model

Simulated Annealing (SA) is subsequently applied to the new WTA model to improve the quality of solutions. SA is a stochastic relaxation algorithm which has been applied and used to solve optimization problems. It derives its name from the analogy of the physical process of thermodynamics and metallurgy. Metal is first heated to a high temperature $T$, thus causing the metal atoms to vibrate vigorously, resulting in a disorderly structure. The metal is then cooled slowly, to allow the atoms rearrange themselves in an orderly fashion, which corresponds to a structure of minimum energy.

The network starts with a high initial temperature $T = 1$ and evolves to a new state after each WTA update. The network energy is calculated for each WTA update and if it is less than that before the update, the new state is readily accepted. If not, the new state is not rejected outright, but there is a certain probability $e^{-\delta E/T}$ of accepting it. After each epoch, the network temperature decreases according to $T = 0.999T$, and so does the probability of accepting a new state of higher energy. Eventually, the temperature decreases to a small value, and the network converges, thereby terminating the SA algorithm.

The algorithm of competitive model incorporating SA can be summarized as follows:

### Algorithm 9.2 Completive Model with SA

1. *Input Gray levels $N$ of image, and its associated histogram, as well as number of classes $C$. Compute distance matrix, according to the following equation, for all gray levels*

$$d_{xy} = (g_x - g_y)^2.$$

2. *Initialized the neural network by randomly assigning one '1' per row, and setting the rest of neurons in the same row to '0'. Do the same for all other rows, while ensuring that there is at least one '1' per column.*

3. *Choose one row randomly and calculate the total input $Net_{x,i}$ to all neurons in that row. Then apply WTA algorithm to update neurons states and calculate the new network energy $E$.*

4. If the energy, $E_{current}$, is less than its previous energy, $E_{previous}$, accept the new state. If not, compute the probability of acceptance,

$$Pr = e^{-(E_{current} - E_{previous})/T},$$

and generate a random number $\alpha$ between 0 and 1. Accept the new state if $\alpha < Pr$, else reject the new state. This constitutes 1 epoch.

5. Decrease temperature $T$ by $T = \eta T$, with $0 < \eta < 1$. Go to step 2 until convergence, such that both (a) and (b) are satisfied
   (a). T reaches below a small value (e.g., 0.1).
   (b). Network state $v$ for previous epoch is the same as current epoch.

## 9.6 Local Minima Escape Algorithm Applied to Competitive Model

The local minima escape, like the SA, can also be used to improve the quality of solutions. It is essentially a combination of the network disturbing technique and CHN's local minima searching property, and was first proposed in (Peng et al., 1993) to improve the quality of solutions of combinatorial optimization problems, in particular the traveling salesman problem (TSP). While the TSP results obtained in (Peng et al., 1993) were promising, it is however, not widely applied to other combinatorial problems, probably because it is still a relatively new technique. This section now investigates the application of LME to the image processing realm, in particular the segmentation process.

The LME technique first perturbs its network parameters ($W$ and $i^b$), using standard white Gaussian noise of varying weights, to obtain a new network $\hat{H}$.

$$\hat{w}_{xi,yj} = (1 + \alpha^w \eta^w_{xi,yj}) w_{xi,yj} + \beta^w \eta^w_{xi,yj}, x, i \leq y, j \quad (9.17\text{a})$$

$$\hat{w}_{xi,yj} = w_{yj,xi}, \quad (9.17\text{b})$$

$$\hat{i}^b_{x,i} = (1 + \alpha^i \eta^i_{xi}) i^b_{x,i} + \beta^i \eta^i_{x,i}, \quad (9.17\text{c})$$

where $\eta^w_{xi,yj}, \eta^i_{x,i}$ are standard Gaussian noises, and $\alpha^w, \beta^w, \alpha^i, \beta^i$ are positive constants which control the strength of disturbance. Note that connection matrix W is still symmetric after disturbance.

Although having different network parameters, the two networks, $H$ and $\hat{H}$, have the same architecture. As such, the neurons in these two networks are one-one related and can be mapped from one to the other easily. The network is initialized as before, and converges eventually under the WTA dynamics. This converged state corresponds to either its global minimum, or one of its local minima (which is often the case).

Assuming the network $H$ is at its local minimum state. Set this local minimum state as the initial state of $\hat{H}$, and apply the WTA dynamics to $\hat{H}$. Upon convergence, this minimum state of $\hat{H}$ is mapped back to the original network $H$ as its current initial state. Allow this new state to converge in $H$ network now. Upon convergence, compute the network energy, and check if this new local minimum state has a lower energy than its former local minimum state. If so, accept the new state, else reject it and retrieve its former local minimum state. This completes one epoch. Repeat the entire procedure until the total number of epochs reaches some pre-set number, or when no new states have been obtained after a certain number of consecutive epochs.

Since the model proposed in this chapter has zero bias, noise is only added to the inter-connection strengths, $W$. As $W_{xi,yj} = \delta_{ij}(1-\delta_{xy})h_x h_y d_{xy}$, adding noise to $d$ has the same effect as noise to $W$. A standard Gaussian noise, with mean 0, standard deviation 1 and variance 1 is added to the distance matrix $d$ of the image, so as to 'disturb' the network.

The LME algorithm, applied to the WTA (proposed Energy) algorithm can be summarized as follows:

### Algorithm 9.3 Competitive Model with LME

1. *Input gray levels $N$ of image, and its associated histogram, as well as number of class $C$. Compute distance matrix, according to the following equation, for all gray levels.*
2. *Initialized the neural network by randomly assigning one '1' per row, and setting the rest of neurons in the same row to '0'. Do the same for all other rows, while ensuring that there is at least one '1' per column.*
3. *Do $N$ times: Choose randomly one row and calculate the total input $Net_{x,i}$ to all neurons in the same row, according to*

$$Net_{xi} = -\sum_{y}^{n} h_x h_y d_{xy} v_{yi}.$$

   *Then apply WTA algorithm.*
4. *Go to step 3 until convergence. This convergent state $V$, is stored temporary and set as the initial state for the disturbed network $\hat{H}$. Calculate the energy $E_{previous}$ of the H network.*
5. *Compute the disturbed distance matrix $\hat{H}$, using*

$$\hat{d}_{xy} = (g_x - g_y)^2 + \mathcal{G}(0,1),$$

   *where $\mathcal{G}$ denotes a Gaussian noise with mean 0, variance 1. Process the competitive dynamics of $\hat{H}$ as H until convergence. Let this local minima be $\hat{V}$.*
6. *This new local minima $\hat{V}$ is now mapped back to H, as its new initial state. Perform the dynamics of H until convergence. Let this new state be $V_{current}$, and calculate the current energy $E_{current}$ of the H network. If $E_{current} < E_{previous}$, accept the new state. Else retrieve its previous minimum state, $V_{previous}$. This constitutes 1 LME epoch.*

7. Repeat above procedure until LME terminates, ie. (a) OR (b) is satisfied:
   (a). The total number of LME epochs reaches some preset number.
   (b). $V_{current} = V_{previous}$, for some consecutive LME epochs.

## 9.7 Simulation Results

### 9.7.1 Error-Correcting

The competitive model demonstrates a high performance of suppressing noise of images. An image with only 4 distinct gray levels (50, 100, 150, 200) is corrupted with a Gaussian noise, thereby resulting in a noisy image with about 40 gray levels, shown as Figure 9.1. A good algorithm will classify the pixels accordingly such that the corrupted pixels with gray levels in the vicinity of each of the 4 distinct levels (50, 100, 150, 200) will be grouped together, thus giving an output image free from noise. For example, the corrupted gray levels 47-53 is likely to be in the same group, while 96-103 will be classified in another group. Hence, the algorithm that performs better will be the one that gives the most number of correctly classified/segmented images.

Images with different number of pixels (200, 2000, 20k), each with varying degrees of white Gaussian noise (Variance = 1, 2, 3, 4, 5, 6) were input to the network. A total of 100 simulations were done for each case. Tables 9.1 and 9.2 compare the results with that of the modeling method used in (Cheng et al., 1996). An example of a corrupted image, a wrongly classified image and the clear, correctly restored image are shown in Figures 9.1, 9.2 and 9.3, respectively.

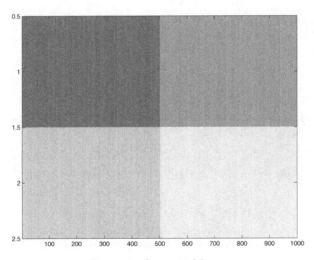

**Fig. 9.1.** Corrupted Image

138    9 Competitive Neural Networks for Image Segmentation

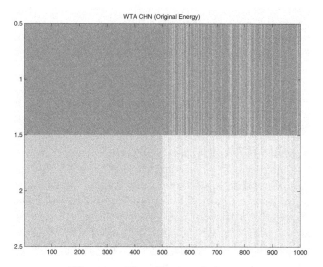

**Fig. 9.2.** Wrongly classified image

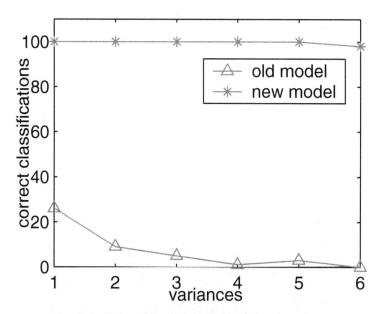

**Fig. 9.3.** Correct classifications for different variances

As observed, the WTA (proposed energy) algorithm performs better in terms of the number correctly segmented images, as well as the percentage error of gray levels and pixels. It is able to give 100% error free image restoration when the pixels are below 2000. When the number of pixels is increased by another factor of 10 to 20k, it is still able to give satisfactory results, with

## 9.7 Simulation Results

**Table 9.1.** The performance of the segmentation model (Cheng et al., 1996)

| Pixels | Noise Variance | Number of completely correct classifications | % error (gray levels) | % error (pixels) |
|---|---|---|---|---|
| 200 | 1 | 33/100 | 40.6208 | 41.0787 |
|  | 2 | 15/100 | 50.8457 | 51.2937 |
|  | 3 | 13/100 | 53.0667 | 53.5505 |
|  | 4 | 4/100  | 58.3102 | 59.1288 |
|  | 5 | 3/100  | 56.8309 | 58.7903 |
|  | 6 | 15/100 | 45.1973 | 47.2101 |
| 2000 | 1 | 26/100 | 42.2615 | 42.8450 |
|  | 2 | 9/100  | 55.6250 | 55.0735 |
|  | 3 | 5/100  | 58.0546 | 59.0965 |
|  | 4 | 1/100  | 58.0443 | 60.9020 |
|  | 5 | 3/100  | 54.6299 | 58.6025 |
|  | 6 | 0/100  | 51.3346 | 55.5435 |
| 20K | 1 | 26/100 | 44.7853 | 44.7154 |
|  | 2 | 9/100  | 55.2891 | 54.4492 |
|  | 3 | 7/100  | 55.1063 | 57.3599 |
|  | 4 | 4/100  | 54.5662 | 59.1239 |
|  | 5 | 2/100  | 54.0060 | 59.3431 |
|  | 6 | 0/100  | 50.3886 | 57.2611 |

approximately 80% correctly segmented images. The reduction in accuracy is due to the presence of more gray levels generated by the Gaussian noise as the number of pixels increases. With a high variance (variance = 6) and a large number of pixels, it becomes rather probable for pixels to be corrupted with noise of deviation up to 25 gray levels. This makes it difficult to correctly classify/segment the image.

The percentage error of gray levels as well as pixels remains at zero, with the exception of high noise variance (variance = 6). The percentage error nevertheless hovers close to zero (0.1%), which implies that the number of wrongly classified pixels or gray levels is very small. Hence, most wrongly segmented images have only a few wrongly classified pixels, which is rather insignificant or noticeable, given the large number of pixels (20k).

The WTA model (Cheng et al., 1996), on the other hand, gives at best 30% correctly segmented images, which occurs only when the noise variance is small (variance=1). Its performance deteriorates rapidly as the noise variance increases, and is unable to restore the original image in most cases when the noise variance is large. As the number of pixels increases, there is a higher probability of generating more gray levels, which thus increases the difficulty of correctly segmenting the image.

**Table 9.2.** The performance of the proposed model

| Pixels | Noise variance | Number of completely correct classifications | % error (gray levels) | % error (pixels) |
|---|---|---|---|---|
| 200 | 1 | 100/100 | 0 | 0 |
|  | 2 | 100/100 | 0 | 0 |
|  | 3 | 100/100 | 0 | 0 |
|  | 4 | 100/100 | 0 | 0 |
|  | 5 | 100/100 | 0 | 0 |
|  | 6 | 100/100 | 0 | 0 |
| 2000 | 1 | 100/100 | 0 | 0 |
|  | 2 | 100/100 | 0 | 0 |
|  | 3 | 100/100 | 0 | 0 |
|  | 4 | 100/100 | 0 | 0 |
|  | 5 | 100/100 | 0 | 0 |
|  | 6 | 98/100 | 0.0151 | 0.0010 |
| 20k | 1 | 100/100 | 0 | 0 |
|  | 2 | 100/100 | 0 | 0 |
|  | 3 | 100/100 | 0 | 0 |
|  | 4 | 100/100 | 0 | 0 |
|  | 5 | 100/100 | 0 | 0 |
|  | 6 | 78/100 | 0.1399 | 0.0013 |

% error (gray levels) = Number (of wrongly classified gray levels) / number (of all gray levels), % error (pixels) = Number (of wrongly classified pixels) / number (of all pixels)

### 9.7.2 Image Segmentation

The neural network is applied to perform the image segmentation task, where the lena image is employed. The definition of a good segmentation is rather subjective. In an attempt to maintain experimental objectivity, the energy values are used as a means of comparison. In other words, for $C$ classes, the lower energy value, the better the segmentation. Since the energy functions are different, the same network state $V$, will give different energy values. Thus, both energy values are calculated for all the simulations, so as to make a better comparison.

Both energy functions were first tested using only the WTA algorithm, before incorporating SA and LME. Different number of classes $C$ (6, 8, 10, 12) were used to segment the image. A total of 20 simulations were done for each case, and the results shown in Table 9.3. The original Lena image is shown in Figure 9.4. The segmentation results for 3 and 6 classes using only WTA are shown in Figures 9.5 and 9.6, respectively.

The results show that the segmentation quality is improved when the number of classes increases, and the energies decrease as well.

9.7 Simulation Results    141

**Fig. 9.4.** Original Lena image

**Fig. 9.5.** Lena image segmented with 3 classes

**Table 9.3.** Image Segmentation using only WTA

| Class | Original energy ($\times 10^4$) | | Equivalent energy ($\times 10^{11}$) | | Proposed energy ($\times 10^{11}$) | |
|---|---|---|---|---|---|---|
|  | Min | Ave | Min | Ave | Min | Ave |
| 6 | 8.2207 | 8.2207 | 36.486 | 36.486 | 32.558 | 32.869 |
| 8 | 4.4871 | 4.4871 | 14.676 | 14.676 | 13.711 | 13.864 |
| 10 | 3.0999 | 3.0999 | 8.3994 | 8.3994 | 6.8463 | 6.8869 |
| 12 | 2.3082 | 2.3322 | 5.6013 | 5.7445 | 4.0353 | 4.0891 |

**Fig. 9.6.** Lena image segmented with 6 classes

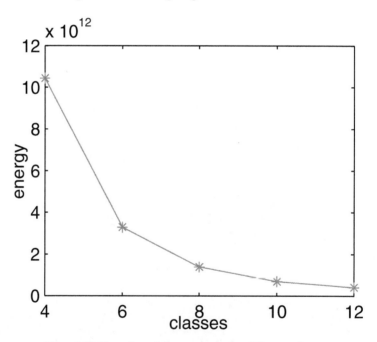

**Fig. 9.7.** Energies of the network for different classes

From Table 9.3, the proposed model gives more satisfactory results. Using the WTA (proposed energy) algorithm, the energy of the network calculated using the proposed energy function is smaller than that of the WTA (original energy) algorithm, for all classes under simulation.

**Table 9.4.** Image Segmentation with SA or LME

| Method | Class | Energy value | |
|---|---|---|---|
| | | Min | Ave |
| WTA with SA | 6 | 3.2524 e+12 | 3.2821 e+12 |
| | 8 | 1.3582 e+12 | 1.3695 e+12 |
| | 10 | 6.8129 e+11 | 6.8658 e+11 |
| | 12 | 4.0269 e+11 | 4.0774 e+11 |
| WTA with LME | 6 | 3.2538 e+12 | 3.2591 e+12 |
| | 8 | 1.3655 e+12 | 1.3723 e+12 |
| | 10 | 6.7925 e+11 | 6.8252 e+11 |
| | 12 | 4.0389 e+11 | 4.0584 e+11 |

Using the SA algorithm, the energy of the network is reduced by $2.0 \times 10^9$ on average, or about 0.4% as shown in Table 9.4. Though a large absolute value, this is however, a small reduction percentage, thus implying that SA does not have a significant effect on the WTA algorithm. This also indicates that the WTA (proposed energy) algorithm is by itself quite efficient, and can give near optimum solutions in most instances. LME is able to reduce the network energy for all cases simulated. It has a percentage energy reduction of about 0.75% for the WTA (proposed energy) algorithm While still a small percentage, it nevertheless shows that it is a more effective technique than with the SA algorithm. As seen from the simulations, SA and LME are able to reduce the energy of the network. LME performs better than SA in terms of the reduction percentage. In addition, it much converges faster than SA. This savings in time is especially important in image processing, as a short processing time would mean that it is possible to extend the algorithm to real-time image processing.

## 9.8 Conclusion

In this chapter, a new modeling method of a competitive neural network based on a winner-takes-all mechanism was presented and its application to image restoration and segmentation was explored. The segmentation task is treated as a constrained optimization problem and is based upon global information of the gray level distribution. The new approach has higher computation efficiency and illustrates better noise suppression as well as image segmentation, as compared to existing approaches.

Two techniques for improving the quality of solutions, simulated annealing (SA) and local escape minima (LME), were incorporated into the WTA algorithm. While these techniques require longer computation time, the results obtained are rather satisfactory. The incorporation of either SA or LME in WTA algorithm would have a much significant advantage in color image segmentation, where good segmentation is more difficult, given the 16 million distinct colors involved in an image.

# 10

# Columnar Competitive Model for Solving Multi-Traveling Salesman Problem

This chapter studies an optimization problem: Multi-Traveling Salesman Problem (MTSP), which is an extension of the well known TSP. A columnar competitive model (CCM) of neural networks incorporating a winner-take-all learning rule is employed to solve the MTSP. Stability conditions of CCM for MTSP are exploited by mathematical analysis. Parameter settings of the network for guaranteeing the network converges to valid solutions are discussed in detail. Simulations are carried out to illustrate the performance of the columnar competitive model compared to heuristic algorithms such as the Tabu Search.

## 10.1 Introduction

There has been an increasing interest in applying the Hopfield neural networks to combinatorial optimization problems, since the original work of Hopfield and Tank (Hopfield and Tank, 1985). Several methods have been proposed to ensure the network converges to valid states. Aiyer et al (Aiyer *et al.*, 1990) have theoretically explained the dynamics of network for traveling salesman problems by analyzing the eigenvalues of the connection matrix. Abe (Abe, 1993) has shown the theoretical relationship between network parameters and solution quality based on the stability conditions of feasible solutions. Chaotic neural network provides another promising approach to solve those problems due to its global search ability and remarkable improvement with less local minima, see in (Nozawa, 1992; Wang and Smith, 1998; Wang and Tian, 2000). Peng et al (Peng *et al.*, 1993) suggested the local minimum escape(LME) algorithm, which improves the local minimum of CHN by combining the network disturbing technique with the Hopfield network's local minimum searching property. In addition, many papers have discussed efficient mapping approaches (Talavan and Yanez, 2002b; Smith, 1999). Talavan and Yanez et al (Talavan and Yanez, 2002b) presented a procedure for parameters settings based on the stability conditions of the network. Cooper et al (Cooper, 2002) developed the higher-order neural networks (HONN) to solve the TSP and study the stability conditions of valid solutions. Brandt et al (Brandt *et al.*, 1988) presented a modified Lyapunov function for mapping the TSP problem. All of those works are noteworthy for the solving of TSP.

MTSP is another optimization problem related with TSP (Potvin *et al.*, 1989; Frederickson *et al.*, 1978; Franca *et al.*, 1995; Samerkae *et al.*, 1999). However, the former is more complex and interesting than the latter. This problem deals with some real world problems where there is a need to account for more then one salesman. Many interesting applications of MTSP have been found. For example, suppose there is a city that is suffering from floods, the government wants to send out some investigators starting from the government building to investigate the disaster areas (all villages and towns in this city) and return to the starting point. The problem is how to find out the nearly equal shortest tour for each investigator. Clearly, it can be mapped to a MTSP. In addition, MTSP can be generalized to a wide variety of routing and scheduling problems, such as the School Bus Problem (Orloff, 1974; Angel *et al.*, 1972) and the Pickup and Delivery Problem (Christofides and Eilon, 1969; Savelsbergh, 1995). In (Svestka and Huckfeldt, 1973), it has shown that MTSP was an appropriate model for the problem of bank messenger scheduling, where a crew of messengers pick up deposits at branch banks and returns them to the central office for processing. Some other problems such as: railway transportation, pipeline laying, routing choosing, computer network topology designing, postman sending mail, torch relay transfer, etc, all can be mapped to the MTSP model.

This chapter discusses a method of solving the MTSP based on an existing maximum neuron model proposed by Y. Takefuji et al (Takefuji *et al.*, 1992), which is also known as a columnar competitive network (Tang *et al.*, 2004). Stability conditions of CCM for MTSP will be explored by mathematical analysis. Parameter settings of the network to guarantee the CCM converges to valid solutions will be discussed.

## 10.2 The MTSP Problem

The description of Multi-Travelling Salesman Problem involves four problems as follows:

*Problem* 1: Given a group of $n$ cities and the distance between any two cities. Suppose there are $m$ visitors starting from a city to visit the group of cities. The objective is to find the nearly equal shortest tour for each visitor such that each city be visited only once and each visitor returns to the starting city.

*Problem* 2: Given a group of $n$ cities and the distance between any two cities. Suppose there are $m$ visitors starting from a city to visit the group of cities. The objective is to find the nearly equal shortest tour for each visitor such that each city be visited only once and each visitor ends at $m$ different cities.

*Problem 3*: Given a group of $n$ cities and the distance between any two cities. Suppose there are $m$ visitors starting from $m$ different cities to visit the group of cities. The objective is to find the nearly equal shortest tour for each visitor such that each city be visited only once and each visitor ends at the same city.

*Problem 4*: Given a group of $n$ cities and the distance between any two cities. Suppose there are $m$ visitors starting from $m$ cities to visit the group of cities. The objective is to find the nearly equal shortest tour for each visitor such that each city is visited only once and each visitor ends at $m$ different cities.

*Problem* 1 is to find $m$ closed tours which start from the the same city, visit each city once, with each tour containing the nearly equal number of city, and return to the starting city with a short total length. *Problem* 2 is to find $m$ routes beginning from the same city and ending with $m$ different cities. *Problem* 3 and *problem* 4 is to find $m$ routes with the $m$ different starting cities. In this chapter, our study is focused on Problem 1. The valid tours for those four problems are illuminated in Fig. 10.1.

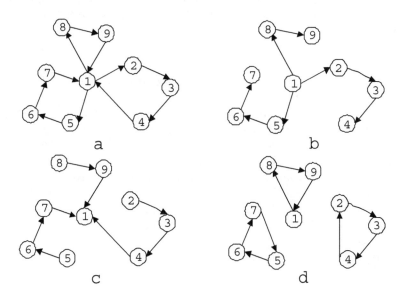

**Fig. 10.1.** Valid tours of various problems. Subplots a, b, c and d are valid tours of problems 1, 2, 3, 4, respectively.

We can describe *problem* 1 with a weighted, non-directed graph $G(V, E)$. The city is represented by the vertex on the graph, denoted as $\nu_i$. The road between each city is represented by the edge, denoted as $e_i$. $w(e_i)$ is the value weighed on edge $e_i$, which represents the real distance of the two cities. $m$ loops should appear in graph $G(V, E)$ to guarantee $m$ closed tours: $C_1, C_2, ..., C_m$, and $\nu_0 \in C_i, i = 1...m$. Then, MTSP can be described as a mathematic model as follows:

$$\min \sum_{j=1}^{m} \sum_{e_i \in (C_j)} \omega(e_i) + D \qquad (10.1)$$

$$\text{subject to } \begin{cases} \nu_o \in C_i(V)(i = 1...m) \\ \bigcup_{i=1}^{m} C_i(V) = G(V), \end{cases} \qquad (10.2)$$

where

$$D = \alpha \left\{ \sum_{j=1}^{m} [n(C_j) - \overline{n}]^2 \right\} \qquad (10.3)$$

In the objective function (1), $\sum_{j=1}^{m} \sum_{e_i \in (C_j)} \omega(e_i)$ is the total length of $m$ loops, $D$ represent the degree of difference among respective loops. In equation (10.3), $n(C_j)$ is the numbers of the vertex in loop $C_j$, $\overline{n}$ is the average vertex number of $m$ loops, $\alpha$ is a scaling factor. It is clear that the difference between loops would be decreased when the value of $D$ changes smaller.

## 10.3 MTSP Mapping and CCM Model

In this section, a neural network model is presented to solve the MTSP. Clearly, there are $m$ sub-tours in a valid solution of the $n$-city and $m$-loop MTSP problems, and each sub-tour starts with the starting city. To map this problem onto the network, an appropriate representation scheme of neurons is required. It is well known, an $n \times n$ square array has been used in $n$-city TSP problem by Hopfield, but it is not suitable for the MTSP problem.

Here, we add $m - 1$ cites to the network, and we call them virtual cities. The virtual cities have the same connections and weights as the starting city. The location of a virtual city is specified by the output state of $m - 1$ neurons in the representation scheme. We define those neurons as the group neurons. Hence, we can distinguish $m$ loops easily from those virtual cities using a $(n + m - 1) \times (n + m - 1)$ square matrix. For example, in a 5-city and 2-loop problem, a feasible solution and the new one after adding the virtual cities are shown in Fig. 10.2. It needs $(5 + 2 - 1) \times (5 + 2 - 1) = 36$ neurons to represent the neurons' output states.

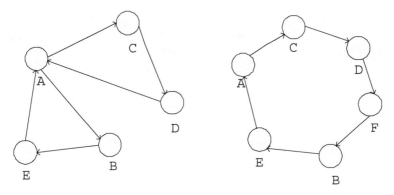

**Fig. 10.2.** The left is the feasible solution tour. The right is the tour after adding the virtual city. F is the virtual city

$$
\begin{array}{c|cccccc}
 & 1 & 2 & 3 & 4 & 5 & 6 \\
\hline
A & 1 & 0 & 0 & 0 & 0 & 0 \\
B & 0 & 0 & 0 & 0 & 1 & 0 \\
C & 0 & 1 & 0 & 0 & 0 & 0 \\
D & 0 & 0 & 1 & 0 & 0 & 0 \\
E & 0 & 0 & 0 & 0 & 0 & 1 \\
F & 0 & 0 & 0 & 1 & 0 & 0 \\
\end{array}
$$

The above output states of neurons represent a multi-tour with 2 tours. $A$ is the starting city to be visited, $C$ is the second and $D$ is the third of the first tour. $F$ is the virtual city which represent city $A$. The second tour starting from $F(A)$, visited $B$, $E$ in sequence. To sum up, the first tour is $A - C - D - F(A)$ and the second is $F(A) - E - D - A(F)$. $E$, $F$ are the group neuron.

In the seminal work of (Hopfield and Tank, 1985), it has been shown that an energy function of the form

$$E = -\frac{1}{2}\mathbf{v}^\top \mathbf{W}\mathbf{v} - (\mathbf{i}^b)^\top \mathbf{v}. \qquad (10.4)$$

can be minimized by the continuous-time neural network with the parameters: $\mathbf{W} = (W_{i,j})_{n\times n}$ is the connection matrix, $n$ is the number of neurons, $\mathbf{v} = (v_{x,i})_{n\times n}$ represents the output sate of the neuron $(x, i)$, and $\mathbf{i}^b$ is the vector form of bias.

To solve MTSP efficiently, we employ a new neural network that has a similar structure as Hopfield network, but obeys a different updating rule: the columnar competitive model (CCM), which incorporates winner-takes-all (WTA) competition column-wise. The WTA mechanism can be described as (Tang et al., 2004): Given a set of $n$ neurons, the input to each neuron is calculated and the neuron with the maximum input value is declared the winner. The winner's output is set to "1" while the remaining neurons will have their values set to "0". The intrinsic competitive nature is in favor of the convergence of CCM to feasible solutions, since it can reduce the number of penalty terms compared to Hopfield's formulation.

The energy function of CCM for MTSP can be written as

$$E(\nu) = \frac{A}{2}\sum_x \sum_i (\nu_{x,i} \sum_{j \neq i} \nu_{x,j}) + \frac{B}{2}\sum_i^m (S_i - \overline{S})^2$$
$$+ \frac{1}{2}\sum_x \sum_{y \neq x} \sum_i d_{xy}\nu_{xi}(\nu_{y,i+1} + \nu_{y,i-1}). \quad (10.5)$$

where $A > 0, B > 0$ are scaling parameters, $d_{xy}$ is the distance between city $x$ and $y$, $\overline{S} = \frac{n}{m}$ is the average number of cities for $m$ tours, $S_i$ is the number of cities that have been visited by team $i$. Let $m_i(i = 1...m-1)$ be the index of the virtual city in the whole city's sequence which composed by $m$ team's tour concatenated end by end, and set $m_0 = 0, m_m = n+m-1$, $n$ is the total number of cites. For example in Fig. 10.1, $m_0 = 0, m_1 = 4, m_2 = 6$. Then $S_i = \sum_x \sum_{i=m_{i-1}}^{m_i-1} \nu_{x,i}$, and

$$\frac{B}{2}\sum_{i=1}^m (S_i - \overline{S})^2$$

$$= \sum_{i=1}^m \frac{B}{2}(S_i^2 - 2\overline{S}S_i + \overline{S}^2)$$

$$= \sum_{i=1}^m \left\{ \frac{B}{2}\sum_x \sum_{i=m_{i-1}}^{m_i-1} \nu_{xi} \cdot \sum_x \sum_{j=m_{i-1}}^{m_i-1} \nu_{xj} - \frac{n}{m}\sum_x \sum_{i=m_{i-1}}^{m_i-1} \nu_{xi} + \frac{Bn^2}{2m^2} \right\}$$

$$= \sum_{i=1}^m \left\{ \frac{B}{2}\sum_x \sum_{i=m_{i-1}}^{m_i-1} \nu_{xi} \cdot \sum_x \sum_{j=m_{i-1}}^{m_i-1} \nu_{xj} \right\} - \frac{n}{m}\sum_x \sum_i \nu_{xi} + \frac{Bn^2}{2m}. \quad (10.6)$$

Thus, comparing (11.3) with (10.4), the connection matrix and input basis are computed as

$$\begin{cases} W_{xi,yj} = -\{A\delta_{xy}(1-\delta_{ij}) + d_{xy}(\delta_{i,j+1} + \delta_{i,j-1}) + B\theta_j\} \\ i^b = \frac{n}{m} \end{cases} \quad (10.7)$$

where

$$\theta_j = \begin{cases} 1, & \text{if } m_{i'-1} \leq j < m_{i'} \\ 0, & \text{otherwise} \end{cases}$$

then the input to neuron (x,i) is calculated as:

$$Net_{x,i} = \sum_y \sum_j (W_{xi,yj} \nu_{yj}) + i^b$$

$$= -\sum_y d_{xy}(\nu_{y,i-1} + \nu_{y,i+1}) - A \sum_{j \neq i} \nu_{x,j} - B \sum_y \sum_{j=m_{i'-1}}^{m_{i'}-1} \nu_{y,j} + \frac{n}{m}$$

$$= -\sum_y d_{xy}(\nu_{y,i-1} + \nu_{y,i+1}) - A \sum_{j \neq i} \nu_{x,j} - BS_{i'} + \frac{n}{m}. \quad (10.8)$$

The columnar competitive model based on winner-take-all (WTA) leaning rule, where the neurons compete with others in each column, and the winner is the one with the largest input. The updating rule of outputs is given by

$$\nu_{x,i}^{t+1} = \begin{cases} 1, & \text{if } Net_{x,i}^t = max\{Net_{1,i}^t, Net_{2,i}^t, ..., Net_{N+m-1,i}^t\} \\ 0, & \text{otherwise} \end{cases} \quad (10.9)$$

For CCM, $\nu_{x,i}$ is updated by the above WTA rule. The whole algorithm is summarized as follows:

**Algorithm 10.1** *The algorithm for solving MTSP by CCM*

1. *Initialize the network, with each neuron having a small value $\nu_{x,i}$.*
2. *Calculated $m_i (i = 0, ..., m)$.*
3. *Select a column(e.g., the first column), compute the input $Net_{x,i}$ of each neuron in this column.*
4. *Apply WTA and update the outputs of the neurons in that column using the WTA updating rule.*
5. *Go to the next column, repeat step 2 until the last column in the network is done. This constitutes one epoch.*
6. *Go to step 2 until the network converges.*

In the next section, by investigating the dynamical stability of the network, we present theoretical results which give an analytical approach in setting the scaling parameters A and B optimally. As a consequence, the convergence to valid solutions is assured.

## 10.4 Valid Solutions and Convergence Analysis of CCM for MTSP

It is well known that the stability of original Hopfield network when it is applied to TSP is guaranteed by the Lyapunov energy function. However, the dynamics of CCM for MTSP is different from the original Hopfield network. In this section, we will investigate the stability of CCM.

The WTA updating rule ensures only one "1" per column, but it is not the case for the rows. The responsibility falls on scaling parameters A and B of the penalty-terms. The following analysis shows that the value of A and B play a predominant role in ensuring the convergence to valid states.

In this section, our efforts are devoted to how to determine the critical value of A and B that can ensure the convergence of the valid solutions for the MSTP based on the stability analysis.

### 10.4.1 Parameters Settings for the CCM Applied to MTSP

Consider the $p$-column of neuron outputs states matrix, suppose row $b$ is an all-zero row and row $a$ is not all-zero. According to equation (10.8), the input to neuron $(a,p)$ and $(b,p)$ is computed as:

$$Net_{a,p} = -A\sum_{j \neq p} \nu_{a,j} - \sum_y d_{ay}(\nu_{y,p-1} + \nu_{y,p+1}) - BS_{p'} + \frac{n}{m}. \quad (10.10)$$

$$Net_{b,p} = -A\sum_{j \neq p} \nu_{b,j} - \sum_y d_{by}(\nu_{y,p-1} + \nu_{y,p+1}) - BS_{p'} + \frac{n}{m}. \quad (10.11)$$

where $0 < S_{p'} < n$. Suppose row $a$ contains $l$ "1" ($1 <= l <= n+m-1$), then

$$Net_{b,p} = -\sum_y d_{by}(\nu_{y,p-1} + \nu_{y,p+1}) - BS_{b'} + \frac{n}{m}$$

$$> -\sum_y d_{by}(\nu_{y,p-1} + \nu_{y,p+1}) - Bn^2 + \frac{n}{m}. \quad (10.12)$$

and

$$Net_{a,p} = -A(l-1) - \sum_y d_{ay}(\nu_{y,p-1} + \nu_{y,p+1}) - BS_{a'} + \frac{n}{m}$$

$$< -A(l-1) - \sum_y d_{ay}(\nu_{y,p-1} + \nu_{y,p+1}) + \frac{n}{m}. \quad (10.13)$$

**Theorem 10.1.** *When $A - Bn^2 > 2d_{max} - d_{min}$, where $d_{max}$ and $d_{min}$ is the maximum and the minimum distance between any two cities, respectively, $n$ is the total number of cities, the neurons' outputs under CCM alway escape from the invalid states.*

*Proof.* It is clear that only one neuron's output per column is set to '1' following the WTA updating rule. Assume the network reaches the following state after some updating:

$$\nu = \begin{pmatrix} \vdots & \vdots & \vdots & \vdots & \vdots \\ 0 & \nu_{b,p} & 0 & 0 & 0 \\ 1 & \nu_{s,p} & 0 & \cdots & 0 \\ 0 & \nu_{t,p} & 1 & \cdots & 0 \\ \vdots & \vdots & \vdots & \ddots & \vdots \\ 0 & \nu_{n,p} & 0 & \cdots & 1 \end{pmatrix}$$

## 10.4 Valid Solutions and Convergence Analysis of CCM for MTSP

According to equations (10.12) and (10.13), the input to each neuron in the $p$-th column is calculated as

$$Net_{b,p} > -(d_{bt} + d_{bs}) - Bn^2 + \frac{n}{m},$$
$$Net_{s,p} < -(A + d_{st}) + \frac{n}{m},$$
$$Net_{t,p} < -(A + d_{st}) + \frac{n}{m},$$
$$Net_{t,p} < -(A + d_{sn} + d_{tn}) + \frac{n}{m}.$$

To ensure the neruon's outputs reach the valid solution in the next updating, the neuron $\nu_{b,p}$ should be the winner, since all the other neurons in row $b$ is zero. In the other word, the input of the neuron $\nu_{b,p}$ should be the maximum of all the inputs in $p$-th column, $Net_{b,p} > Net_{a,p}, a \neq b$. Therefore it is ensured by

$$d_{bt} + d_{bs} + Bn^2 - \frac{n}{m} < A + d_{st} - \frac{n}{m},$$
$$d_{bt} + d_{bs} + Bn^2 - \frac{n}{m} < A + d_{sn} + d_{tn} - \frac{n}{m}.$$

It is well known that

$$d_{st} < d_{sn} + d_{tn}$$

then it follows that

$$A - Bn^2 > d_{bt} + d_{bs} - d_{st}. \tag{10.14}$$

which can be ensured by $A - Bn^2 > 2d_{max} - d_{min}$. The neurons' outputs always escape from invalid states.

### 10.4.2 Dynamical Stability Analysis

Suppose $E^t$ and $E^{t+1}$ are two states before and after WTA update respectively. Consider $p$-th column, let neuron $(a, p)$ be the only active neuron before updating, and neuron $(b, p)$ be the winning neuron, and a row contains $l$ "1", then

$$\nu_{x,p}^t = \begin{cases} 1, & x = a \\ 0, & x \neq a \end{cases} \quad \text{and} \quad \nu_{x,p}^{t+1} = \begin{cases} 1, & x = b \\ 0, & x \neq b \end{cases}$$

The energy function (10.5) can be broken into three terms $E_p$, $E_q$ and $E_o$, that is, $E = E_p + E_q + E_o$. $E_p$ stands for the energy of the columns $p-1$, $p$ and $p+1$ of the rows $a$ and $b$. $E_q$ stands for the energy of the groups. $E_o$ stands for the energy of the rest columns and rows. Then $E_p$ is computed by

$$E_p = \frac{A}{2}\left(\sum_i \nu_{a,i} \sum_j \nu_{b,j}\right) + \sum_x \sum_y d_{xy}\nu_{x,p}(\nu_{y,p+1} + \nu_{y,p-1}). \tag{10.15}$$

$E_o$ is computed as

$$E_o = \frac{A}{2}\sum_{x \neq a,b}\sum_i (\nu_{x,i}\sum_{j \neq i}\nu_{x,j}) + \frac{1}{2}\sum_x\sum_y \sum_{i \neq p-1,p,p+1} d_{xy}\nu_{x,i}(\nu_{y,i-1} + \nu_{y,i+1})$$
$$+ \frac{1}{2}\sum_x\sum_y d_{xy}\nu_{x,p-1}\nu_{x,p-2} + \frac{1}{2}\sum_x\sum_y d_{xy}\nu_{x,p+1}\nu_{x,p+2}. \tag{10.16}$$

And $E_q$ is calculated as

$$E_q = \frac{B}{2}\sum_{i=1}^{m} S_i^2 - \frac{n(n+m-1)}{m} + \frac{Bn^2}{2m}. \qquad (10.17)$$

Now, we investigate the change of $E$ under the WTA learning rule of the CCM.

**Theorem 10.2.** *Let $A - Bn^2 > 2d_{max} - d_{min}$, where $d_{max}$ and $d_{min}$ is the maximum and the minimum distance respectively, $n$ is the total city number, then the CCM is always convergent to valid states.*

*Proof.* Three cases are studied in the follow analysis.

**Case 1: $(a, p)$ and $(b, p)$ are both not group neurons**

In this case $0 < a, b < n$, it can be seen that only $E_p$ will be affected by the state of column $p$. According to equation (10.15), $E_p^t$ and $E_p^{t+1}$ is computed by

$$E_p^t = -\frac{A}{2}l(l-1) - \sum_y d_{ay}(\nu_{y,p-1} + \nu_{y,p+1}). \qquad (10.18)$$

$$E_p^{t+1} = -\frac{A}{2}(l-1)(l-2) - \sum_y d_{by}(\nu_{y,p-1} + \nu_{y,p+1}). \qquad (10.19)$$

Then

$$E_p^{t+1} - E_p^t = -A(l-1) + \sum_y d_{by}(\nu_{y,p-1} + \nu_{y,p+1})$$
$$- \sum_y d_{ay}(\nu_{y,p-1} + \nu_{y,p+1}).$$

At the same time, the input to neuron $(a, p)$ and $(b, p)$ before updating are computed as follows:

$$Net_{a,p}^t = -A(l-1) - \sum_y d_{ay}(\nu_{y,p-1} + \nu_{y,p+1}) - BS_{p'} + \frac{n}{m}.$$

$$Net_{b,p}^t = -\sum_y d_{by}(\nu_{y,p-1} + \nu_{y,p+1}) - BS_{p'} + \frac{n}{m}.$$

According to the discussion of the above subsection, when

$$A - Bn^2 > 2d_{max} - d_{min}$$

is guaranteed, then
$$Net_{b,p} > Net_{a,p}(a \neq b)$$
can be ensured. This implies that

## 10.4 Valid Solutions and Convergence Analysis of CCM for MTSP

$$-A(l-1) + \sum_y d_{by}(\nu_{y,p-1} + \nu_{y,p+1}) - \sum_y d_{ay}(\nu_{y,p-1} + \nu_{y,p+1}) < 0. \quad (10.20)$$

So we can clearly deduce that

$$E_p^{t+1} - E_p^t < 0.$$

Thus, $E^{t+1} - E^t < 0$.

**Case 2: $(a, p)$ is a group neuron while $(b, p)$ is not**

In this case $n < a < n+m$ while $0 < b < n$, it can be concluded that not only $E_p$ but also $E_q$ would be changed before and after the WTA updating rule in $p$-th column. $(a, p)$ in a group of neurons, can be active before updating and nonactive after updating. This implies that two connected groups that are distinguished by neuron $(a, p)$ before updating merge into a single group after updating. Suppose $s_g^t$ and $s_h^t$ represent the city's number of those two connected groups before updating, and $s_p^{t+1}$ stands for the city's number of this merged group after updating. Then, $s_g^t + s_h^t = s_p^{t+1}$. According to equation (10.17), $E_q^t$ and $E_q^{t+1}$ are computed as:

$$E_q^t = \frac{B}{2}(s_g^t)^2 + \frac{B}{2}(s_h^t)^2 + \frac{B}{2}\sum_{i \neq g,h}^m S_i^2 - \frac{n(n+m-1)}{m} + \frac{Bn^2}{2m}. \quad (10.21)$$

$$E_q^{t+1} = \frac{B}{2}(s_p^{t+1})^2 + \frac{B}{2}\sum_{i \neq g,h}^m S_i^2 - \frac{n(n+m-1)}{m} + \frac{Bn^2}{2m}. \quad (10.22)$$

Now, we consider the change of $E$:

$$E_p^{t+1} + E_q^{t+1} - E_p^t - E_q^t = -A(l-1) + \sum_y d_{by}(\nu_{y,p-1} + \nu_{y,p+1})$$
$$- \sum_y d_{ay}(\nu_{y,p-1} + \nu_{y,p+1}) + Bs_g^t s_h^t. \quad (10.23)$$

According to the discussion of the above subsection (10.12) and (10.13), when $A - Bn^2 > 2d_{max} - d_{min}$, $Net_{b,p} > Net_{a,p}$ are ensured, it implies that

$$-\sum_y d_{by}(\nu_{y,p-1} + \nu_{y,p+1}) - Bn^2 + \frac{n}{m}$$
$$> -A(l-1) - \sum_y d_{ay}(\nu_{y,p-1} + \nu_{y,p+1}) + \frac{n}{m}. \quad (10.24)$$

that is

$$-A(l-1) + \sum_y d_{by}(\nu_{y,p-1} + \nu_{y,p+1}) - \sum_y d_{ay}(\nu_{y,p-1} + \nu_{y,p+1}) < -Bn^2. \quad (10.25)$$

Applying (10.25) to (10.23), we get that

$$E_p^{t+1} + E_q^{t+1} - E_p^t - E_q^t < -Bn^2 + Bs_g^t s_h^t$$
$$< -Bns_h^t + Bs_g^t s_h^t$$
$$< -Bs_h^t(n - s_g^t) < 0. \qquad (10.26)$$

It also implies that $E^{t+1} - E^t < 0$.

**Case 3:** $(b, p)$ **is group neuron while** $(a, p)$ **is not**

In this case $n < b < n + m$ while $0 < a < n$. As in case 2, both $E_p$ and $E_q$ are changed with the change of neuron's state in column $p$. $(b, p)$ is group neuron, it can be active after updating while nonactive before updating. This implies that one group which neuron $(a, p)$ be contained in before updating would be divided into two groups after updating. Suppose $s_g^{t+1}$ and $s_h^{t+1}$ represent the city number of these two connected groups after updating respectively, and $s_p^t$ stand for the city number of the original group before updating. Then, $s_g^{t+1} + s_h^{t+1} = s_p^t$.

As in case 2, $E_q^t$ and $E_q^{t+1}$ is computed as following, respectively.

$$E_q^t = \frac{B}{2}(s_p^t)^2 + \frac{B}{2} \sum_{i=1, i \neq g, h}^{m} S_i^2 - \frac{n(n+m-1)}{m} + \frac{Bn^2}{2m}.$$

$$E_q^{t+1} = \frac{B}{2}[(s_g^{t+1})^2 + (s_h^{t+1})^2] + \frac{B}{2} \sum_{i \neq g, h}^{m} S_i^2 - \frac{n(n+m-1)}{m} + \frac{Bn^2}{2m}.$$

Now we consider the change of $E$:

$$E_p^{t+1} + E_q^{t+1} - E_p^t - E_q^t = -A(l-1) + \sum_y d_{by}(\nu_{y,p-1} + \nu_{y,p+1})$$
$$- \sum_y d_{ay}(\nu_{y,p-1} + \nu_{y,p+1}) - Bs_g^{t+1} s_h^{t+1}.$$

From $A - Bn^2 > 2d_{max} - d_{min}$ and (10.20), it follows that

$$E_p^{t+1} + E_q^{t+1} - E_p^t - E_q^t < -Bs_g^{t+1} s_h^{t+1} < 0$$

since $B > 0, s_h^{t+1} > 1, s_g^{t+1} > 1$. This also implies that $E^{t+1} - E^t < 0$.

In conclusion, the energy always decreases during the process of WTA updating if $A - Bn^2 > 2d_{max} - d_{min}$. In other words, the CCM model is always convergent under the WTA updating rule.

## 10.5 Simulation Results

An application of MTSP is studied in this section. Suppose there is a city that is suffering from floods. The government wants to send out 3 investigators starting from the government building to investigate the disaster areas (all

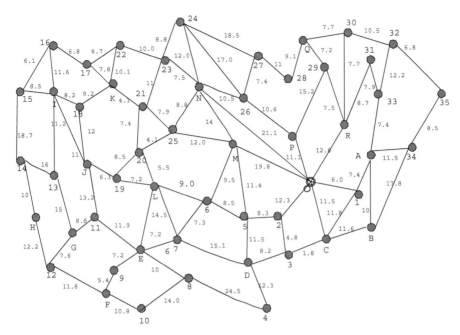

**Fig. 10.3.** The city network.

the towns in this city) and return to the starting position. The city's road graph is given in Fig. 10.3. The aim is to find the nearly equal shortest tour for each investigator.

There are 53 cities in this network, denoted as $A - R$ and $1 - 35$, $O$ is the site of government. The arcs represent the road among cities. The length of each road is given by the number on the graph.

It is simple to say that the distance triangularity of the given network does not hold. In order to ensure the distance triangularity of the network, the shortest path distances between two vertexes are used in our simulations. If $v_i$ and $v_j$ are not adjacent, then we can add a virtual arc between $v_i$ and $v_j$, denoted as $w_{ij}$, and associate the length of the shortest path from $v_i$ to $v_j$ with the value of $w_{ij}$. After this process, the distance triangularity of network holds, then this example can be solved by the method introduced in above sections. After getting the solutions, the original path should be resumed. The method used to resume the result path is shown in the following example.

For example (as shown in Fig. 10.4), if $b$ and $c$ were not adjacent in the result path $(\cdots a \quad b \quad c \cdots)$, and the shortest path between $b$ and $c$ is $b\,d\,c$, then the resumed path is $(\cdots a \quad b \quad d \quad c \cdots)$.

In this section we will validate the criteria of the previous two theorems by computer simulations. Here we adopt the algorithm introduced in Section 3. In this work, 500 simulations have been performed with different scale of parameters $A$ and $B$.

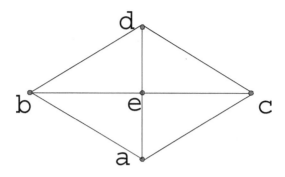

**Fig. 10.4.** The road example

In these simulations, $MAX = 132.8$, $MIN = 1.8$ and $n = 53$. Three case are studied in the simulation based on the value of $A$ and $B$: (1) $A - Bn^2 < 2MAX - MIN$, set $A = 300$ AND $B = 0.05$; (2) $A - Bn^2 \doteq 2MAX - MIN$, set $A = 400$ AND $B = 0.05$; (3) $A - Bn^2 > 2MAX - MIN$, chose $A = 400$ AND $B = 0.01$. The energy change of the system is studied in detail. The energy decreases in all instance of simulation when $A - Bn^2 > 2MAX - MIN$ are guaranteed. We show some good results of energy change in Fig. 10.5. At the same time, the simulation results of case (3) are given in Table 10.1.

**Table 10.1.** The resulting paths

|         | tour path | length(km) |
|---------|-----------|------------|
| GROUP 1 | O→2→5→6→7→L→19→J→11 →E→9→F→10 →8→4→D→3→C→O | 175.4 |
| GROUP 2 | O→1→A→B→34→35→32→33→31→30→R→29 →Q→28→27→24→26→P→O | 180.5 |
| GROUP 3 | O→M→25→20→21→K→18→I→13→G→12→H→14→15 →16→17→22→23→N→O | 197.4 |

Among the several significant developments of general local search methods (Pirlot, 1996) such as simulated annealing and Tabu search, Tabu search appears to be one of the more powerful methods (Habib et al., 2001; Rego, 1998). To evaluate the efficacy of the model we presented, a heuristic algorithm: a Tabu search algorithm with a 4-exchange reference structure is considered as a comparison with our work.

Table 10.2 shows the simulation results for CCM model and the 4-exchange Tabu Search algorithms. The simulation result shows that the CCM works more effective than the Tabu search for solving MTSP in CPU's, while the results obtained by the Tabu search are better. It clearly shows that the balance effect of optimal solution attained by CCM is greatly improved.

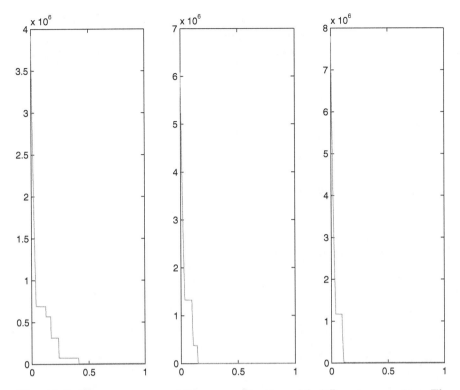

**Fig. 10.5.** The convergence of the energy function with different parameters. The left panel $A = 300, B = 0.05, A - Bn^2 < 2MAX - MINX$, the middle panel $A = 400, B = 0.05, A - Bn^2 \doteq 2MAX - MIN$, the right panel $A = 400, B = 0.01, A - Bn^2 > 2MAX - MIN$

**Table 10.2.** The computational result of CCM for MTSP with tabu search method

| n | m | CCM | | | | | TS | | | | |
|---|---|---|---|---|---|---|---|---|---|---|---|
| | | Max | Min | Dis | BL | CPU | Max | Min | Dis | BL | CPU |
| 53 | 3 | 197.4 | 175.4 | 553.3 | 0.67 | 10.23 | 200.5 | 169.2 | 545.1 | 8.90 | 110.53 |
| | 4 | 130.9 | 121.3 | 511.7 | 1.25 | 15.49 | 156.2 | 98.8 | 493.5 | 23.25 | 113.57 |
| | 5 | 111.9 | 93.6 | 499.2 | 2 | 22.32 | 133.6 | 67.8 | 487.8 | 47 | 109.23 |

Max: the maximal cost of the $m$ closed tour; Min: the least cost of the $m$ closed tour; Dis: the total cost of the resuled solution; BL: the degree of balance; CPU: the CPU time.

## 10.6 Conclusions

In this chapter, MTSP is described by a mathematical model. Based on the description, a columnar competitive model was employed to solve MTSP. The stability condition of CCM for MTSP was exploited. According to the theoretical analysis, the critical values of the network parameters were found. The simulation results shown that WTA updating rule makes CCM an efficient and fast computational model for solving MTSP.

# 11

# Improving Local Minima of Columnar Competitive Model for TSPs

[1]A new approach to solve the TSP based on columnar competitive model (CCM) has been proposed recently. This method performed much better than the original Hopfield network in terms of both the number and the quality of valid solutions. However, the local minima problem is still an open issue. This chapter studies the performance of the CCM and aims to improve its local minima problem. Mathematical analysis shows that, in principle, the CCM is hardly able to escape from local minimum states. An alternative version of the CCM is presented based on a modified energy function so as to enhance the capability of CCM. A new algorithm is proposed here by combining the alternative model and the original one, which enables the network to find lower energy level when trapped in local minima, so as to reach the optimal or near-optimal state quickly. Simulations are carried out to illustrate the performance of the proposed method.

## 11.1 Introduction

Since the Hopfield network was first used (Hopfield and Tank, 1985) to solve optimization problems, mainly the traveling salesman problem (TSP), much research has been aimed at applying Hopfield neural networks to solve combinatorial optimization problems. It is well known that there are some drawbacks with Hopfield networks when it is applied to optimization problems: invalidity of the obtained solutions, trial-and-error parameter settings, and low computation efficiency and so on (Smith, 1999; Tan et al., 2005). Recently, Tang and Tan et al (Tang et al., 2004) proposed a new approach to solve TSP with a Hopfield network based columnar competitive model (CCM), which is also termed as a maximum neuron model by Y. Takefuji

---

[1] Reuse of the materials of "Improving local minima of columnar competitive model for TSPs", 53(6), 2006, 1353–1362, IEEE Trans. on Circuits and Systems-I, with permission from IEEE.

et al (Takefuji et al., 1992). This method incorporating a winner-takes-all (WTA) learning rule guarantees convergence to valid states and suppression of spurious states. Theoretical analysis and simulation results illustrated that this model offered the advantage of fast convergence with low computational effort, as well as performing much better than the Hopfield network in terms of both the number and the quality of valid solutions, thereby allowing it to solve large scale computational problems effectively. However, local minima are still a problem for the CCM. Moreover, with the increase in the network size, the local minima problem becomes significantly worse. In order to avoid the local minima problem, many researchers have adopted different heuristic search techniques, such as chaotic network (Chen and Aihara, 1995; Hasegawa et al., 2002b; Wang and Smith, 1998), and local minima escape (LME) algorithms (Peng et al., 1993; Papageorgiou et al., 1998; Martin-Valdivia et al., 2000), which can be used to find a new state or lower energy level whenever the network is trapped to a local minimum. In this chapter, we propose a new algorithm based on the columnar competitive model to solve lager-scale traveling salesman problems, which allows the network to escape from local minima and converge to the global optimal or near-optimal state quickly.

## 11.2 Performance Analysis for CCM

Let $n$ be the number of cities and let $d_{x,y}$ be the distance between the cities $x$ and $y$, $x, y \in 1, ..., n$. A tour of the TSP can be represented by a $n \times n$ permutation matrix, where each row and each column is corresponds respectively to a particular city and order in the tour. Let $H = \{V \in [0,1]^{n \times n}\}$ and $H_C = \{V \in \{0,1\}^{n \times n}\}$ be a unit hypercube of $R^{n \times n}$ and its corner set, respectively. Given a fixed $V \in H$, define $S_i^V \equiv \sum_x v_{x,i}$ and $S_x^V \equiv \sum_i v_{x,i}$ as the sum of rows and columns. Then the valid tour set for TSP is

$$H_T = \{V \in H_c | S_x = 1, S_i = 1, \forall x, i \in \{1, ..., n\}\} \quad (11.1)$$

In the seminal work of Hopfield, it has been shown that an energy function of the form

$$E = -\frac{1}{2} \mathbf{v}^T \mathbf{W} \mathbf{v} - (\mathbf{i}^b)^T \mathbf{v}. \quad (11.2)$$

can be minimized by the continuous-time neural network if $W$ is symmetric (Hopfield and Tank, 1985), with the parameters: $\mathbf{W} = (W_{i,j})$ is the connection matrix, $i, j = 1, ..., n$ and $n$ is the number of neurons, $\mathbf{i}^b$ is the vector of bias.

The columnar competitive model has a similar connection structure as Hopfield network, but obeys a different updating rule which incorporates winner-takes-all (WTA) column-wise. The intrinsic competitive nature is in favor of the convergence of CCM to feasible solutions, since it can reduce the number of penalty terms compared to Hopfield's formulation.

The associated energy function of CCM for TSP can be written as

## 11.2 Performance Analysis for CCM

$$E(V) = \frac{k}{2}\sum_x\sum_i(\nu_{x,i}\sum_{j\neq i}\nu_{x,j})$$

$$+\frac{1}{2}\sum_x\sum_{y\neq x}\sum_i d_{xy}\nu_{xi}(\nu_{y,i+1}+\nu_{y,i-1}). \tag{11.3}$$

where $k > 0$ is a scaling parameter and $d_{xy}$ is the distance between cities $x$ and $y$. Comparing (11.2) and (11.3), the connection matrix and the external input of the network are computed as follows,

$$W_{xi,yj} = -\{k\delta_{xy}(1-\delta_{ij}) + d_{xy}(\delta_{i,j+1} + \delta_{i,j-1})\}, \tag{11.4}$$
$$\mathbf{i}^b = 0, \tag{11.5}$$

where $\delta_{i,j}$ is the Kronecker's delta.

In the following, we present some analytical results for the CCM.

**Theorem 11.1.** *Given* $\forall V \in H_T$, $\exists C > 0$, *when* $k > C$, *then the evolution would be stopped whenever the network defined by (11.3)-(10.9) is trapped to state $V$.*

*Proof.* Since $V \in H_T$,

$$S_i^V \equiv \sum_x \nu_{x,i} = 1, \tag{11.6}$$

$$S_x^V \equiv \sum_i \nu_{x,i} = 1. \tag{11.7}$$

To investigate the evolution of CCM in this case, we consider the input of the network for any column, assumed to be column $i$. Suppose the state at iteration $t$ to be

$$V^t = \begin{pmatrix} \cdots & \vdots & & 0 & \nu_{x,i} & 0 & \vdots & \cdots \\ \cdots & 0 & & 1 & \nu_{s,i} & 0 & 0 & \cdots \\ \cdots & \vdots & & 0 & \nu_{x,i} & 0 & \vdots & \cdots \\ \cdots & 0 & & 0 & \nu_{p,i} & 0 & 0 & \cdots \\ \cdots & \vdots & & 0 & \nu_{x,i} & 0 & \vdots & \cdots \\ \cdots & 0 & & 0 & \nu_{t,i} & 1 & 0 & \cdots \\ \cdots & \vdots & & & 0 & \nu_{x,i} & 0 & \vdots & \cdots \end{pmatrix}$$

and let

$$\nu_{x,i}^t = \begin{cases} 1, & \text{if } x = p \\ 0, & \text{otherwise.} \end{cases} \tag{11.8}$$

Then the input to each neuron in the $i$-th column can be calculated as

$$Net_{p,i}^t = -(d_{ps} + d_{pt}),$$

$$Net^t_{x(x\neq p),i} = -(d_{xs} + d_{xt} + k).$$

Thus,
$$\exists C_i = d_{xs} + d_{xt} - d_{ps} - d_{pt}$$

when $k > C_i$
$$Net^t_{p,i} > Net^t_{x,i}(x = 1..n, x \neq p)$$

according to equation (10.9), the outputs of the $i-th$ columnar are

$$v^{t+1}_{x,i} = v^t_{x,i} = \begin{cases} 1, & \text{if } x = p \\ 0, & \text{otherwise.} \end{cases} \quad (11.9)$$

Clearly, the competition of the neurons in $i$-th columnar will be stopped. Hence
$$\exists C = \max\{C_1, C_2, ..., C_n\},$$

when $k > C$, the evolution of the CCM would be stopped whenever the network is trapped into state $V$, which is obviously a local minimum of the energy function.

This completes the proof.

To gain a deeper insight into the performance of CCM for TSP, we employ a simple example, which consists of four cities as shown in Fig. 11.1.

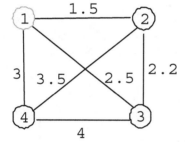

**Fig. 11.1.** The example. 1 is the start city.

Seen from Fig. 11.1, clearly there are 6 valid tours:

$T_1 : 1 \to 2 \to 3 \to 4 \to 1$, $T_2 : 1 \to 2 \to 4 \to 3 \to 1$, $T_3 : 1 \to 3 \to 2 \to 4 \to 1$
$T_4 : 1 \to 4 \to 3 \to 2 \to 1$, $T_5 : 1 \to 3 \to 4 \to 2 \to 1$, $T_6 : 1 \to 4 \to 2 \to 3 \to 1$.

Let $S(T_i)(i = 1 \ldots 6)$ denote the length of tour $T_i$, then

$$S(T_1) = S(T_4) = 10.7, \ S(T_2) = S(T_5) = 11.5, \ S(T_3) = S(T_6) = 11.2.$$

Obviously, $T_1$ and $T_4$ are the global optimal solution, and the rest are local minima. The competition will stop when the network converges to any one of the local minimum, such as $T_2, T_3, T_5$ or $T_6$. Moreover, the final solution is randomly selected from these six solutions, since the initial value of $V$ is obtained randomly.

The following table gives the experimental results of 100 runs, which reports the probability of $T_1, T_2, T_3, T_4, T_5, T_6$ that the CMM can converge to: the solutions distribute in the valid state's space nearly uniformly.

**Table 11.1.** The probability of the convergence to every valid solution

| Solution | $T_1$ | $T_2$ | $T_3$ | $T_4$ | $T_5$ | $T_6$ |
|---|---|---|---|---|---|---|
| Convergence (%) | 17 | 17 | 16 | 18 | 17 | 15 |

The above analysis shows that the local minima problem is still a restriction for CCM when it is applied to the TSP, especially when the number of cities becomes large. The selection of parameter values cannot guarantee the network converging to the global minima, though it can lead the network arriving at the valid solution quickly. In the next section, an algorithm is presented to tackle this problem, which permits the network to escape from local minima and converge to optimal or near-optimal states effectively.

## 11.3 An Improving for Columnar Competitive Model

To improve the performance of the original Hopfield network, Wilson and Pawley (Wilson and Pawley, 1988b) reexamined Hopfields formulation for the TSP and encountered difficulties in converging to valid tours. They found that the solutions represented by local minima may correspond to invalid tours. To remedy this problem, Brandt et al. (Brandt *et al.*, 1988) proposed a modified energy function, which gives better convergence to valid tours than Hopfield and Tanks formulation, but the tour quality is not as good as the original one (Protzel *et al.*, 1994). To study the performance of the columnar competitive model, the energy formulation is also an essential factor that should be taken into account. In this section, by examining the competitive updating rule, the condition that ensures escaping from local minima is derived. Subsequently, an alternative representation of the CCM satisfying this condition is proposed.

### 11.3.1 Some Preliminary Knowledge

From the view point of mathematical programming, the TSP can be described as a quadratic 0-1 programming problem with linear constraints,

$$\min \quad E^{obj}(V) \tag{11.10}$$

$$\text{subject to} \quad \begin{cases} S_i = \sum_{x=1}^{n} v_{xi} & \forall i \in \{i = 1, ..., n\} \\ S_x = \sum_{i=1}^{n} v_{xi} & \forall x \in \{x = 1, ..., n\}, \end{cases} \tag{11.11}$$

where $v_{xi} \in \{0, 1\}$ and $E^{obj}(V)$ is the total tour length. Then energy function to be minimized by the network is

$$E(V) = E^{obj}(V) + E^{cns}(V) \tag{11.12}$$

where $E^{cns}(V)$ are the constraints described by (11.11). The input of the network also can be written as the sum of two portions, accordingly,

$$Net_{x,i}(V) = Net_{x,i}^{obj}(V) + Net_{x,i}^{cns}(V) \tag{11.13}$$

where $Net_{x,i}^{obj}(V)$ is the input produced by (11.10) and $Net_{x,i}^{cns}(V)$ is by (11.11). Moreover, from the perspective of the original neural network, we can write $Net_{x,i}^{cns}(V)$ as

$$Net_{x,i}^{cns}(V) = Net_{x,i}^{cnsw}(V) + Net_{x,i}^{cnsi}(V) \tag{11.14}$$

$Net_{x,i}^{cnsw}(V)$ is the portion of input brought from the connection matrix $W$ and $Net_{x,i}^{cnsi}(V)$ results from the external input $i^b$ of the network.

Apparently, the expressions of $Net_{x,i}^{obj}(V)$, $Net_{x,i}^{cnsw}(V)$ and $Net_{x,i}^{cnsi}(V)$ are different for different neural representations. According to the representation of equation (11.3) presented by (Tang et al., 2004), it holds that

$$Net_{x,i}^{obj}(V) = -\sum_{y} d_{xy}(\nu_{y,i-1} + \nu_{y,i+1}),$$

$$Net_{x,i}^{cnsw}(V) = -k \sum_{j \neq i} \nu_{x,j},$$

$$Net_{x,i}^{cnsi}(V) = 0.$$

Hence, for $V \in H_T$ it is obvious that

$$Net_{x,i}^{cnsw}(V) = \begin{cases} -k, & \text{if } v_{x,i} = 0 \\ 0, & \text{if } v_{x,i} = 1 \end{cases}$$

The following theorem presents the condition ensuring the CCM to escape from local minima.

**Theorem 11.2.** *Give $\forall V \in H_T$ and $V$ is not the global optimal solution, if*

$$Net_{x,i}^{cnsw}(V) = \begin{cases} -k, & \text{if } v_{x,i} = 0 \\ -k, & \text{if } v_{x,i} = 1 \end{cases} \tag{11.15}$$

*are guaranteed, then the CCM will escape from the local minimum state $V$.*

## 11.3 An Improving for Columnar Competitive Model

*Proof.* When

$$Net_{x,i}^{cnsw}(V) = \begin{cases} -k, & \text{if } v_{x,i} = 0 \\ -k, & \text{if } v_{x,i} = 1 \end{cases} \quad (11.16)$$

are guaranteed, the input to the neurons in $i$-th column can be computed as, according to (11.13)-(11.14)

$$Net_{x,i}(V) = Net_{x,i}^{obj}(V) + Net_{x,i}^{cnsw}(V) + Net_{x,i}^{cnsi}(V)$$
$$= -\sum_y d_{xy}(v_{y,i-1} + v_{y,i+1}) - k + Net_{x,i}^{cnsi}(V), \quad (11.17)$$

where $Net_{x,i}^{cnsi}(V)$ is a constant.

Since $V \in H_T$, there is only one active neuron in $i-th$ column. Suppose $(p, i)$ is the active neuron, $(r, i)$ one of the inactive neuron, then their input difference is computed by

$$Net_{r,i}(V) - Net_{p,i}(V) = \sum_y d_{py}(v_{y,i-1} + v_{y,i+1}) - \sum_y d_{ry}(v_{r,i-1} + v_{r,i+1})$$
$$= (d_{pa} + d_{pb}) - (d_{ra} + d_{rb}) \quad (11.18)$$

It shows that the inputs difference between any two neurons in column $i$ is independent of the value of $k$, but only depending on the distance terms. The neuron which has the shortest path to city $a$ and $b$ will thus win the competition.

To ensure the network escape from the local minimum $V$, the input to the active neuron $v_{p,i}$ ($v_{p,i} = 1$) should not be the maximal one, i.e.,

$$Net_{p,i}(V) < \max\{Net_{1,i}(V), Net_{2,i}(V), ..., Net_{n,i}(V)\}, \quad (11.19)$$

considering the updating rule of CCM.

Since $V$ is not the global optimal solution, a column $j$ should exist, in which the the input to the active neuron $(p, j)$ is not maximal. That is to say there exists a neuron $(r, j)$ in formulation (11.19) such that, $i = j$, $Net_{r,j}(V) - Net_{p,j}(V) > 0$. Therefore, the network will escape from $V$. This completes the proof.

In the sequel, a new neural representation following the condition of Theorem 11.2 will be introduced.

### 11.3.2 A Modified Neural Representation for CCM

Consider a modified energy function formulation as follows

$$E'(V) = \frac{k}{2} \sum_x \sum_i \left( v_{x,i}(1 - \sum_j v_{x,j}) \right)$$
$$+ \frac{1}{2} \sum_x \sum_{y \neq x} \sum_i d_{xy} v_{xi}(v_{y,i+1} + v_{y,i-1}). \quad (11.20)$$

where $k > 0$ is a scaling parameter and $d_{xy}$ is the distance between cities $x$ and $y$. Comparing (11.2) and (11.3), the connection matrix and the external input of the network are computed as follows,

$$W'_{xi,yj} = -\{k\delta_{xy} + d_{xy}(\delta_{i,j+1} + \delta_{i,j-1})\}, \quad (11.21)$$

$$\mathbf{i'^b} = -k, \quad (11.22)$$

where $\delta_{i,j}$ is the Kronecker's delta. The input to neuron $(x, i)$ is calculated by

$$Net'_{x,i} = \sum_y \sum_j (W'_{xi,yj} \nu_{yj}) + \mathbf{i'^b}$$
$$= -\sum_y d_{xy}(\nu_{y,i-1} + \nu_{y,i+1}) - k \sum_j \nu_{x,j} - k. \quad (11.23)$$

For $V \in H_T$, it is derived that

$$Net'^{cnsw}_{x,i}(V) = -k \sum_j \nu_{x,j} = -k. \quad (11.24)$$

This shows that the condition in Theorem 11.2 can be satisfied by the modified energy function (11.20). Thus, the CCM using this energy function will escape from the local minimum state when the network is trapped into a valid solution. On the basis of this modified representation, an improved algorithm is proposed in the next subsection.

### 11.3.3 The Improvement for Columnar Competitive Model

Theorem 11.1 has shown that the CCM, denoted as $C$, defined by (11.3)-(10.9) operates as a local minima searching algorithm. Given any initial state, it will stabilize at a local minimum state. While the modified CCM defined by (11.20)-(11.23), denoted as $C$ will cause the network to escape from the local minimum state and bring the network to a new state, which can be used as a new initial state of $C$. Then the network can proceed further and stabilize at a new local minimum state. Hence, to attain the optimal solution, we can alternately use $C$ and $C$. Starting from this idea, we develop a local minima escape algorithm for CCM. This algorithm is a realization of combining the network disturbing technique and the local minima searching property of CCM.

The connections and the biases for both networks are as follows:

$$C: \begin{cases} W_{xi,yj} = -\{k\delta_{xy}(1 - \delta_{ij}) + d_{xy}(\delta_{i,j+1} + \delta_{i,j-1})\} \\ \mathbf{i^b} = 0 \end{cases} \quad (11.25)$$

and

$$C': \begin{cases} W'_{xi,yj} = -\{k\delta_{xy} + d_{xy}(\delta_{i,j+1} + \delta_{i,j-1})\} \\ \mathbf{i'^b} = -k \end{cases} \quad (11.26)$$

## 11.3 An Improving for Columnar Competitive Model

The networks $C'$ and $C$ have the same architecture but different parameters. Therefore, the neurons of the two networks are one-to-one related and the states of the two networks can be easily mapped to each other. Based on the two networks, the algorithm performs such computations as

$$W_{xi,yj}(V) = \begin{cases} -\{k\delta_{xy}(1-\delta_{ij}) + d_{xy}(\delta_{i,j+1} + \delta_{i,j-1})\}, & \text{if } V \notin H_T \\ -\{k\delta_{xy} + d_{xy}(\delta_{i,j+1} + \delta_{i,j-1})\}, & \text{if } V \in H_T \end{cases} \quad (11.27)$$

and

$$i^b = \begin{cases} 0, & \text{if } V \notin H_T \\ -k, & \text{if } V \in H_T \end{cases} \quad (11.28)$$

Then the input to each neuron (x,i) is calculated as:

$$Net_{x,i}(V) = \begin{cases} -\sum_y d_{xy}(\nu_{y,i-1} + \nu_{y,i+1}) - k\sum_{j \neq i} \nu_{x,j}, & \text{if } V \notin H_T \\ -\sum_y d_{xy}(\nu_{y,i-1} + \nu_{y,i+1}) - k\sum_j \nu_{x,j} - k, & \text{if } V \in H_T \end{cases} \quad (11.29)$$

the updating rule of outputs is the same as the rule given by equation (10.9).

Fig. 11.2 shows the simple block diagram of the improved CCM, where $C_{xi}, x, i = 1, \cdots, n$ are the recurrent connections from $v_{xi}, x, i = 1, \cdots, n$, and all have the same structure as $C_{11}$. In order to decide when to switch between $C$ and $C'$, $n$ single layer perceptrons, $M_1, M_2, ..., M_n$ with identical architecture, such as the weight matrix $W$, the bias $b$ and the transfer function $f$ are defined by

$$W = [1, 1, ..., 1], \qquad b = 0,$$

and

$$f(x) = \begin{cases} 1, & \text{if } x > 0 \\ 0, & \text{otherwise} \end{cases}$$

The input $P_i$ is the vector equal to $(v_{i1}, \ldots, v_{in})$, $i = 1, \ldots, n$. Then the output of $M_i$ can be computed as

$$a_i = f(W_i P_i - b) = \begin{cases} 0, & \text{if } P_i(j) = 0 \ (\forall j = 1, 2, ..., n) \\ 1, & \text{otherwise} \end{cases} \quad (11.30)$$

Subsequently, the logic element $L$ performs the "and" operation,

$$u = \begin{cases} 1, & \text{if } a_i = 1 (i = 1, 2, ..., n) \\ 0, & \text{otherwise} \end{cases}$$

It is clearly shown that the value of $u$ will be equal to 1 if and only if the output $V$ of the network is a valid solution. In this case, the network takes the parameters of $C$.

With the two networks $C$ and $C'$ defined above, now the new algorithm can be expressed as follows: Assuming the network $C$ is at one of its local minima, by disturbing its network parameters according to (11.26), a new network $C'$

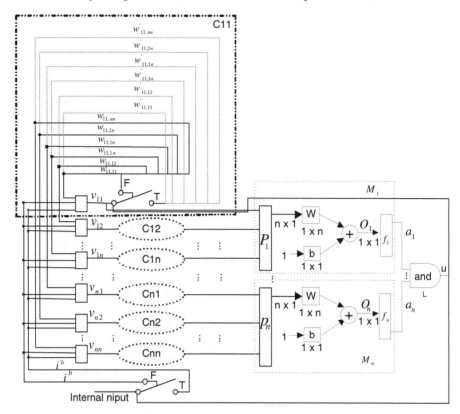

**Fig. 11.2.** Block diagram of the improved CCM

is created. Keep a copy of the current local minima state of $C$ and then set if as the initial state of $C$. Then the network $C$ will proceed and lead the system to escape from the local minimum. The new state of $C$ will be mapped back to $C$ as its new initial state. Then check whether the network $C$, starting from the net state, can converge to a new local minima state which is at a lower energy level than the present local minimum state. If it can, the new local minima state of $C$ is accepted; this completes an iteration. The next iteration will follow the same procedure. The algorithm will terminate if it reaches a preset number or if a new local minima state has been accepted for a certain number of consecutive iterations.

In summarizing the algorithm, the following notations are used:
Notations:
$V^{Obi}$: the best solution at present;
$E^{obj}$: the system's energy when the network reach the state $V^{Obi}$;
$RepeatCount$: the current repeat count of the network;
$NECount$: the repeat count of the network when the no better solution are found;

*MaxRCount*: the maximal value of *RepeatCount*;
*MaxNECount*: the maximal value of *NECount*.

### Algorithm 11.1 Escape Local Minima

1. Initialize the network with a starting state $V^{init} \in H_T$, this means each neuron of the network has a value $v_{xi} = 0$ or $v_{xi} = 1$.
2. Select a column $i$ (e.g., the first column $i = 1$). Set $E^{obj} = E^{init}$, $RepeatCount = 0$, $ENCount = 0$.
3. Search for valid solution: $V^{val}$
    a) Set $V^{val} := V^{init}$, if $V^{val} \in H_T$ then goto step 3.
    b) Compute the input $Net_{x,i}$ of each neuron in $i$-th column using $Net_{x,i} = -\sum_y d_{xy}(\nu_{y,i-1} + \nu_{y,i+1}) - k\sum_{j \neq i} \nu_{x,j}$ shown in (11.29).
    c) Apply WTA and update the outputs of the neurons in $i$-th column using equation (10.9), then the network reaches a new state $V^{cur}$, set $V^{init} := V^{cur}$.
    d) Go to the next column, set $i := ((i+1) \mod n)$ and then repeat step 3.
4. Bring the network to escape from a local minimum state
    a) $RepeatCount := RepeatCount + 1$.
    b) If $E^{obj} < E^{val}$ then $V^{obj} := V^{val}$; $E(V^{obj}) := E(V^{val})$ else $ENCount := ENCount + 1$.
    c) If $((RepeatCount > MaxRCount)$ or $(ENCount > MaxENCount))$ then go to step 5.
    d) Compute the input $Net_{x,i}$ of each neuron in $i$-th column using $Net_{x,i} = -\sum_y d_{xy}(\nu_{y,i-1} + \nu_{y,i+1}) - k\sum_j \nu_{x,j} - k$ shown in (11.29).
    e) Apply WTA and update the outputs of the neurons in $i$-th column using equation (10.9), then the network reach a new state $V^{cur}$, set $V^{init} := V^{cur}$.
    f) Set $i := ((i+1) \mod n)$ go to step 3.
5. Stop updating and output the final solution $V^{obj}$ and the corresponding energy: $E(V^{obj})$.

## 11.4 Simulation Results

To verify the theoretical results and the effectiveness of the improved CCM in improving local minima when applied to TSP, two experiments were carried out. These programs were coded in MATLAB 6.5 run in an IMB PC with Pentium 4 2.66 GHz and 256 RAM.

The first experiment is on three 10-city problems where the city coordinates are shown in Table II. The first data set is the one used in (Hopfield and Tank, 1985) and (Wilson and Pawley, 1988a) and the other sets are randomly generated within the unit box. The optimal tours for these data sets are known to be $(1,4,5,6,7,8,9,10,2,3)$, $(1,4,8,6,2,3,5,10,9,7)$ and

$(1, 10, 3, 9, 4, 6, 5, 2, 7, 8)$, respectively, with the corresponding minimal tour lengths being 2.6907, 2.1851 and 2.6872.

**Table 11.2.** City Coordinates for The Three 10-City Problems

| i | X | Y | X | Y | X | Y |
|---|---|---|---|---|---|---|
| 1 | 0.4000 | 0.4439 | 0.4235 | 0.3798 | 0.3050 | 0.6435 |
| 2 | 0.2439 | 0.1463 | 0.5155 | 0.7833 | 0.8744 | 0.3200 |
| 3 | 0.1707 | 0.2293 | 0.3340 | 0.6808 | 0.0150 | 0.9601 |
| 4 | 0.2293 | 0.7610 | 0.4329 | 0.4611 | 0.7680 | 0.7266 |
| 5 | 0.5171 | 0.9414 | 0.2259 | 0.5678 | 0.9708 | 0.4120 |
| 6 | 0.8732 | 0.6536 | 0.5798 | 0.7942 | 0.9901 | 0.7446 |
| 7 | 0.6878 | 0.5219 | 0.7604 | 0.0592 | 0.7889 | 0.2679 |
| 8 | 0.8488 | 0.3609 | 0.5298 | 0.6029 | 0.4387 | 0.4399 |
| 9 | 0.6683 | 0.2536 | 0.6405 | 0.0503 | 0.4983 | 0.9334 |
| 10 | 0.6195 | 0.2634 | 0.2091 | 0.4154 | 0.2140 | 0.6833 |

To examine into the relationship between the value of $k$ and the convergence of CCM and the improved CCM, 500 simulations were performed for each problem. The simulation results are given in Tables III and IV.

**Table 11.3.** The Performance of CCM Applied to 10-City TSP

| k | Problem 1 | | | Problem 2 | | | Problem 3 | | |
|---|---|---|---|---|---|---|---|---|---|
|   | Valid | Invalid | Ave | Valid | Invalid | Ave | Valid | Invalid | Ave |
| 0.2 | 0 | 500 | - | 0 | 500 | - | 0 | 500 | - |
| 0.6 | 346 | 154 | 3.0451 | 221 | 279 | 2.4763 | 112 | 388 | 3.1510 |
| 0.8 | 450 | 50 | 3.1999 | 386 | 114 | 2.5676 | 230 | 270 | 3.2430 |
| 1.1 | 489 | 11 | 3.2484 | 473 | 26 | 2.6782 | 362 | 38 | 3.3664 |
| 1.5 | 496 | 4 | 3.2956 | 500 | 0 | 2.7212 | 414 | 86 | 3.5241 |
| 1.7 | 500 | 0 | 3.3055 | 500 | 0 | 2.7914 | 472 | 28 | 3.6154 |
| 2.2 | 500 | 0 | 4.2484 | 500 | 0 | 3.5796 | 500 | 0 | 4.6110 |
| 3.0 | 500 | 0 | 4.3477 | 500 | 0 | 3.6237 | 500 | 0 | 4.7318 |
| 10.0 | 500 | 0 | 4.6345 | 500 | 0 | 3.8642 | 500 | 0 | 5.1039 |

"Ave" means the average length.

The simulation results show that both the CCM and the improved CCM can converge to the valid solution with a percentage 100%, when $k > 2.2$. But the average length of the valid tour for the 10-city problems 1, 2 and 3, can be reduced by 32.52%, 38.62% and 39.67%, by the improvement to CCM. That is to say that the improved CCM is more effective than CCM when concerned with the tour length of the final solution.

In terms of the solution quality, when $k = 3$, the improved CCM found the minimal tour 45, 21 and 31 times for the 500 runs, this is, 9%, 4.2% and 6.2%

**Table 11.4.** The Performance of Improved CCM Applied to 10-City TSP

| k | Problem 1 | | | Problem 2 | | | Problem 3 | | |
|---|---|---|---|---|---|---|---|---|---|
| | Valid | Invalid | Ave | Valid | Invalid | Ave | Valid | Invalid | Ave |
| 0.2 | 0 | 500 | - | 0 | 500 | - | 0 | 500 | - |
| 0.6 | 345 | 155 | 3.0461 | 299 | 211 | 3.1016 | 150 | 350 | 2.9290 |
| 0.8 | 460 | 40 | 3.0623 | 400 | 100 | 2.2677 | 250 | 250 | 3.1824 |
| 1.1 | 495 | 5 | 2.9770 | 460 | 40 | 2.5080 | 400 | 100 | 3.4126 |
| 1.5 | 497 | 3 | 2.7929 | 500 | 0 | 2.3170 | 439 | 61 | 3.2289 |
| 1.7 | 500 | 0 | 2.7792 | 500 | 0 | 2.2392 | 489 | 11 | 3.1202 |
| 2.2 | 500 | 0 | 2.7769 | 500 | 0 | 2.2435 | 500 | 0 | 2.9003 |
| 3.0 | 500 | 0 | 2.7865 | 500 | 0 | 2.2767 | 500 | 0 | 2.9098 |
| 10.0 | 500 | 0 | 2.8154 | 500 | 0 | 2.2736 | 500 | 0 | 2.9055 |

"Ave" means the average length.

solutions attained by the improved CCM for these three 10-city problems were optimal solution; in comparison, the CCM found the minimal tour 6, 2 and 9 times. Moreover, the improved CCM found the GOOD solution 417, 491 and 435 times for the three data sets, respectively, while the CCM did 57, 56 and 34 times. All the solutions found by improved CCM are ACCEPT, while CCM did only 26.4%, 22.2% and 19.4%. The GOOD indicates the tour distance within 110% of the optimum distance. The ACCEPT indicates the tour distance within 130% of the optimum distance. Fig. 11.3 shows the histograms of the tour lengths found by the two algorithms with parameter $k = 3$.

On the other hand, regarding the convergence rate, the CCM is much faster then the improved CCM, as shown in Fig. 11.4. Table 11.5 summarizes the overall performance of these two models.

**Table 11.5.** Comparison performance for 10-city problems

| Model | Valid (%) | Minima (%) | GOOD (%) | ACCEPT (%) | Convergence |
|---|---|---|---|---|---|
| CCM | 100, 100, 100 | 9, 4.2, 6.2 | 83.4, 98.2, 87 | 100, 100, 100 | fast |
| Improved CCM | 100, 100, 100 | 1.2, 0.4, 1.8 | 11.4, 11.2, 6.8 | 26.4, 22.2, 19.4 | slow |

The theoretical results are also validated upon a 51-city example with data from a TSP data library collected by Reinelt (Reinelt, 1991). This problem is believed to have an optimal tour with a length of 426.

For a total of 10 runs, the average tour length obtained by CCM is 1019, which is 139.2% longer than the optimal tour. While, when using the improved CCM, the average tour length is 536, which is 26.5% longer than the optimal tour; The shortest tour is 519, 21.8% longer than the optimal tour. A typical tour obtained is shown if Fig. 11.5.

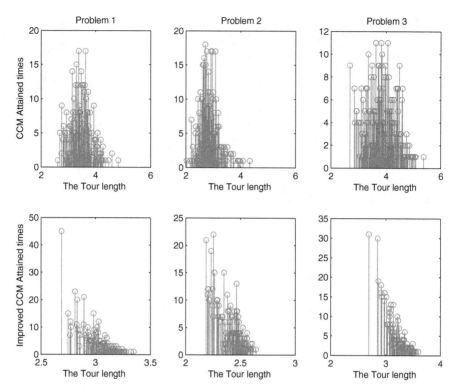

**Fig. 11.3.** Tour length histograms for 10-city problems 1, 2, and 3 produced by CCM (top panels) and Improved CCM (bottom panels).

## 11.5 Conclusions

We presented a method to improve the local minimum problem of the CCM. Mathematical analysis was carried out to show the local searching property of the CCM. Subsequently, we derived the condition that drives the network to escape from local minimum states. Consequently, the network is shown to converge to global optimal or near-optimal states. Simulations have been carried out to illustrate this improvement. From the experimental results, it can be seen that the improved CCM is more effective than the CCM and the original Hopfield model. The improvement is especially significant when the problem scale is small. It is not difficult to see that when the problem scale is larger, further improvement becomes difficult when the networks energy decreases to a certain level, since local minimum problem becomes more serious with larger-scale TSPs.

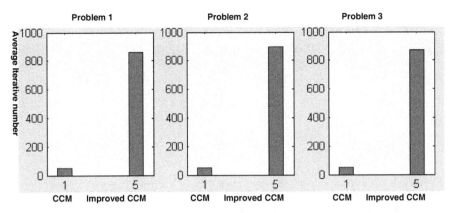

**Fig. 11.4.** Average iteration number histograms for problems 1, 2, and 3 produced by CCM and the improved CCM

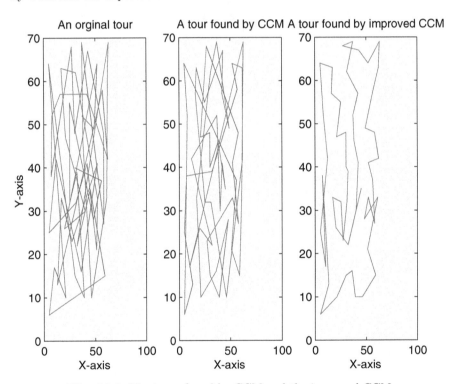

**Fig. 11.5.** The tours found by CCM and the improved CCM

# 12

# A New Algorithm for Finding the Shortest Paths Using PCNN

Pulse coupled neural networks (PCNN), based on the phenomena of synchronous pulse bursts in the animal visual cortex, are different from traditional artificial neural networks. Caulfield and Kinser have presented the idea of utilizing the autowave in PCNN to find the solution of the maze problem. This chapter studies the performance of the autowave in PCNN and apples PCNN to optimization problems, such as the shortest path problem. A multi-output model of pulse coupled neural networks (MPCNN) is studied. A new algorithm for finding the shortest path problem using MPCNN is presented. Simulations are carried out to illustrate the performance of the proposed method.

## 12.1 Introduction

Given a connected graph $G = (V, E)$, a weight $w_{ij}$ for each edge connected $v_i$ with $v_j$, and a fixed vertex $s$ in $V$, to find the shortest paths from $s$ to other nodes in $V$. This is the single-source shortest path problem. It is a classical problem with diverse applications, such as vehicle routing in transportation (Orloff, 1974), traffic routing in communication networks (Ephremids and S.Verdu, 1989), pickup and delivery system (Savelsbergh, 1995), website page searching in internet information retrieval system (Ricca and Tonella, 2001), and so on.

PCNNs is a result of research efforted on the development of artificial neuron model that is capable of emulating the behavior of cortical neurons observed in the visual cortices of animal (Gray and Singer, 1989)-(R. Eckhorn and Dicke, 1990). According to the phenomena of synchronous pulse burst in the cat visual cortex, Eckhorn et al (R. Eckhorn and Dicke, 1990) developed the linking field network. Johnson introduced PCNNs based on the linking model et al (Johnson and Ritter, 1993) and (Johnson, 1993). Since Johnson's work (Johnson, 1994) in 1994, interesting has been increasing in using PCNNs for various applications, such as target recognition (Ranganath and Kuntimad, 1999), image processing (G. Xiaodong and Liming, 2005), motion matching (Xiaofu and Minai, 2004), pattern recognition (Muresan, 2003) and optimization (Caulfield and Kinser, 1999; Junying et al., 2004).

It is well known that many combinatorial optimization problems, such as the well known traveling salesman problem (TSP) (Bai et al., 2006)-(Likas and Paschos, 2002), can be formulated as a shortest path problem. It is believed that neural networks are a viable method for solving these problems (Park, 1989; Zuo and Fan, n.d.). Since the original work of Hopfield (Hopfield and Tank, 1985) in 1985, many researches have aimed at applying the Hopfield neural networks to combinatorial optimization problems (Hopfield and Tank, 1985; Tang et al., 2004; Qu et al., n.d.; Cheng et al., 1996). The major drawbacks of Hopfield networks are: (1). Invalidity of obtained solutions. (2). Trial-and-error setting value process of the networks parameters. (3). Low computational efficiency. As a comparison, the spatiotemporal dynamics of PCNNs provide good computational capability for optimization problems. Caulfield and Kinser have present the idea of utilizing autowaves in PCNNs to find the solution of the maze problem (Caulfield and Kinser, 1999).

In this chapter, a novel modified model of PCNNs : Multi-Output model of Pulse Coupled Neural Networks (MPCNNs) is presented to solve the shortest path problem. Some conditions for the exploring of autowave in MPCNNs are exploited by mathematical analysis, which can guarantee the waves explore from the single "start" node to the multi "end" nodes. The shortest paths from a single "start" node to multi destination nodes can be found simultaneously.

## 12.2 PCNNs Neuron Model

A typical neuron of PCNNs consists of three parts: the receptive fields, the modulation fields, and the pulse generator. The neuron receives input signals from other neurons and external sources through the receptive fields. The receptive fields can be divided into two channels: one is the feeding inputs and the other is the linking inputs. A typical neuron's model of PCNNs is shown in Fig. 17.2.

Suppose, $N$ is the total number of iterations and $n(n = 0...N-1)$ is the current iteration, then the receptive fields can be described by the following equations:

$$F_{ij}(n) = e^{-\alpha_F} F_{ij}(n-1) + V_F \sum_{kl} M_{ij,kl} Y_{kl}(n-1) + I_{ij} \quad (12.1)$$

$$L_{ij}(n) = e^{-\alpha_L} L_{ij}(n-1) + V_L \sum_{kl} W_{ij,kl} Y_{kl}(n-1) + J_{ij} \quad (12.2)$$

where the $(i,j)$ pair is the position of neuron. $F$ and $L$ are feeding inputs and linking inputs, respectively. $Y$ is the pulse output, $\alpha_F$ and $\alpha_L$ are time constants for feeding and linking, $V_F$ and $V_L$ are normalizing constants, $M$ and $W$ are the synaptic weights, $I_{ij}$ and $J_{ij}$ are external inputs.

## 12.2 PCNNs Neuron Model

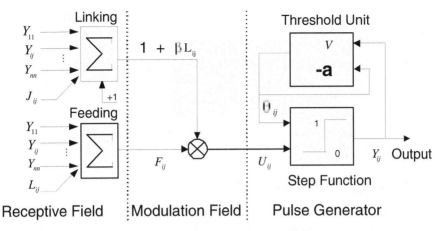

**Fig. 12.1.** PCNNs Neuron Model

The modulation fields generate the internal activity of each neuron, is modeled as follows:

$$U_{ij}(n) = F_{ij}(n)(1 + \beta L_{ij}(n)) \tag{12.3}$$

where $\beta$ is strength of the linking.

The pulse generator receives the result of total internal activity $U_{ij}$ and determines the firing events, can be modeled as

$$Y_{ij}(n) = \text{step}(U_{ij}(n) - \theta_{ij}(n-1)) = \begin{cases} 1, & \text{if } U_{ij}(n) > \theta_{ij}(n-1) \\ 0, & \text{otherwise} \end{cases} \tag{12.4}$$

$$\theta_{ij}(n) = e^{-\alpha_\theta}\theta_{ij}(n-1) + V_\theta Y_{ij}(n-1) \tag{12.5}$$

where $\theta_{ij}(n)$ represents the dynamic threshold of the neuron $(i, j)$. $\alpha_\theta$ and $V_\theta$ are the time constant and the normalization constant, respectively. If $U_{ij}$ is greater than the threshold, the output of neuron $(i, j)$ turns into 1, neuron $(i, j)$ fires, then $Y_{ij}$ feedbacks to make $\theta_{ij}$ rise over $U_{ij}$ immediately, then the output of neuron $(i, j)$ turn into 0. This produces a pulse output. It is clear that the pulse generator is responsible for the modeling of the refractory period of spiking.

The basic idea in solving optimization problems is in utilizing the autowave in PCNNs. In the next section, we propose a Multi-Output model of PCNNs which can mimic the autowaves in PCNNs and find the shortest paths from a single "start" node to multi destination nodes simultaneously.

## 12.3 The Multi-Output Model of Pulse Coupled Neural Networks (MPCNNs)

In this section, we will introduce the designation of Multi-Output model of pulse coupled neural networks (MPCNNs), in which a linear attenuated threshold is used. The reason for the dynamic threshold is designed to be a linearly decreasing function rather than exponentially decreasing function is that such a function would make the autowaves travel uniformly and simplify the digital implementation of the network.

### 12.3.1 The Design of MPCNNs

The proposed neuron's model of MPCNNs is shown in Fig.17.2. Each neuron $i$ in MPCNNs has two outputs: $Y_i$ and $M_i$. $Y_i$ produces the pulse output as that in original PCNNs, while $M_i$ works as an indicator, which indicates the fired state of neuron $i$. $M_i$ is set to be "1" when neuron $i$ fired and "0" if neuron $i$ not fired. $Y_i$ and $M_i$ in the network is designed as the following functions:

$$Y_i(t) = \text{step}(U_i(t) - \theta_i(t)) = \begin{cases} 1, & \text{if } U_i(t) >= \theta_i(t), \\ 0, & \text{otherwise,} \end{cases} \quad (12.6)$$

$$M_i(t) = \begin{cases} 1, & \text{if neuron } i \text{ have fired,} \\ 0, & \text{if neuron } i \text{ have not fired,} \end{cases} \quad (12.7)$$

i=1, 2, ..., N, where $U_i(t)$ and $\theta_i(t)$ are the internal activity and threshold at time $t$, respectively. $N$ is the total neuron's number in the networks.

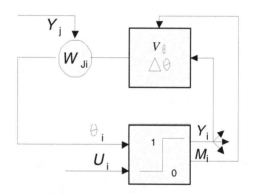

**Fig. 12.2.** The Neuron's Model of MPCNNs

The threshold of neuron $i$ ($i = 1, 2, ..., N$) is designed as equation (12.8). It is clear that the threshold of neuron $i$ is decrease linearly respected to a constant value $\Delta\theta$.

## 12.3 The Multi-Output Model of Pulse Coupled Neural Networks (MPCNNs)

$$\theta_i(t) = \begin{cases} min\{W_{ji} - \Delta\theta, (\theta_i(t - \Delta T) - \Delta\theta)\}, & \text{if } M_i(t - \Delta T) = 0 \text{ and} \\ & Y_j(t - \Delta T) = 1 \text{ and } W_{ji} \neq 0, \forall j \\ & (\text{neuron } j \text{ fired in } t - \Delta T \text{ and} \\ & i \text{ have not fired before } t - \Delta T) \\ V_\theta, & \text{if } Y_i(t - \Delta T) = 1, \\ & (\text{neuron } i \text{ fired in } t - \Delta T) \\ \theta_i(t - \Delta T) - \Delta\theta, & \text{otherwise} \\ & (\text{no neuron fired in } t - \Delta T). \end{cases}$$

where $V_\theta$ is the threshold with a larger value. $W_{ji}$ is the linking strength from neuron $j$ to neuron $i$. If there have no linking connection between neuron $j$ and neuron $i$, then $W_{ji} = 0$.

The internal activity $U_i$ of neuron $i$ determines the firing events. It can be designed as a constant in MPCNNs, as shown in follows

$$U_i = C \quad (i = 1, 2, ..., N) \text{ and } C \text{ is s constant.} \tag{12.8}$$

Now, we have accomplished the design of MPCNNs by Equation (12.6) - (12.8). Some performances of the travelling of autowaves in MPCNNs will be discussed in the next subsection.

### 12.3.2 Performance Analysis of the Travelling of Autowaves in MPCNNs

To study the performances of the travelling of autowaves in MPCNNs, three calculation of crucial datas are involved. There are: (1). $T_i^1$ : neuron $i$'s firing periods. (2). $T_{ij}^1$ : the periods from the time when neuron $i$ fired to the time neuron $j$ ($j = 1, 2, ..., N, j \neq i$) fired. (3). $T^1$ : the periods from the time when the first neuron fired to the time when the last neuron fired. Before calculating these three data, some basic notions of the autowaves are first introduced.

We define the single fired pass as a minimal continuous fired period, in which each neuron is fired at least once. The first neuron fired in a single fired pass is called the "source" of the autowave. On the contrary, the last neuron fired in a single fired pass is called the "end" of the autowave. If each neuron fired only once in a single fired pass, then it is called the normal single fired pass. It is clear that the autowave would explore from the "source", visiting each neuron only once, and reach to the "end" step by step in the normal single fired pass.

(1). *The Calculation of* $T_i^1$

Suppose, after $t_i^1 - \Delta T$ times iterations, neuron $i$ fire at time $t_i^1$, the output of neuron $i$ at time $t_i^1$ is

$$Y_i(t_i^1) = 1,$$

then

$$V_\theta > U_i >= \theta_i(t_i^1)$$

In the next iteration, $t = t_i^1 + \Delta T$, the threshold and output of neuron $i$ are shown in equation (12.9) and (12.10), respectively, according to equation (12.6) and (12.8).

$$\theta_i(t + \Delta T) = V_\theta \qquad (12.9)$$
$$Y_i(t + \Delta T) = Step(U_i - \theta_i(T + \Delta T)) = Step(U_i - V_\theta) = 0 \qquad (12.10)$$

In the following iterations, the threshold will be attenuated with value $\Delta\theta$, the output will remain at 0, until when the value of threshold decreased down to the level of $U_i$, supposed at time $t_i^2$, then

$$T_i^1 = t_i^2 - t_i^1 = \frac{V_\theta - U_i}{\Delta\theta}\Delta T = \frac{V_\theta - C}{\Delta\theta}\Delta T \qquad (12.11)$$

(2). The Calculation of $T_{ij}^1$

Suppose, neuron $i$ fires at time $t_i^1$ and neuron $j$ fires at time $t_j^1$, and there have $k$ neuron fired from time $t_i^1$ to $t_j^1$, then

$$T_{ij}^1 = t_j^1 - t_i^1$$
$$= \frac{\min\{P_{ij}\} - C}{\Delta\theta}\Delta T$$
$$< \frac{(k+1)W_{max} - C}{\Delta\theta}\Delta T, k \in \{1, 2, ..., N-1\} \qquad (12.12)$$

where $\min\{P_{ij}\}$ is the minimal path length from neuron $i$ to neuron $j$, and $W_{max}$ is the maximal one of all the linking strength, this is

$$W_{max} = \max\{W_{ij}\}, 0 < i, j < N,$$

and $N$ is the total neuron numbers.

(3). The Calculation of $T^1$

Suppose, neuron 1 is the first fired neuron and neuron $n$ is the last fired neuron, neuron 1 fired at time $t_1^1$ and neuron $n$ fired at time $t_n^1$. The shortest path length from neuron 1 to neuron $n$ is denoted as $\min\{P_{1n}\}$, then

$$T^1 = t_n^1 - t_1^1 = \frac{\min\{P_{1n}\} - C}{\Delta\theta}\Delta T < \frac{NW_{max} - C}{\Delta\theta}\Delta T \qquad (12.13)$$

With $T_i^1$, $T_{ij}^1$ and $T^1$, the following theorem is introduced.

**Theorem 12.1.** *Suppose, neuron start is the first fired neuron, $N$ is the neuron's number and $W_{max}$ is the maximal one of the linking strengths in MPC-NNs, if the following conditions*

(I) $\begin{cases} C >= \theta_{start}(0) \\ C < \theta_j(0) = V_\theta, \quad j = 1, 2, ...N \text{ and } j \neq start \end{cases}$ \hfill (12.14)

(II) $V_\theta > NW_{max}$ \hfill (12.15)

\hfill (12.16)

## 12.4 The Algorithm for Solving the Shortest Path Problems using MPCNNs

hold, then the autowaves travel from neuron start, visit each neuron in the network step by step, end to the last neuron, each neuron fires only once

*Proof.* When $(I)$ holds, the outputs of the neurons at time $t = 0$ are shown as (12.17) according to equation (12.6).

$$Y_i(0) = \text{step}(U_i(0) - \theta_i(0)) = \text{step}(C - \theta_i(0)) = \begin{cases} 1, & \text{if } i = start \\ 0, & \text{otherwise} \end{cases} \quad (12.17)$$

only neuron *start* fires at time $t = 0$, so the autowaves travel starting from neuron *start*.

When $(II)$ holds,

$$T_i^1 > T^1 > T_{ij}^1 \quad (i, j = 1, 2, ..., N), \quad (12.18)$$

equation (12.18) would be satisfied according to (12.11), (12.12) and (12.13). This is a sufficient condition for the form of the normal single fired pass. So the process of the firstly $N$ neuron's firing in MPCNNs form a normal single fired pass. This is also say that the autowaves of MPCNNs would explore from the "source", visiting each neuron only once, and reach to the "end" step by step in the normal single fired pass.

This completes the proof of Theorem 12.1. This theorem presents a sufficient condition for MPCNNs to optimization problems. In next section, a method for finding the shortest paths using MPCNNs will be discussed in detail.

## 12.4 The Algorithm for Solving the Shortest Path Problems using MPCNNs

In Caulfield and Kinser's work, PCNNs are structured in such a way that each point in the geometric maze figure corresponds to a neuron in the network, and the autowave in the PCNNs travels from each neuron to its neighborhood neuron(s) along the maze from iteration to iteration of the network. While in our work, each neuron in the network corresponds to a node in the graph, and each directed connection between neurons in the network corresponds to the directed edge between the nodes in the graph, the autowave travels along the connection between two neurons.

The algorithm of discrete-time MPCNNs for finding the shortest paths in the graph is expressed in the following steps. The continuous-time MPCNN is the case of $\Delta T \to 0$.

**Algorithm 12.1** *The algorithm for the discrete-time MPCNN*
    // *start* : the starting neuron to fire;
    // $N$ : the total number of neurons;

// $k$ : the current iteration;
// $l$ : the number of neurons that have fired;
// $RouteRecord[i,j]$ : a matrix to record the routes, $RouteRecord[i,j] = 1$ if the autowave travel from neruon $i$ to neuron $j$, else $RouteRecord[i,j] = 0$;
// $PreNode(j)$ : the neuron representing the parent of neuron $j$ ;
// $FiredNeuron$ : the neuron that is fired in the pre-iteration
// $W_{max}$ : the maximal one of the linking strength;

1. Initialize the network, set the iteration number $k = 0$, and the number of neurons that have fired $l = 0$, and the record of route $RouteRecord[i,j] = 0$, $(i,j = 1,2,...,N)$. Set $PreNode(i) = 0$ for $\forall\ i = \{1,2,...,N\}$, and $FiredNeuron = 0$.
2. Set

$$\begin{cases} U_i = C, & i = 1,2,...,N \\ \theta_{start}(0) < C \\ \theta_i(0) > C, & i = 1,2,...,N \text{ and } i \neq start, \\ V_\theta > NW_{max} \end{cases}$$

then the neuron start fires first, as shown in the following

$$\begin{cases} Y_{start}(0) = 1 \\ Y_i(0) = 0 \end{cases} \text{ and } \begin{cases} M_{start}(0) = 1 \\ M_i(0) = 0 \end{cases} \quad (i = 1,2,...,N \text{ and } i \neq start),$$

autowaves are generated and neuron start is the source of the autowaves.
3. For each neuron $j$
   a) Calculate $\theta_j(t)$ according to equation (12.8).
   b) if $W(FiredNeuron, j) < Theta(j)$, set $PreNode(j) = FiredNeuron$.
   c) Calculate $Y_j(t)$ and $M_j(t)$ according to equation (12.6) and (12.7), respectively.
   d) if $Y_j(t) == 1$
      i. Set $FiredNeuron = j$ and $l = l + 1$.
      ii. Record the route, set $RouteRecord[PreNode(j), j] = 1$.
   e) Set $k = k + 1$
4. Repeat step 3 until all destination neurons in the network are in fired state.

This algorithm is a nondeterministic method, which guarantees the globally shortest path solution. The shortest paths from a single "start" node to all multi destination nodes are searched simultaneously in a parallel manner by running the network only once.

## 12.5 Simulation Results

To verify the the theoretical results and the effectiveness of the discrete-time MPCNNs when it be applied to solve the shortest paths problem, several experiments have been carried out. These programs, which were coded in

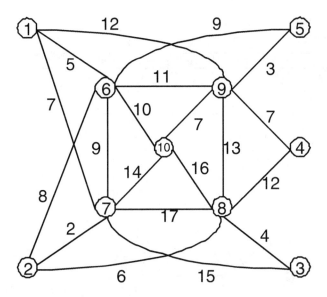

**Fig. 12.3.** A symmetric weighted graph.

MATLAB 6.5, were run in an IBM-compatible PC, Pentium 4 2.66 GHz and 256 MB RAM.

The first experiment is based on a symmetric weighted graph with 10 nodes and 20 edges, as shown as Fig. 12.3. The maximal weight of all edges is: $W_{max} = 17$. In this experiment, $V_\theta$ is taken the value of 200 and $\Delta\theta$ is set to be 1. The internal actives and the initial values of the thresholds are set to be:

$$U_i = C = 0 \quad \text{and} \quad \begin{cases} \theta_{start}(0) < 0 \\ \theta_i(0) = V_\theta, \text{ if } i \neq start. \end{cases} \quad (12.19)$$

Table 12.1 presents the results of the computer simulation. Clearly, all the paths found by the algorithm are globally shortest paths.

The finding process from the "start" node 1 to the others with the time increasing are shown in Fig. 12.4.

Theorem 12.1 has shown that a larger value of $V_\theta$ should be required to guarantee the exploring of autowaves in MPCNNs. To test the relationship between the value of $V_\theta$ and the exploring of autowaves in MPCNNs, 8 simulations will be performed. The "start" node is set to be 1 in each simulation. Different values of $V_\theta$ are taken in each running. In this experiment, $\Delta\theta$ is also set to be 1. The initial value of the internal actives of each neuron $U_i$, and the the initial value of the threshold of each neuron $\theta_i(0)$ is set to be the same value with the above experiment as shown in equation (12.19). Simulation results are given in Table 12.2.

The simulation results in Table 12.2 show that when $V_\theta >= 22$, the resulted solutions are the globally solutions. This is also shows that $V_\theta > NW_{max}$ is

**Table 12.1.** Results of MPCNN for the graph in Fig. 12.3

| Start | Shortest Path | Iters. | Global Sols. |
|---|---|---|---|
| 1 | 1→6→5; 1→6→10; 1→7→2→8→3; 1→9→4 | 23 | YES |
| 2 | 2→6→5; 2→7→1; 2→7→10; 2→8→3; 2→8→4; 2→8→9 | 21 | YES |
| 3 | 3→8→2→6; 3→8→2→7→1; 3→8→4; 3→8→9→5; 3→8→10 | 23 | YES |
| 4 | 4→8→2; 4→8→3; 4→8→2→7; 4→9→5; 4→9→6; 4→9→10 | 29 | YES |
| 5 | 5→6→1; 5→6→2; 5→6→7; 5→9→4; 5→9→8→3; 5→9→10 | 23 | YES |
| 6 | 6→1; 6→2→8→3; 6→5; 6→7; 6→9→4; 6→10 | 21 | YES |
| 7 | 7→1→9→5; 7→2→8→3; 7→2→8→4; 7→6; 7→10 | 25 | YES |
| 8 | 8→2→6; 8→2→7→1; 8→3; 8→4; 8→9→5; 8→10 | 18 | YES |
| 9 | 9→1→7; 9→4; 9→5; 9→6→2; 9→8→3; 9→10 | 21 | YES |
| 10 | 10→6→1; 10→7→2; 10→8→3; 10→9→4; 10→9→5 | 22 | YES |

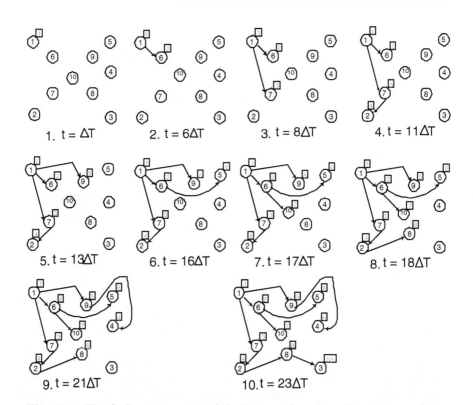

**Fig. 12.4.** The finding processing of the instance from "start" node 1 to others.

a sufficient condition for the globally solutions of the shortest paths problem, but not the necessary conditions.

**Table 12.2.** Simulation Results with the different value of $V_\theta$

| $V_\theta$ | Fired Sequence of Neuron | Spiking Times | Iters. Times | Global Sols. |
|---|---|---|---|---|
| 5 | 1→ 6→ 7→ 2→ 3→ 4→ 5→ 8→ 9→ 10 | 10 | 6 | NO |
| 10 | 1→ 6→ 7→ 2→ 3→ 4→ 5→ 8→ 9→ 10 | 10 | 11 | NO |
| 15 | 1→ 6→ 7→ 2→ 9→ 3→ 4→ 5→ 8→ 10 | 10 | 16 | NO |
| 20 | 1→ 6→ 7→ 2→ 9→ 5→ 10→ 8→ 3→ 4 | 10 | 21 | NO |
| 22 | 1→ 6→ 7→ 2→ 9→ 5→ 10→ 8→ 4→ 3 | 10 | 24 | YES |
| 50 | 1→ 6→ 7→ 2→ 9→ 5→ 10→ 8→ 4→ 3 | 10 | 24 | YES |
| 170 | 1→ 6→ 7→ 2→ 9→ 5→ 10→ 8→ 4→ 3 | 10 | 24 | YES |
| 300 | 1→ 6→ 7→ 2→ 9→ 5→ 10→ 8→ 4→ 3 | 10 | 24 | YES |

Another simulation was made to test the relationship between the number of iterations and the value of $\Delta\theta$, also based on the graph shown in Fig.12.3. Three case of the "start" node: 1, 5 and 10 are involved in this experiment, and all the initial conditions is set to the same value as in the first experiment, expect the value of $\Delta\theta$. The variation of the number of iterations with the different value of $\Delta\theta$ are shown in Table 12.3.

**Table 12.3.** Simulation Results with the different value of $V_\theta$

| Start Node | $\Delta\theta$ | Iterations Times: $k$ | Globally Solution |
|---|---|---|---|
| 1 | $0 < \Delta\theta <= 2.2$ | $k >= 12$ | YES |
|   | $\Delta\theta > 2.2$ | $k <= 12$ | NO |
| 5 | $0 < \Delta\theta < 40$ | $k >= 7$ | YES |
|   | $\Delta\theta >= 40$ | $k < 7$ | NO |
| 10 | $0 < \Delta\theta <= 66$ | $k >= 4$ | YES |
|   | $\Delta\theta > 67$ | $k < 4$ | NO |
| For All | $0 < \Delta\theta <= 2.2$ | - | YES |
|   | $\Delta\theta > 2.2$ | - | NO |

Table 12.3 shows that a small value of $\Delta\theta$ should be required to guarantee the globally solution of the shortest paths problem, while the number of iteration of algorithm will be increased with the decreasing of $\Delta\theta$.

We also apply our algorithm to finding the shortest paths from arbitrarily specified "start" node to all other nodes in randomly generated undirected and symmetric graphs. The graphs are generated in such a way that the $N$ nodes are uniformly distributed in a square with the edge length $D$, and each two nodes is connected with an edge if their distance is no more than $30\%D$, the weight of the edges which connected the two nodes is just the Euclidean distance between the two nodes. An MPCNNs associated with the graph is constructed to find the shortest paths for each such generated graph from

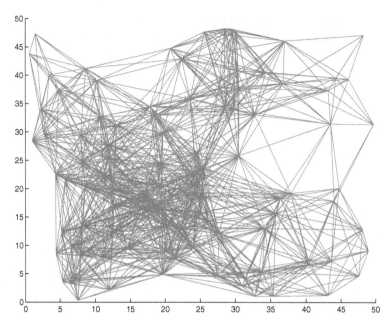

**Fig. 12.5.** A symmetric weighted graph randomly generated with $N = 100$ in a square of edge length $D = 50$

any specified "start" node to all of the other $N - 1$ nodes. Fig. 12.5 shows a generated graph with $D = 50$ and $N = 100$.

The shortest paths searched by the new algorithm from three different "start" nodes to the others for the randomly generated graph are shown in Fig. 12.6.

## 12.6 Conclusions

In this chapter, we present a new approach based on multi-output pulse coupled neural networks (MPCNNs) to solve the shortest path problem. Our works are based on a symmetric weighted graph. This is a nondeterministic method which would guarantee the globally solutions. The shortest paths from a single "start" node to all multi destination nodes are search simultaneously parallelly by running the network only once. The highly parallel computational nature of the network results in faster discovery of the shortest paths if the network is realized with VLSI. It is easy to extend the proposed MPC-NNs to other cases when the topology has some slight changes. The approach also can be apply to many real time applications of shortest paths problems such as website page searching in an internet information retrieval system and traffic routing in communication network.

12.6 Conclusions 189

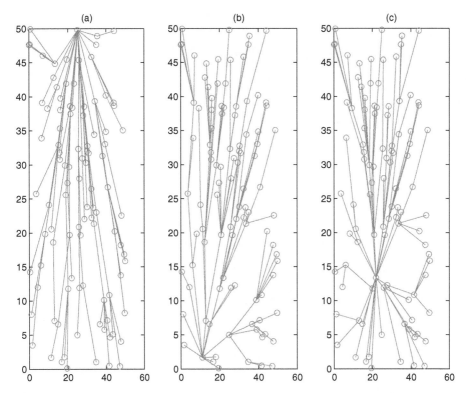

**Fig. 12.6.** (a), (b) and (c) show the shortest paths form differently specified "start" nodes to all others

As shown in this study, the discrete-time MPCNN is an iterative procedure in which the output from one iteration stimulates the iteration. In the case of continuous-time network, this is still the same if the time step is small enough.

# 13

# Qualitative Analysis for Neural Networks with LT Transfer Functions

## 13.1 Introduction

[1]Dynamical properties of recurrent neural networks are of crucial importance in their applications. Stability analysis of recurrent neural networks has attracted extensive interests in recent years as the growing literatures illustrate (Cohen and Grossberg, 1983; Forti, 1994; Tang et al., 2002; Yi et al., 1999; Yi et al., 2003). So far one of the major results is focused on behavior of monostable networks, i.e., the networks are forced to possess a unique stationary states under some conditions. However, these monostable networks are computationally restrictive in applications as well as in modeling biological behaviors.

The activation functions play an important role in binding the neural dynamics of recurrent neural networks. Various activation functions have been employed in neural networks, see e.g., the sigmoid function and limiter function, which formulate the Hopfield-Tank model (Hopfield and Tank, 1985), cellular neural networks (CNN) (Chua and Yang, 1988) and Linsker's model (Feng et al., 1996). Both transfer functions confine any inputs to bounded outputs. The linear threshold (LT) transfer function permits the networks to have unbounded outputs, thus more complex dynamics arise in the networks, e.g. limit cycles or chaos. Most recently, the LT neural networks which model cortical neurons have attracted growing interest in both theoretical analysis and applications (Hahnloser, 1998; Hahnloser et al., 2000; Wersing et al., 2001a; Wersing et al., 2001b). The LT networks have been observed to exhibit one important property, i.e., *multistability*, which allows the networks to possess multiple equilibria ("steady states") coexisting under certain synaptic weights and external inputs. Multistability endows the LT networks with distinguished application potentials in decision making, digital selection and

---

[1] Reuse of the materials of "Qualitative analysis for recurrent neural networks with linear threshold transfer functions", 52(5), 2005, 1003–1012, IEEE Trans. on Circuits and Systems-I, with permission from IEEE.

analogy amplification (Hahnloser, 1998; Hahnloser et al., 2000). Therefore, qualitative analysis for the LT networks deserves more effort and attention for both theoretical investigations and applications.

In this chapter, recurrent neural networks described by the following system of nonlinear differential equations are studied:

$$\dot{x}_i(t) = -x_i(t) + \sum_{j=1}^{n} w_{ij}\sigma(x_j(t)) + h_i, \ i = 1, \cdots, n, \tag{13.1}$$

or its equivalent form

$$\dot{x}(t) = -x(t) + W\sigma(x(t)) + h, \tag{13.2}$$

where $x_i$ denotes the activity of neuron $i$, $x = (x_1, \cdots, x_n)^T \in \Re^n$ and $h \in \Re^n$ denotes external inputs. $W = (w_{ij})_{n \times n}$ is a constant real matrix, each entry of which denotes the synaptic weight representing the strength of the synaptic connection from neuron $j$ to neuron $i$. $\sigma(x)$ is the LT transfer function, i.e., $\sigma(x) = \max(0, x)$. A vector that satisfies

$$-x^* + W\sigma(x^*) + h \equiv 0$$

is called a *stationary state* or an *equilibrium* of the network.

In the recent paper (Yi et al., 2003), the authors studied three basic properties of multistable networks and obtained the corresponding conditions of boundedness, global attractivity and complete convergence. This chapter is focused on the analysis of global attractivity and equilibria properties for LT networks and propose new results for boundedness and global attractivity. As compared with the condition that ensures global attractivity in (Yi et al., 2003), the condition obtained here can be seen as a weaker one.

The rest of this chapter is organized as follows. The equilibria properties and distributions of two-dimensional networks are addressed in Section 13.2. Sufficient and necessary conditions for multiple equilibria coexisting are derived in Section 13.3. In Section 13.4 we give a boundedness and global attractivity analysis for the network (13.2). Conditions will be derived to ensure the boundedness and global attractivity of the network. Under such conditions, the network possesses certain compact set which attracts all the trajectories of the network. Simulation examples illustrate the theory developed in Section 13.5. Finally, conclusions are drawn in Section 13.6.

## 13.2 Equilibria and Their Properties

The LT network is one of the three typical and classical models of neural network and no references are found to show qualitative properties of its 2-coupled system.

## 13.2 Equilibria and Their Properties

In this section, the equilibria and their properties for two-dimensional networks are studied. Analysis for dynamics of the planar systems will provide us a fundamental but nontrivial investigation into the neurodynamics of the LT networks. Generally, a degenerate system of differential equations will have an infinite number of equilibria. It is assumed that the system of the networks is nondegenerate, i.e., its Jacobian matrix is invertible.

Since at any time $t$, each LT neuron is either firing (active) or silent (inactive), the whole ensemble of the LT neurons can be divided into a partition $P^+(t)$ of neurons with positive states $x_i(t) \geq 0$ for $i \in P^+(t)$, and a partition $P^-(t)$ of neurons with negative states $x_i(t) < 0$ for $i \in P^-(t)$. Clearly $P^+(t) \cup P^-(t) = [1, \cdots, N]$ and there are in total $2^N$ different possible partitions for $x$. The qualitative properties of the LT network's equilibria can then be investigated in each corresponding partition.

Define a matrix $W^e$ such that

$$W^e x(t) = W \sigma(t). \tag{13.3}$$

The matrix $W^e$ is called an *effective recurrence matrix* (Hahnloser, 1998), which describes the connectivity of the operating network, where only active neurons are dynamically relevant. The matrix $W^e$ is simply constructed out of $W$ such that $w_{ij}^e = w_{ij}$ for $j \in P^+$ and $w_{ij}^e = 0$ for $j \in P^-$.

Define a matrix $\Phi = \text{diag}(\phi_1, \cdots, \phi_N)$, where $\phi_i = 1$ for $i \in P^+$ and $\phi_i = 0$ for $i \in P^-$. The matrix $\Phi$ carries the digital information about which neurons are active. According to the definition of $\Phi$, it holds that

$$W^e = W\Phi. \tag{13.4}$$

This relation holds for any particular partition of $x$.

Consequently, the nonlinear dynamics of the LT neural network is linearized as

$$\dot{x}(t) = (W^e - I)x(t) + h, \tag{13.5}$$

where $I$ denotes the identity matrix. It implies that for any partition, the LT network is accurately described by a linear system, but nonlinearities arise when the partitions are switching. This explains the observations that digital selection and analogy amplification coexist in a LT type of silicon circuit (Hahnloser et al., 2000).

Let $J$ denote the Jacobian matrix of the LT network, then $J = (W^e - I)$. Define $\delta = \det J$, $\tau = \text{trace } A$ and $\Delta = \tau^2 - 4\delta$. The dynamical properties of the network are determined by the relationship among $\delta$, $\tau$ and $\Delta$: a) If $\delta < 0$ then $x^*$ is a saddle. b) If $\delta > 0$ and $\Delta \geq 0$ then $x^*$ is a node; it is stable if $\tau < 0$ and unstable if $\tau > 0$. c) If $\delta > 0$, $\Delta < 0$ and $\tau \neq 0$, then $x^*$ is a focus; it is stable if $\tau < 0$ and unstable if $\tau > 0$. d) If $\delta > 0$ and $\tau = 0$ then $x^*$ is a center (Perko, 2001).

Now the equilibrium is computed by

$$x^* = -J^{-1}h. \tag{13.6}$$

It is a true equilibrium provided that $x^*$ lies in the partition dictated by its recurrence matrix $W^e$, i.e., $x_i \geq 0$ for $i \in P^+$ and $x_i < 0$ for $i \in P^-$. The equilibrium is unique for a given partition $P$.

In the following, a fundamental theory about the two-dimensional LT networks is proposed.

**Theorem 13.1.** *Suppose the plane is divided into four quadrants, i.e., $\Re^2 = D_1 \cup D_2 \cup D_3 \cup D_4$, where*

$$D_1 = \{(x_1, x_2) \in \Re^2 : x_1 \geq 0, x_2 \geq 0\},$$
$$D_2 = \{(x_1, x_2) \in \Re^2 : x_1 < 0, x_2 \geq 0\},$$
$$D_3 = \{(x_1, x_2) \in \Re^2 : x_1 < 0, x_2 < 0\},$$
$$D_4 = \{(x_1, x_2) \in \Re^2 : x_1 \geq 0, x_2 < 0\}.$$

*The qualitative properties and distributions of the planar LT network equilibria and their corresponding parameter conditions are described in Table 14.1.*

**Table 13.1.** Properties and distributions of the equilibria

| $D_1$ | | $(w_{11}-1)(w_{22}-1) < w_{12}w_{21}$ | | saddle | I(1) |
|---|---|---|---|---|---|
| | | $(w_{22}-1)h_1 \geq w_{12}h_2$ | | | |
| | | $(w_{11}-1)h_2 \geq w_{21}h_1$ | | | |
| | $(w_{11}-1)(w_{22}-1) > w_{12}w_{21}$ | $(w_{11}-w_{22})^2 \geq -4w_{12}w_{21}$ | $w_{11}+w_{22} < 2$ | s-node | I(2) |
| | $(w_{22}-1)h_1 \leq w_{12}h_2$ | | $w_{11}+w_{22} > 2$ | uns-node | I(3) |
| | $(w_{11}-1)h_2 \leq w_{21}h_1$ | $(w_{11}-w_{22})^2 < -4w_{12}w_{21}$ | $w_{11}+w_{22} < 2$ | s-focus | I(4) |
| | | | $w_{11}+w_{22} > 2$ | uns-focus | I(5) |
| | | $w_{11}+w_{22}-2=0$ | | center | I(6) |
| $D_2$ | | $w_{22} < 1, h_1 < \frac{w_{12}h_2}{w_{22}-1}, h_2 \geq 0$ | | s-node | II(1) |
| | | $w_{22} > 1, h_1 < \frac{w_{12}h_2}{w_{22}-1}, h_2 \leq 0$ | | saddle | II(2) |
| $D_3$ | | $h_1 < 0, h_2 < 0$ | | s-node | III |
| $D_4$ | | $w_{11} < 1, h_1 \geq 0, h_2 < \frac{w_{21}h_1}{w_{11}-1}$ | | s-node | IV(1) |
| | | $w_{11} > 1, h_1 \leq 0, h_2 < \frac{w_{21}h_1}{w_{11}-1}$ | | saddle | IV(2) |

s: stable, uns: unstable

*Proof.* Let the neuron states $x \in \Re^2$ be divided into the four partitions $D_1, D_2, D_3$ and $D_4$. Thus four exhaustive cases are considered in what follows.

Case 1: $x^* = (x_1^*, x_2^*) \in D_1$.

The digital information about this equilibrium is $\Phi = \text{diag}(1,1)$, then the effective recurrence matrix $W^e$ is equal to $W$. Hence $J = W - I$. Obviously,

## 13.2 Equilibria and Their Properties

$$\delta = \det(W - I) = (w_{11} - 1)(w_{22} - 1) - w_{12}w_{21},$$
$$\tau = w_{11} + w_{22} - 2,$$
$$\Delta = (w_{11} - w_{22})^2 + 4w_{12}w_{21}.$$

The equilibrium is computed by

$$x^* = -J^{-1}h = \frac{1}{w_{12}w_{21} - (w_{11} - 1)(w_{22} - 1)} \begin{bmatrix} (w_{22} - 1)h_1 - w_{12}h_2 \\ -w_{21}h_1 + (w_{11} - 1)h_2 \end{bmatrix}.$$

It is a true equilibrium provided that

(1a) $\begin{cases} (w_{22} - 1)h_1 - w_{12}h_2 \geq 0 \\ (w_{11} - 1)h_2 - w_{21}h_1 \geq 0 \\ w_{12}w_{21} - (w_{11} - 1)(w_{22} - 1) > 0 \end{cases}$

or

(1b) $\begin{cases} (w_{22} - 1)h_1 - w_{12}h_2 \leq 0 \\ (w_{11} - 1)h_2 - w_{21}h_1 \leq 0 \\ w_{12}w_{21} - (w_{11} - 1)(w_{22} - 1) < 0. \end{cases}$

In Case (1a), the equilibrium is a saddle since $\delta < 0$. In Case (1b), $\delta > 0$, and therefore three exhaustive subcases are needed to be considered.

Subcase i : $\Delta \geq 0$. It is known that the equilibrium is a node for $\delta > 0$ and $\Delta \geq 0$. Thus the following conditions are obtained.

$$\begin{cases} (w_{11} - 1)(w_{22} - 1) > w_{12}w_{21} \\ (w_{22} - 1)h_1 \leq w_{12}h_2 \\ (w_{11} - 1)h_2 \leq w_{21}h_1 \\ (w_{11} - w_{22})^2 + 4w_{12}w_{21} \geq 0. \end{cases}$$

Furthermore, the node is ensured stable or unstable by $w_{11} + w_{22} - 2 < 0$ or $w_{11} + w_{22} - 2 > 0$, respectively.

Subcase ii : $\Delta < 0, \tau \neq 0$. It is known that the equilibrium is a focus under these conditions, which can be expressed equivalently as

$$\begin{cases} (w_{11} - 1)(w_{22} - 1) > w_{12}w_{21} \\ (w_{22} - 1)h_1 \leq w_{12}h_2 \\ (w_{11} - 1)h_2 \leq w_{21}h_1 \\ (w_{11} - w_{22})^2 + 4w_{12}w_{21} < 0. \end{cases}$$

Furthermore, a stable focus is ensured by $w_{11} + w_{22} - 2 < 0$, an unstable focus by $w_{11} + w_{22} - 2 > 0$.

Subcase iii : $\tau = 0$. It is known that the equilibrium is a center. The conditions are obtained as follows:

$$\begin{cases} (w_{11} - 1)(w_{22} - 1) > w_{12}w_{21} \\ (w_{22} - 1)h_1 \leq w_{12}h_2 \\ (w_{11} - 1)h_2 \leq w_{21}h_1 \\ w_{11} + w_{22} - 2 = 0. \end{cases}$$

Case 2: $x^* = (x_1^*, x_2^*) \in D_2$.

Since $\Phi = \text{diag}(0, 1)$, it leads to

$$J = W^+ - I = W\Phi - I = \begin{bmatrix} -1 & w_{12} \\ 0 & w_{22} - 1 \end{bmatrix}.$$

Then

$$\delta = -(w_{22} - 1),$$
$$\tau = w_{22} - 2,$$
$$\Delta = (w_{22} - 2)^2 + 4(w_{22} - 1) = w_{22}^2 \geq 0.$$

The equilibrium is computed by

$$x^* = -J^{-1}h = \frac{1}{1 - w_{22}} \begin{bmatrix} w_{12}h_2 + (1 - w_{22})h_1 \\ h_2 \end{bmatrix}.$$

It is a true equilibrium provided that

$$(2a) \quad \begin{cases} w_{22} < 1 \\ h_1 < \frac{w_{12}h_2}{w_{22}-1} \\ h_2 \geq 0 \end{cases}$$

or

$$(2b) \quad \begin{cases} w_{22} > 1 \\ h_1 < \frac{w_{12}h_2}{w_{22}-1} \\ h_2 \leq 0. \end{cases}$$

Analogously as in Case 1, it is proven that in Case (2a) the equilibrium $x^*$ is a stable node because $\delta > 0$, $\tau < 0$ and $\Delta \geq 0$ are always satisfied. In Case (2b), $x^*$ is a saddle because $\delta < 0$.

Case 3: $x^* = (x_1^*, x_2^*) \in D_3$.

Obviously $\Phi$ becomes a zero matrix and the equilibrium $x^* = h$. Then $x^* \in D_2$ exists if and only if $h_1 < 0$ and $h_2 < 0$. This equilibrium is always a stable node.

Case 4: $x^* = (x_1^*, x_2^*) \in D_4$.

Since $\Phi = \text{diag}(1, 0)$, the Jacobi matrix becomes

$$J = W\Phi - I = \begin{bmatrix} w_{11} - 1 & 0 \\ w_{21} & -1 \end{bmatrix}.$$

Then

$$\delta = 1 - w_{11},$$
$$\tau = w_{11} - 2,$$
$$\Delta = T^2 - 4 * \delta = (w_{11} - 2)^2 + 4(w_{11} - 1) = w_{11}^2 \geq 0.$$

The equilibrium is computed by

$$x^* = -J^{-1}h = \frac{1}{1-w_{11}}\left[\begin{array}{c} h_1 \\ w_{21}h_1 + (1-w_{11})h_2 \end{array}\right].$$

It is a true equilibrium provided that

$$(4a) \quad \begin{cases} w_{11} < 1 \\ h_1 \geq 0 \\ h_2 < \frac{w_{21}h_1}{w_{11}-1} \end{cases}$$

or

$$(4b) \quad \begin{cases} w_{11} > 1 \\ h_1 \leq 0 \\ h_2 < \frac{w_{21}h_1}{w_{11}-1}. \end{cases}$$

In Case (4a), $x^*$ is a stable node since $\delta > 0$, $\tau < 0$ and $\Delta \geq 0$ are always satisfied. In Case (4b), $x^*$ is a saddle because $\delta < 0$.

The proof is completed.

## 13.3 Coexistence of Multiple Equilibria

As discussed above, Table 14.1 reveals the relations between the parameters (synaptic weights $w_{ij}$'s and external inputs $h1, h_2$) and the features of equilibria of the network. The equilibria can coexist when the parameters satisfy some conditions. Therefore, the results of Table 14.1 allow us to derive the sufficient and necessary conditions for the coexistence of 2, 3 or 4 equilibria by solving the intersections of the parameter space defined by the inequalities. At the same time the properties of the equilibria are obtained in this way.

Though the work of solving the intersections (more than 30 groups) is somewhat tedious, it is exhaustive and explicit to describe the distributions and properties of the equilibria of the system in different parameter spaces.

Let $D_i$ denote the set formulated by the parameter space which is corresponding to the class index $i$ in the last column of Table 14.1. The following theorems are derived.

**Theorem 13.2.** *The planar system has exactly 4 equilibria, i.e., unstable node or unstable focus (I), saddle (II), stable node (III) and saddle (IV), coexisting in four quadrants if and only if the parameters belong to the intersection* $D_{I(3)} \cap D_{II(2)} \cap D_{III} \cap D_{IV(2)}$ *or* $D_{I(5)} \cap D_{II(2)} \cap D_{III} \cap D_{IV(2)}$.

*Proof.* It is noted that the conditions are sufficient and necessary for the coexisting of 4 equilibria of the planar system. It is equivalent to solve the intersections among any four groups of conditions from $I(1)-IV(2)$. To satisfy $D_{I(3)}$, it is required $-\frac{1}{4}(w_{11} - w_{22})^2 \leq w_{12}w_{21} < (w_{11} - 1)(w_{22} - 1)$. To ensure its validity, an additional inequality is necessary, i.e., $(w_{11} - 1)(w_{22} - $

1) $> -\frac{1}{4}(w_{11}-w_{22})^2$. According to $II(2)$ and $IV(2)$, $w_{11} > 1, w_{22} > 1$ and $h_1 < \min\{\frac{w_{12}h_2}{w_{22}-1},0\}, h_2 < \min\{\frac{w_{21}h_1}{w_{11}-1},0\}$. After a sequence of simple inequality computations, finally the conditions of $D_{I(3)} \cap D_{II(2)} \cap D_{III} \cap D_{IV(2)}$ are acquired:

$$\begin{cases} (w_{11}-1)(w_{22}-1) > -\frac{1}{4}(w_{11}-w_{22})^2 \\ -\frac{1}{4}(w_{11}-w_{22})^2 \le w_{12}w_{21} < (w_{11}-1)(w_{22}-1) \\ w_{11} > 1, w_{22} > 1 \\ h_1 < \min\{\frac{w_{12}h_2}{w_{22}-1},0\} \\ h_2 < \min\{\frac{w_{21}h_1}{w_{11}-1},0\}. \end{cases}$$

It is analogous to derive the conditions of $D_{I(5)} \cap D_{II(2)} \cap D_{III} \cap D_{IV(2)}$:

$$\begin{cases} w_{12}w_{21} < \min\{(w_{11}-1)(w_{22}-1), -\frac{1}{4}(w_{11}-w_{22})^2\} \\ w_{11} > 1, w_{22} > 1 \\ h_1 < \min\{\frac{w_{12}h_2}{w_{22}-1},0\} \\ h_2 < \min\{\frac{w_{21}h_1}{w_{11}-1},0\}. \end{cases}$$

Any other intersections of 4 groups of conditions are empty. This completes the proof.

In the next, the analysis is restricted to discuss an important case where only nonnegative external inputs are fed in. Analogously, the following two theorems are obtained with ignoring the proofs for space limit.

**Theorem 13.3.** *The nondegenerate network has at least 3 equilibria if the network parameters belong to any of the intersection: $D_{I(1)} \cap D_{II(1)} \cap D_{IV(1)}$.*

The above intersection is given by

$$D_{I(1)} \cap D_{II(1)} \cap D_{IV(1)} : \begin{cases} 1 + w_{12}w_{21}/(w_{22}-1) < w_{11} < 1, \\ w_{22} < 1, w_{12} < 0, w_{21} < 0, \\ 0 \le h_1 < w_{12}h_2/(w_{22}-1), \\ 0 \le h_2 < w_{21}h_1/(w_{11}-1). \end{cases}$$

**Theorem 13.4.** *The nondegenerate network has at least 2 equilibria if the network parameters fall into any of the intersections: $D_{I(1)} \cap D_{II(1)}, D_{I(1)} \cap D_{IV(1)}, D_{II(1)} \cap D_{IV(1)}$.*

The above intersections are given by, respectively:

$$D_{I(1)} \cap D_{II(1)} : \begin{cases} (w_{11}-1)(w_{22}-1) < w_{12}w_{21}, \\ (w_{22}-1)h_1 > w_{12}h_2, \\ (w_{11}-1)h_2 \ge w_{21}h_1, \\ w_{22} < 1, h_2 \ge 0. \end{cases}$$

$$D_{I(1)} \cap D_{IV(1)} : \begin{cases} (w_{22}-1)h_1 \geq w_{12}h_2, \\ w_{11} < 1, \ w_{22} > 1 + w_{12}w_{21}/(w_{11}-1), \\ h_1 \geq 0, \ h_2 < w_{21}h_1/(w_{11}-1), \end{cases}$$

$$D_{II(1)} \cap D_{IV(1)} : \begin{cases} w_{11} < 1, \ w_{22} < 1, \\ w_{12} < 0, \ w_{21} < 0, \\ 0 \leq h_1 < w_{12}h_2/(w_{22}-1), \\ 0 \leq h_2 < w_{21}h_1/(w_{11}-1). \end{cases}$$

The following corollary illustrates the multistability conditions for the two-dimensional networks.

**Corollary 13.5.** *The nondegenerate planar system has at most 2 stable equilibria. The sufficient and necessary conditions for coexistence of 2 stable equilibria are that the parameters fall in any of the intersections $D_{I(2)} \cap D_{III}, D_{I(4)} \cap D_{III}, D_{I(6)} \cap D_{III}$ and $D_{II(1)} \cap D_{IV(1)}$.*

For a straightforward illustration of the relations between the coexistence of equilibria with the network parameters, Figure 13.1 is drawn in the parameter plane of $w_{11} - w_{22}$ according to Table 1. This plane is divided by the lines $w_{11} = 1$ and $w_{22} = 1$ and the lines L1: $w_{22} = w_{11} + 2\sqrt{|w_{12}w_{21}|}$, L2: $w_{22} = w_{11} - 2\sqrt{|w_{12}w_{21}|}$ and L3: $w_{11} + w_{22} = 2$. Regarding the equilibrium of I(3), for example, the two arrows above L3 and L1 indicate that both $w_{11} + w_{22} > 2$ and $(w_{11} - w_{22})^2 \geq -4w_{12}w_{21}$ should be met. It is shown that the equilibrium (III) always exists regardless of the weights since there are no such inequality constraints, as long as the external inputs are negative.

It should be noted that if $w_{12}w_{21} \geq 0$ then L1 and L2 will coalesce and the regions for I(4) and I(5) will disappear.

## 13.4 Boundedness and Global Attractivity

In this section, the conditions which ensure the boundedness of the networks are derived. Moreover, a compact set which globally attracts all the trajectories of the network is obtained in an explicit expression.

Let the matrix norm and the vector norm be computed by $\|A\| = (\sum_{i=1}^{m}\sum_{j=1}^{n} a_{ij}^2)^{\frac{1}{2}} = (tr(A^T A))^{\frac{1}{2}}$, and $\|x\| = (\sum_{i=1}^{n} x_i^2)^{\frac{1}{2}} = (x^T x)^{\frac{1}{2}}$ for all $A \in \Re^{n \times n}$ and $x \in \Re^n$. Obviously, $\|Ax\| \leq \|A\|\|x\|$.

**Definition 13.6.** *Consider the recurrent neural network described by a system of differential equations $\dot{x} = f(t,x)$, where $f(t,x) \in C(I \times \Re^n, \Re^n)$, the network is ultimately bounded for bound B, if there exists a $B > 0$ and a $T(t_0) > 0$ such that for every trajectory of the system, $\|x(t,t_0,x_0)\| < B$ for all $t \geq t_0 + T$, where B is independent of the particular trajectory.*

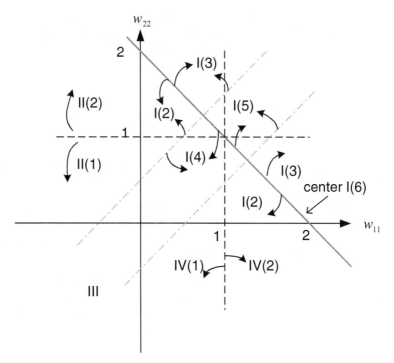

**Fig. 13.1.** Equilibria distribution in $w_{11}, w_{22}$ plane.

**Definition 13.7.** *Let $S$ be a compact subset of $\Re^n$. Let $S_\epsilon$ denote the $\epsilon$-neighborhood set of $S$, $\forall \epsilon > 0$. The compact set $S$ is said to globally attract the network if all trajectories of the network ultimately enter and remain in $S_\epsilon$.*

For any matrix $A = (a_{ij})_{n \times n}$, define $A^+ = (a_{ij}^+)$ such that

$$a_{ij}^+ = \begin{cases} a_{ii}, i = j, \\ \max(0, a_{ij}), i \neq j, \end{cases}$$

and define $A^- = (a_{ij}^-)$ such that

$$a_{ij}^- = \begin{cases} a_{ii}, i = j, \\ \min(0, a_{ij}), i \neq j. \end{cases}$$

Let $V$ and $\Omega$ denote the effective recurrence matrices of $W^+$ and $W^-$ for the network (13.2) respectively, i.e., $V = (W^+)^e$, $\Omega = (W^-)^e$. Let $\|M\| = \max(\|V\|, \|\Omega\|)$.

**Theorem 13.8.** *Suppose that $\|M\| < 1$. Then the network (13.2) is ultimately bounded. Furthermore, the compact set*

$$S = \left\{ x \mid \|x\| \leq \frac{\|h\|}{1 - \|M\|} \right\}$$

*globally attracts the network (13.2).*

## 13.4 Boundedness and Global Attractivity

*Proof.* The trajectory $x(t)$ of network (13.2) satisfies

$$x(t) = x(t_0)e^{-(t-t_0)} + \int_{t_0}^{t} e^{-(t-s)}[W\sigma(x(s)) + h]ds, \qquad (13.7)$$

where $t_0 \in \Re^1$. Since $\sigma(\cdot) \geq 0$, from the above equation, it is obtained that

$$x(t) \leq x(t_0)e^{-(t-t_0)} + \int_{t_0}^{t} e^{-(t-s)}[W^+\sigma(x(s)) + h]ds. \qquad (13.8)$$

Since $V = (W^+)^e$, it follows from equation (13.3) that $W^+\sigma(x) = Vx$. Therefore,

$$x(t) \leq x(t_0)e^{-(t-t_0)} + \int_{t_0}^{t} e^{-(t-s)}[Vx(s) + h]ds. \qquad (13.9)$$

On the other hand, $W^-\sigma(x) = \Omega x$. It follows from equation (15.13) that

$$x(t) \geq x(t_0)e^{-(t-t_0)} + \int_{t_0}^{t} e^{-(t-s)}[W^-\sigma(x(s)) + h]ds$$

$$= x(t_0)e^{-(t-t_0)} + \int_{t_0}^{t} e^{-(t-s)}[\Omega x(s) + h]ds. \qquad (13.10)$$

From equations (13.9) and (13.10), it is derived that

$$\|x(t)\| \leq \|x(t_0)\|e^{-(t-t_0)} + \int_{t_0}^{t} e^{-(t-s)}\|h\|ds$$

$$+ \max\left(\int_{t_0}^{t} e^{-(t-s)}\|V\|\|x(s)\|ds, \int_{t_0}^{t} e^{-(t-s)}\|\Omega\|\|x(s)\|ds\right).$$

Let $\|M\| = \max(\|V\|, \|\Omega\|)$. Then

$$\|x(t)\| \leq \|x(t_0)\|e^{-(t-t_0)} + \int_{t_0}^{t} e^{-(t-s)}\|h\|ds + \int_{t_0}^{t} e^{-(t-s)}\|M\|\|(x(s))\|ds$$

$$= e^{-(t-t_0)}\|x(t_0)\| + (1 - e^{-(t-t_0)})\|h\| + \int_{t_0}^{t} e^{-(t-s)}\|M\|\|x(s)\|ds,$$

that is,

$$e^t\|x(t)\| \leq e^{t_0}\|x(t_0)\| + (e^t - e^{t_0})\|h\| + \int_{t_0}^{t} \|M\|e^s\|x(s)\|ds. \qquad (13.11)$$

Let $y(t) = e^t\|x(t)\|$. Thus

$$y(t) \leq y(t_0) + (e^t - e^{t_0})\|h\| + \int_{t_0}^{t} \|M\|y(s)ds. \qquad (13.12)$$

By Gronwall's inequality it is obtained

$$y(t) \le y(t_0) + (e^t - e^{t_0})\|h\| + \int_{t_0}^t \left(y(t_0) + (e^s - e^{t_0})\|h\|\right) \|M\| e^{\|M\|(t-s)} ds$$

$$= y(t_0) + (e^t - e^{t_0})\|h\| + \int_{t_0}^t \left(y(t_0) - e^{t_0}\|h\|\right) \|M\| e^{\|M\|(t-s)} ds$$

$$+ \int_{t_0}^t \|h\| \|M\| e^{\|M\|t} e^{(1-\|M\|)s} ds$$

$$= y(t_0) + (e^t - e^{t_0})\|h\| + \left(y(t_0) - e^{t_0}\|h\|\right)(e^{\|M\|(t-t_0)} - 1)$$

$$+ \frac{\|h\| \|M\| e^{\|M\|t}}{1 - \|M\|} \left(e^{(1-\|M\|)t} - e^{(1-\|M\|)t_0}\right)$$

$$= e^t \|h\| + \left(y(t_0) - e^{t_0}\|h\|\right) e^{\|M\|(t-t_0)}$$

$$+ \frac{\|h\| \|M\|}{1 - \|M\|} (e^t - e^{\|M\|t + (1-\|M\|)t_0}). \tag{13.13}$$

Thus

$$e^{-t} y(t) \le \|h\| + \left(y(t_0) - e^{t_0}\|h\|\right) e^{-\|M\|t_0} e^{-(1-\|M\|)t}$$

$$+ \frac{\|h\| \|M\|}{1 - \|M\|} (1 - e^{-(1-\|M\|)(t-t_0)}), \tag{13.14}$$

that is,

$$\|x(t)\| \le \|h\| + \left(e^{-t_0} y(t_0) - \|h\|\right) e^{-(1-\|M\|)(t-t_0)} + \frac{\|h\| \|M\|}{1 - \|M\|} (1 - e^{-(1-\|M\|)(t-t_0)})$$

$$= \|h\| + \frac{\|h\| \|M\|}{1 - \|M\|} + \left(x(t_0) - \|h\| - \frac{\|h\| \|M\|}{1 - \|M\|}\right) e^{-(1-\|M\|)(t-t_0)}$$

$$= \frac{\|h\|}{1 - \|M\|} + \left(x(t_0) - \frac{\|h\|}{1 - \|M\|}\right) e^{-(1-\|M\|)(t-t_0)}. \tag{13.15}$$

Since $\|M\| < 1$,

$$\lim_{t \to +\infty} \|x(t)\| \le \frac{\|h\|}{1 - \|M\|}.$$

It implies that the network is ultimately bounded and the compact set $S$ globally attracts any trajectory of the network (13.2). This completes the proof.

Notice that $V - (W^+)^e$, $\Omega = (W^-)^e$. Recalling equation (13.4),

$$V = W^+ \Phi, \tag{13.16}$$
$$\Omega = W^- \Phi, \tag{13.17}$$

where $\Phi$ is the digital information matrix as defined in Section II.

Consider an extreme case where all neurons are active, $\Phi$ becomes the identity matrix I. It always holds that

$$\|V\| = \|W^+\Phi\| \leq \|W^+\mathrm{I}\| = \|W^+\|,$$
$$\|\Omega\| = \|W^-\Phi\| \leq \|W^+\mathrm{I}\| = \|W^-\|.$$

It implies that the all firing states of neurons leads to the maximal norms of $V$ and $\Omega$.

Let $\|\bar{M}\| = \max(\|W^+\|, \|W^-\|)$. Obviously, $\|M\| \leq \|\bar{M}\|$. Furthermore,

$$\frac{\|h\|}{1-\|M\|} \leq \frac{\|h\|}{1-\|\bar{M}\|}.$$

Then the following corollary is obtained, which avoids the computation of effective recurrence matrix in Theorem 13.8.

**Corollary 13.9.** *Suppose $\|\bar{M}\| < 1$. Then the network (13.2) is ultimately bounded. Furthermore, the compact set*

$$S = \left\{ x \mid \|x\| \leq \frac{\|h\|}{1-\|\bar{M}\|} \right\}$$

*globally attracts the network (13.2).*

*Remarks:* A multistable dynamical system is one in which multiple attractors coexist. Although multistability is not exclusive to systems with non-saturating transfer functions, it appears that only the multistability analysis of LT networks has found significant meaning in neural network areas (Hahnloser, 1998; Wersing et al., 2001b; Yi et al., 2003). In contrast to saturating transfer functions e.g., the sigmoid function and limiter function, the distinction of the LT transfer function lies in that in the actual computations of the recurrent network, the upper saturation may not be involved, since the activation is dynamically bounded due to the effective net interactions. In addition, the LT networks can perform both digital selection and analog amplification (Hahnloser et al., 2000). Another important characteristics of LT transfer functions is that the lack of an upper saturation can be used to exclude spurious ambiguous states (Wersing et al., 2001b).

## 13.5 Simulation Examples

In this section, some simulation examples are presented to illustrate the theorems obtained in this work.

Example 1 : Consider the following two-dimensional neural network,

$$\begin{bmatrix} \dot{x}_1(t) \\ \dot{x}_2(t) \end{bmatrix} = -\begin{bmatrix} x_1(t) \\ x_2(t) \end{bmatrix} + \begin{bmatrix} 2 & -2 \\ 2 & 0 \end{bmatrix} \begin{bmatrix} \sigma(x_1(t)) \\ \sigma(x_2(t)) \end{bmatrix} + \begin{bmatrix} -1 \\ -4 \end{bmatrix} \quad (13.18)$$

for $t \geq 0$. Obviously,

$$W = \begin{bmatrix} 2 & -2 \\ 2 & 0 \end{bmatrix}, h = \begin{bmatrix} -1 \\ -4 \end{bmatrix}.$$

It is verified that the network parameters belong to the intersection $D_{I(6)} \cap D_{III}$. By Corollary 13.5, the network possess exactly two stable stationary states: one center in quadrant I and one stable node in quadrant III. Figure 13.2 illustrates the simulation results for two stable equilibria coexisting in the planar system.

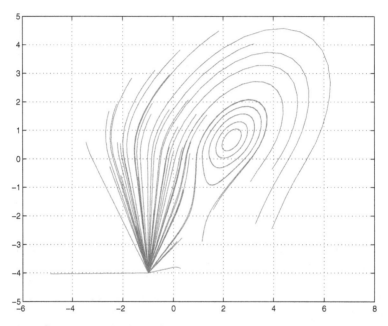

**Fig. 13.2.** Center and stable node coexisting in quadrants I and III of network (13.18)

Example 2 : Suppose the network in Example 1 is now constructed by the following parameters:

$$W = \begin{bmatrix} 0.2 & -2 \\ -1 & 0.6 \end{bmatrix}, h = \begin{bmatrix} 4 \\ 3 \end{bmatrix}.$$

The connection weights and external inputs fall in the intersection $D_{I(1)} \cap D_{II(1)} \cap D_{IV(1)}$, which meet Theorem 13.3 but not Theorem 13.2. It implies that the planar system has exactly three equilibria: one saddle in quadrant I and two stable nodes in quadrants II and IV, respectively. See figure 13.3 for an illustration.

Example 3 : Consider the network (13.18) with the network parameters as follows:

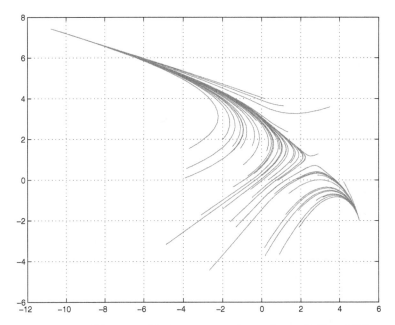

**Fig. 13.3.** Three equilibria coexisting in quadrants I, II and IV

$$W = \begin{bmatrix} 2 & 1 \\ -1 & 3 \end{bmatrix}, h = \begin{bmatrix} -2 \\ -1 \end{bmatrix}.$$

The parameters fall into the intersection $D_{I(5)} \cap D_{II(2)} \cap D_{III} \cap D_{IV(2)}$. According to Theorem 13.2, the network possesses exactly 4 equilibria, that is, unstable focus (I), saddle (II), stable node (III) and saddle (IV). Figure 13.4 illustrates the simulation results.

Example 4 : Consider a three-dimensional neural network

$$\dot{x}(t) = -x(t) + W\sigma(x(t)) + h, \tag{13.19}$$

where

$$W = \begin{bmatrix} 0.4 & 0.6 & -0.5 \\ 0.5 & 0 & 0 \\ -0.6 & 0 & 0.3 \end{bmatrix}, \quad h = \begin{bmatrix} 1 \\ 1 \\ 1 \end{bmatrix}.$$

It is easy to see that $\|W\| > 1$ but $\|W^+\| = \|W^-\| = 0.9274 < 1$. By Theorem 13.8, the network is bounded and all the trajectories of the network are globally attracted into a compact set

$$S = \{x : \|x\| \leq 23.8574\}.$$

Figure 13.5 shows the trajectories of the network starting from randomly 50 initial points in the three-dimensional space. Figures 13.6-13.8 show the projections of the trajectories onto the phase planes $(x_1, x_2)$, $(x_2, x_3)$ and $(x_1, x_3)$, respectively.

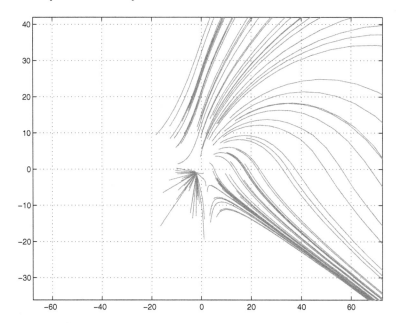

**Fig. 13.4.** Four equilibria coexisting in four quadrants

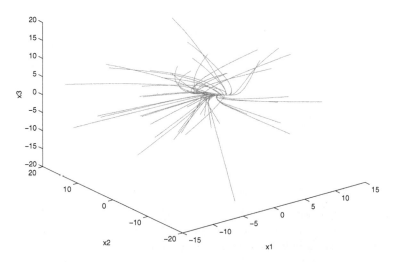

**Fig. 13.5.** Global attractivity of the network (14.43)

## 13.6 Conclusion

In this chapter, the dynamical properties of recurrent neural networks with LT transfer functions have been analyzed, such as geometrical properties of

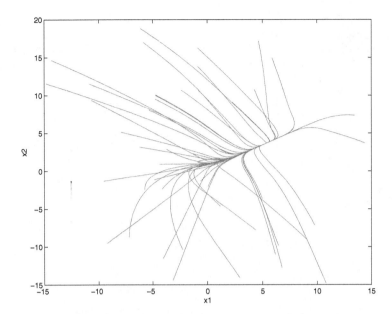

**Fig. 13.6.** Projection on $(x_1, x_2)$ phase plane

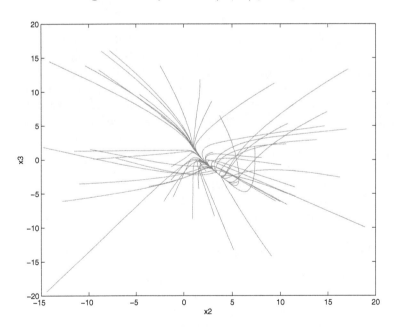

**Fig. 13.7.** Projection on $(x_2, x_3)$ phase plane

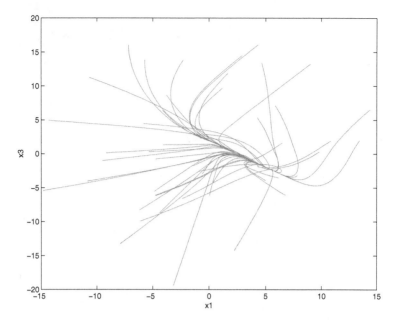

**Fig. 13.8.** Projection on $(x_1, x_3)$ phase plane

equilibria, global attractivitity and multistability. Firstly, the qualitative properties and distributions of the equilibria of general two-dimensional networks were investigated. The relations between the dynamical properties and the network parameters are revealed clearly and conditions for the coexistence of multiple equilibria are therefore established. Consequently, multistability was explained in a clear manner, though it was based on a two-dimensional system. Secondly, the condition for the global attractivity was derived, which is an important result in ensuring the bounded dynamics of LT networks. The theories developed herein are applicable to both symmetric and asymmetric connection weights, and thus they can have more implications in practice. In addition, our theoretical analysis confirmed the results obtained in (Hahnloser et al., 2000) that both digital selection and analog amplification coexist.

The LT network is actually nonlinear because the LT transfer function $\sigma(x)$ is nonlinear (although it is piecewise linear). The different shape of $\sigma(x)$ in different interval melts simple dynamical behaviors into a complicated one, so its qualitative analysis is technical. Nonlinearity may cause periodic oscillation and chaotic oscillation, and various types of nonlinear oscillators have been studied (Guckenheimer and Holmes, 1983; Haken, 2002; Amari and Ksabov, 1997). The 2-coupled LT network is a special nonlinear system and might undergo those complicated phenomena, but its qualitative analysis at equilibria and global stability are rather basic and uneasy work before proceeding further. Our work not only clarifies the issues of equilibria and stability but also provides a foundation for further investigation to its periodic oscillation and chaotic oscillation.

The main purpose of this chapter is the theoretical analysis of LT network dynamics, of which the stability result has significant meaning in developing practical applications of such networks, for example, in feature binding and sensory segmentation (Wersing *et al.*, 2001a). It is still a difficult task to develop a design method or learning algorithm of a LT network for a target behavior, and therefore it would be a meaningful topic for the future research.

# 14

# Analysis of Cyclic Dynamics for Networks of Linear Threshold Neurons

[1]A network of neurons with linear threshold (LT) transfer functions is a prominent model to emulate the behavior of cortical neurons. The analysis of dynamic properties for LT networks has attracted growing interest, such as multistability and boundedness. However, not much is known about how the connection strength and external inputs are related to oscillatory behaviors. Periodic oscillation is an important characteristic that relates to nondivergence, which shows that the network is still bounded despite the existence of unstable modes. By concentrating on a general parameterized two-cell network, theoretical results for geometrical properties and existence of periodic orbits are presented. Although it is restricted to two-dimensional systems, the analysis provides a useful tool to analyze cyclic dynamics of some specific LT networks of higher dimension. As an application, it is extended to an important class of biologically motivated networks of large scale, i.e., the winner-take-all model using local excitation and global inhibition.

## 14.1 Introduction

Networks of linear threshold (LT) neurons with nonsaturating transfer functions have attracted growing interest, as seen in the recent literature (Ben-Yishai et al., 1995; Hahnloser, 1998; Wersing et al., 2001b; Xie et al., 2002; Yi et al., 2003). The prominent properties of an LT network are embodied in at least two categories. Firstly, it is believed to be more biological plausible, e.g., in modeling cortical neurons which rarely operate close to saturation (Douglas et al., 1995). Secondly, the LT network is suitable for performing multistability analysis. A multistable network is not as computationally restrictive as compared to a monostable network, typical of which is found in the area of decision

---

[1] Reuse of the materials of "Analysis of cyclic dynamics for networks of linear threshold neurons", 17(1), 2005, 97–114, Neural Computation, with permission from MIT Press.

making. Its computational abilities have been explored in a wide range of applications. For example, a class of multistable WTA networks was presented in (Hahnloser, 1998) and an LT network composed of competitive layers was employed to accomplish perceptual grouping (Wersing et al., 2001a). An LT network was also designed using microelectronic circuits, which demonstrated both analog amplification and digital selection (Hahnloser et al., 2000).

The transfer function plays an important factor in bounding the network dynamics. Unlike the sigmoid function in Hopfield networks (Hopfield, 1984) and the piecewise linear sigmoid function in cellular networks (Kennedy and Chua, 1988), both of which ensure bounded dynamics and at least one equilibrium, nonsaturating transfer function may give rise to unbounded dynamics and even nonexistence of equilibrium (Forti and Tesi, 1995). The existence and uniqueness of equilibrium of LT networks have been investigated (Hadeler and Kuhn, 1987; Feng and Hadeler, 1996). A stability condition such that local inhibition is sufficient to achieve nondivergence of LT networks was established (Wersing et al., 2001b). A wide scope of dynamics analysis, including boundedness, global attractivity and complete convergence was shown in (Yi et al., 2003).

However, in the dynamical analysis of the LT networks, little attention has been paid to theoretically analyzing geometrical properties of both equilibria and periodic oscillations. It is often difficult to perform theoretical analysis for the cyclic dynamics of a two-dimensional system, and is even impossible to study that of high dimensional systems. It was reported that slowing down the global inhibition produced periodic oscillations in a WTA network and the epileptic network approached a limit cycle by computer simulations (Hahnloser, 1998). Nevertheless, there is a lack of theoretical proof about the existence of periodic orbits. It also remains unclear what factors affect the amplitude and period of the oscillations. This chapter provides an in-depth analysis on these issues based upon parameterized networks with two linear threshold neurons. Generally, the analytical result is only applicable to two-cell networks; however, it can be extended to a special class of high dimensional networks, such as generalized WTA network using local excitation and global inhibition. The principle analysis of the planar system would provide a useful framework to study the dynamical behaviors and provide an insight into understanding LT networks.

## 14.2 Preliminaries

The network with $n$ linear threshold neurons is described by the additive recurrent model,

$$\dot{x}(t) = -x(t) + W\sigma(x(t)) + h, \qquad (14.1)$$

where $x, h \in \mathbf{R}^n$ are the neural states and external inputs, $W = (w_{ij})_{n \times n}$ the synaptic connection strengths. $\sigma(x) = (\sigma(x_i))$ is called the linear threshold function defined by $\sigma(x_i) = \max\{0, x_i\}, i = 1, \cdots, n$.

Due to the simple nonlinearity of threshold, network (14.1) can be divided into pieces of linear systems and there are $2^n$ in total of such partitions for $n$ neurons. In each partition, we are able to study the qualitative behavior of the dynamics in details. In general, equation (14.1) can be linearized as

$$\dot{x}(t) = -x(t) + W^+ x(t) + h, \tag{14.2}$$

and it must be made corresponding to each partition. For a given partition, $W^+$ is simply computed by $w_{ij}^+ = w_{ij}$ for $x_j \geq 0$ and $w_{ij}^+ = 0$ for $x_j < 0$, for all $i, j$. Then it is easy to compute the equilibria of (14.2) by solving linear equation $Jx^* + h = 0$, where $J = (W^+ - I)$ is the Jacobian, and $I$ is the identity matrix. If $J$ is singular, the network may possess an infinite number of equilibria, for example, a line attractor (necessary conditions for line attractors have been studied in (Hahnloser et al., 2003)). If $J$ is nonsingular, a unique equilibrium exists, which is explicitly given by $x^* = -J^{-1}h$. It should be noted that if $x^*$ lies in this region, it is called a *true equilibrium*, otherwise it is called a *virtual equilibrium*.

## 14.3 Geometrical Properties of Equilibria

For a planar system, the phase plane is divided into four quadrants, i.e., $\mathbf{R}^2 = D_1 \cup D_2 \cup D_3 \cup D_4$, where $D_1 = \{(x_1, x_2) : x_1 \geq 0, x_2 \geq 0\}$, $D_2 = \{(x_1, x_2) : x_1 < 0, x_2 \geq 0\}$, $D_3 = \{(x_1, x_2) : x_1 < 0, x_2 < 0\}$, $D_4 = \{(x_1, x_2) : x_1 \geq 0, x_2 < 0\}$. Obviously, $D_1$ is a linear region which is unsaturated. $D_2$ and $D_4$ are partially saturated but $D_3$ is saturated. The Jacobians of the two-cell network in each region are formulated explicitly,

$$\begin{pmatrix} w_{11} - 1 & w_{12} \\ w_{21} & w_{22} - 1 \end{pmatrix}, \begin{pmatrix} -1 & w_{12} \\ 0 & w_{22} - 1 \end{pmatrix}, \begin{pmatrix} -1 & 0 \\ 0 & -1 \end{pmatrix}, \begin{pmatrix} w_{11} - 1 & 0 \\ w_{21} & -1 \end{pmatrix},$$

adding the subscript $1, \cdots, 4$ when it is necessary to indicate their corresponding regions.

Define $\delta = \det J$, $\mu = \operatorname{trace} J$ and $\Delta = \mu^2 - 4\delta$. The dynamical properties of the network are determined by the relations among $\delta$, $\mu$ and $\Delta$ according to the linear system theory in $\mathbf{R}^2$ (Perko, 2001): (i) If $\delta < 0$ then $x^*$ is a saddle, (ii) If $\delta > 0$ and $\Delta \geq 0$ then $x^*$ is a node which is stable if $\mu < 0$ and unstable if $\mu > 0$, (iii) If $\delta > 0$, $\Delta < 0$ and $\mu \neq 0$, then $x^*$ is a focus which is stable if $\mu < 0$ and unstable if $\mu > 0$, (iv) If $\delta > 0$ and $\mu = 0$ then $x^*$ is a center. Based on this theory, for the sake of concision, the following theory is presented without detailed analysis.

**Theorem 14.1.** *A distribution of equilibria of the two-cell LT network (14.1) and their geometrical properties are given for various cases of connection strengths and external inputs in Table 1.*

Table 14.1. Properties and distributions of the equilibria

| dist. | conditions | | | propt. |
|---|---|---|---|---|
| $D_1$ | $(w_{11}-1)(w_{22}-1) < w_{12}w_{21}$ | | | saddle |
| | $(w_{22}-1)h_1 \geq w_{12}h_2$ | | | |
| | $(w_{11}-1)h_2 \geq w_{21}h_1$ | | | |
| | $(w_{11}-1)(w_{22}-1) > w_{12}w_{21}$ | $(w_{11}-w_{22})^2 \geq -4w_{12}w_{21}$ | $d < 2$ | s-node |
| | $(w_{22}-1)h_1 \leq w_{12}h_2$ | | $d > 2$ | u-node |
| | $(w_{11}-1)h_2 \leq w_{21}h_1$ | | | |
| | | $(w_{11}-w_{22})^2 < -4w_{12}w_{21}$ | $d < 2$ | s-focus |
| | | | $d > 2$ | u-focus |
| | | | $d = 2$ | center |
| $D_2$ | $w_{22} < 1,\ h_1 < w_{12}h_2/(w_{22}-1),\ h_2 \geq 0$ | | | s-node |
| | $w_{22} > 1,\ h_1 < w_{12}h_2/(w_{22}-1),\ h_2 \leq 0$ | | | saddle |
| $D_3$ | $h_1 < 0, h_2 < 0$ | | | s-node |
| $D_4$ | $w_{11} < 1,\ h_1 \geq 0,\ h_2 < w_{21}h_1/(w_{11}-1)$ | | | s-node |
| | $w_{11} > 1,\ h_1 \leq 0,\ h_2 < w_{21}h_1/(w_{11}-1)$ | | | saddle |

$d=w_{11}+w_{22}$, s: stable, u: unstable.

## 14.4 Neural States in $D_1$ and $D_2$

### 14.4.1 Phase Analysis for Center Type Equilibrium in $D_1$

In $D_1$, the Jacobian matrix reads

$$J = \begin{pmatrix} w_{11} - 1 & w_{12} \\ w_{22} & w_{22} - 1 \end{pmatrix},$$

and its eigenvalues are $\lambda_{1,2} = \pm \omega i$. By defining $\tilde{x}_1(t) = x_1(t) - x_1^*, \tilde{x}_2(t) = x_2(t) - x_2^*$ for $t \geq 0$, the linear system (14.1) becomes $\dot{\tilde{x}} = J\tilde{x}$. Note that $w_{12}w_{21} \neq 0$. The linear system has a pair of complex eigenvectors corresponding to $\lambda_{1,2}$,

$$e_{1,2} = \begin{pmatrix} 1 \\ \frac{-w_{11}+1}{w_{12}} \end{pmatrix} \pm i \begin{pmatrix} 0 \\ \frac{\omega}{w_{12}} \end{pmatrix},$$

or

$$e_{1,2} = \begin{pmatrix} \frac{w_{11}-1}{w_{21}} \\ 1 \end{pmatrix} \pm i \begin{pmatrix} \frac{\omega}{w_{21}} \\ 0 \end{pmatrix}.$$

Denote $u = \text{Re}(e)$ and $v = \text{Im}(e)$, according to linear system theory (Perko, 2001), $J$ can be reduced to the Jordan canonical form by defining a invertible matrix $P = (v, u)$. We choose the first pair of eigenvectors for our derivation. Adopting the second pair of eigenvectors will lead to the same results.

Therefore, we can define a linear transformation $y(t) = P^{-1}\tilde{x}(t)$ for all $t \geq 0$, where

$$P = \begin{pmatrix} 0 & 1 \\ \frac{\omega}{w_{12}} & \frac{-w_{11}+1}{w_{12}} \end{pmatrix},$$

such that

then we have
$$\dot{y} = \begin{pmatrix} 0 & -\omega \\ \omega & 0 \end{pmatrix} y. \tag{14.3}$$

It follows that
$$y(t) = \begin{pmatrix} \cos\omega t & -\sin\omega t \\ \sin\omega t & \cos\omega t \end{pmatrix} y(0). \tag{14.4}$$

Hence,
$$y_1(t)^2 + y_2(t)^2 = y_1(0)^2 + y_2(0)^2. \tag{14.5}$$

Recall that $y(t) = P^{-1}\widetilde{x}(t)$, i.e.,
$$\begin{cases} y_1(t) = \frac{w_{11}-1}{\omega}\widetilde{x}_1(t) + \frac{w_{12}}{\omega}\widetilde{x}_2(t), \\ y_2(t) = \widetilde{x}_1(t), \end{cases} \tag{14.6}$$

substituting it back into (14.5), we thus obtain the trajectory equation in $D_1$,
$$(\frac{w_{11}-1}{\omega}\widetilde{x}_1(t) + \frac{w_{12}}{\omega}\widetilde{x}_2(t))^2 + \widetilde{x}_1(t)^2 = (\frac{w_{11}-1}{\omega}\widetilde{x}_1(0) + \frac{w_{12}}{\omega}\widetilde{x}_2(0))^2 + \widetilde{x}_1(0)^2, \tag{14.7}$$

that is,
$$(1 + \frac{(w_{11}-1)^2}{\omega^2})\widetilde{x}_1(t)^2 + \frac{2(w_{11}-1)w_{12}}{\omega^2}\widetilde{x}_1(t)\widetilde{x}_2(t) + \frac{w_{12}^2}{\omega^2}\widetilde{x}_2(t)^2$$
$$= (\frac{w_{11}-1}{\omega}\widetilde{x}_1(0) + \frac{w_{12}}{\omega}\widetilde{x}_2(0))^2 + \widetilde{x}_1(0)^2. \tag{14.8}$$

### 14.4.2 Phase Analysis in $D_2$

In $D_2$, the Jacobian matrix is given as
$$J = \begin{pmatrix} -1 & w_{12} \\ 0 & w_{22}-1 \end{pmatrix}.$$

By defining $\widetilde{x}_1(t) = x_1(t) - x_1^*, \widetilde{x}_2(t) = x_2(t) - x_2^*$ for $t \geq 0$, the linear system (14.1) becomes $\dot{\widetilde{x}} = J\widetilde{x}$. Since $J$ may have one or two distinct eigenvalues when $w_{22}$ takes different values, thus two different cases need to be considered.

**Case 1:** $w_{22} \neq 0$. In this case the Jacobian matrix has two distinct eigenvalues, $\lambda_1 = -1, \lambda_2 = w_{22} - 1$. The corresponding linear-independent eigenvectors are given below,
$$e_1 = \begin{pmatrix} 1 \\ 0 \end{pmatrix}, e_2 = \begin{pmatrix} \frac{w_{12}}{w_{22}} \\ 1 \end{pmatrix}.$$

According to the linear system theory, $J$ can be reduced to the Jordan canonical form by defining an invertible matrix $P = (e_1, e_2)$. Therefore, we can define a linear transformation $y(t) = P^{-1}\widetilde{x}(t)$ for all $t \geq 0$, where

$$P = \begin{pmatrix} 1 & \frac{w_{12}}{w_{22}} \\ 0 & 1 \end{pmatrix}.$$

Then

$$y = P^{-1}\tilde{x} = \begin{pmatrix} 1 & -\frac{w_{12}}{w_{22}} \\ 0 & 1 \end{pmatrix} \tilde{x} = \begin{pmatrix} \tilde{x}_1 - \frac{w_{12}}{w_{22}}\tilde{x}_2 \\ \tilde{x}_2 \end{pmatrix}. \tag{14.9}$$

It follows that

$$\dot{y} = P^{-1}JPy = \begin{pmatrix} -1 & 0 \\ 0 & w_{22}-1 \end{pmatrix} y.$$

It is solved as

$$\begin{cases} y_1(t) = \exp(-t)y_1(0), \\ y_2(t) = \exp((w_{22}-1)t)y_2(0). \end{cases} \tag{14.10}$$

Thus we have

$$y_1(t)^{(w_{22}-1)} y_2(t) = y_1(0)^{(w_{22}-1)} y_2(0). \tag{14.11}$$

Substitute (14.9) into (14.11), we get

$$(\tilde{x}_1(t) - \frac{w_{12}}{w_{22}}\tilde{x}_2(t))^{(w_{22}-1)} \tilde{x}_2(t) = (\tilde{x}_1(0) - \frac{w_{12}}{w_{22}}\tilde{x}_2(0))^{(w_{22}-1)} \tilde{x}_2(0). \tag{14.12}$$

**Case 2:** $w_{22} = 0$.

The eigenvalues are $\lambda_{1,2} = -1$. The Jacobian matrix has only one linear independent vector $e_1 = (1,0)^T$. A generalized eigenvector $e_2 = (0, \frac{1}{w_{12}})^T$ is obtained such that $Je_2 = (-1)e_2 + e_1$. Then we can define a similarity transformation $y = P^{-1}\tilde{x}$, where

$$P = \begin{pmatrix} 1 & 0 \\ 0 & \frac{1}{w_{12}} \end{pmatrix}. \tag{14.13}$$

Then we have

$$y = P^{-1}\tilde{x} = \begin{pmatrix} 1 & 0 \\ 0 & w_{12} \end{pmatrix} \tilde{x} = \begin{pmatrix} \tilde{x}_1 \\ w_{12}\tilde{x}_2 \end{pmatrix}, \tag{14.14}$$

and

$$\dot{y} = P^{-1}JPy = \begin{pmatrix} -1 & 1 \\ 0 & -1 \end{pmatrix} y. \tag{14.15}$$

Its solutions are given by

$$y(t) = e^{-t} \begin{pmatrix} 1 & t \\ 0 & 1 \end{pmatrix} y(0). \tag{14.16}$$

Substitute (14.14) into (14.16), it is obtained that

$$\begin{cases} \tilde{x}_1(t) = e^{-t}(\tilde{x}_1(0) + tw_{12}\tilde{x}_2(0)), \\ \tilde{x}_2(t) = e^{-t}\tilde{x}_2(0), \end{cases} \tag{14.17}$$

or equivalently,

$$\frac{\tilde{x}_1(t)}{\tilde{x}_2(t)} = \frac{\tilde{x}_1(0)}{\tilde{x}_2(0)} + tw_{12}. \tag{14.18}$$

From the above equation, it is derived that

$$t = \frac{1}{w_{12}}(\frac{\tilde{x}_1(t)}{\tilde{x}_2(t)} - \frac{\tilde{x}_1(0)}{\tilde{x}_2(0)}), \tag{14.19}$$

substitute it back into (14.17), we get

$$\ln \tilde{x}_2(t) + \frac{1}{w_{12}}\frac{\tilde{x}_1(t)}{\tilde{x}_2(t)} = \ln \tilde{x}_2(0) + \frac{1}{w_{12}}\frac{\tilde{x}_1(0)}{\tilde{x}_2(0)}, \tag{14.20}$$

for $\tilde{x}_2(0) > 0$.

### 14.4.3 Neural States Computed in Temporal Domain

The orbit of the system in $D_1$ is computed by

$$(x(t)-x^e) = \exp(\frac{\tau}{2}t)\begin{pmatrix} \frac{w_{11}-w_{22}}{2\omega}\sin\omega t + \cos\omega t & \frac{w_{12}}{\omega}\sin\omega t \\ \frac{w_{21}}{\omega}\sin\omega t & \frac{w_{22}-w_{11}}{2\omega}\sin\omega t + \cos\omega t \end{pmatrix}(x(0)-x^e). \tag{14.21}$$

The orbit of the system in $D_2$ is computed by

$$(x(t) - x^e) = \begin{pmatrix} \exp(-t) & \frac{w_{12}}{w_{22}}(\exp((w_{22}-1)t) - \exp(-t)) \\ 0 & \exp((w_{22}-1)t) \end{pmatrix}(x(0) - x^e), (w_{22} \neq 0) \tag{14.22}$$

or else

$$(x(t) - x^e) = \begin{pmatrix} \exp(-t) & w_{12}t\exp(-t) \\ 0 & w_{22}t\exp(-t) + \exp(-t) \end{pmatrix}(x(0) - x^e), (w_{22} = 0). \tag{14.23}$$

## 14.5 Rotated Vector Fields

Since within each region, the system (14.1) is linear and the properties of its equilibria are determined by the eigenvalues of the Jacobian matrix in each region. Thus the linear system cannot exhibit a limit cycle in each region, but must cross boundaries of regions if it exists. Therefore switching between different regions must happen on the limit cycle. A necessary condition for the existence of periodic orbits is that there exists a rotated vector field in the dynamics (14.1). It is observed that the most complex and interesting phenomena occurs in the unsaturated region $D_1$. In the sequel, we assume excitatory inputs ($h = (h_1, h_2) > 0$) in network (14.1).

According to the results from the last section, the two-cell LT network has only one nonzero equilibrium that is a focus or a center in $D_1$, if it satisfies

$$\begin{cases} (w_{11}-1)(w_{22}-1) > w_{12}w_{21}, \\ (w_{22}-1)h_1 < w_{12}h_2, \\ (w_{11}-1)h_2 < w_{21}h_1, \\ (w_{11}-w_{22})^2 + 4w_{12}w_{21} < 0. \end{cases} \quad (14.24)$$

Thus the network has a *rotated vector field* prescribed by (14.24) in the linear region.

From the Jacobian matrix $J_1$, it is noted that $\tau = w_{11} + w_{22} - 2$, $\delta = (w_{11}-1)(w_{22}-1) - w_{12}w_{21} > 0$, and $\Delta = \tau^2 - 4\delta < 0$. Define $\omega = \frac{1}{2}\sqrt{|\Delta|}$, then the eigenvalues of the linear system are given as

$$\lambda = \frac{\tau}{2} \pm \omega i. \quad (14.25)$$

If $\tau \neq 0$, the equilibrium is a topological focus. If $\tau = 0$, then $\omega = \sqrt{\delta}$, $\lambda = \pm \omega i$, and the equilibrium is a topological center.

Let $L_1$ and $L_2$ be the two lines that satisfy $\dot{x}_1 = 0$ and $\dot{x}_2 = 0$, respectively, i.e.,

$$\begin{cases} L_1 : (w_{11}-1)x_1 + w_{12}x_2 + h_1 = 0, \\ L_2 : w_{21}x_1 + (w_{22}-1)x_2 + h_2 = 0. \end{cases}$$

Let $k_1$ and $k_2$ denote their corresponding slopes, $k_1 = \frac{1-w_{11}}{w_{12}}$ and $k_2 = \frac{w_{21}}{1-w_{22}}$. $L_1$ and $L_2$ cross at the equilibrium $x^*$ and suppose $L_1$ and $L_2$ intersect $x_2$-axis at points $s(0,s)$ and $p(0,p)$, respectively, where $s = -\frac{h_1}{w_{12}}$ and $p = \frac{h_2}{1-w_{22}}$. Note that it always holds that $p < s$, as derived from the conditions in (14.24). Consequently, the vector field of the network prescribed by (14.24) has one of the forms shown in Fig. 14.1.

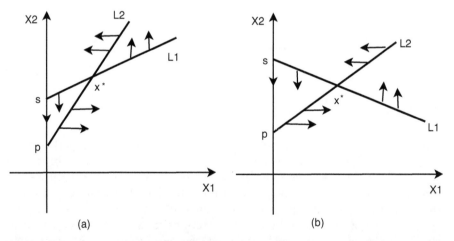

**Fig. 14.1.** Rotated vector fields prescribed by (14.24). (a) is under the conditions $w_{12} < 0, 0 < k_1 < k_2$, and (b) $w_{12} < 0, k_1 < 0 < k_2$. If $w_{12} > 0$, the arrows from $L_1$ and $L_2$ are reversed and $p < s < 0$ for both (a) and (b).

## 14.6 Existence and Boundary of Periodic Orbits

According to the results of the previous Theorem, the equilibrium $x^*$ is an unstable focus or a center, if it satisfies

$$\begin{cases} w_{12}w_{21} < -\frac{1}{4}(w_{11} - w_{22})^2, \\ (w_{22} - 1)h_1 < w_{12}h_2, \\ (w_{11} - 1)h_2 < w_{21}h_1, \\ w_{11} + w_{22} > 2(\text{unstable focus}), \ w_{11} + w_{22} = 2(\text{center}). \end{cases} \qquad (14.26)$$

If network (14.1) has a topological center in $D_1$, then it always exhibits oscillations around the equilibrium in $D_1$. However, the outmost of the periodic orbits needs further study. When $x^*$ is an unstable focus, the network may tend to be unstable and it is nontrivial to determine if periodic orbits exist.

**Remark 1:** *Poincaré-Bendixson Theorem* (Zhang et al., 1992) tells that if there exists an annular region $\Omega \subset \mathbf{R}^2$, containing no equilibrium, such that both orbits from its outer boundary $\gamma_2$ and from its inner boundary $\gamma_1$ enter into $\Omega$, then there exists a (may not be unique) periodic orbit $\gamma$ in $\Omega$. If the vector field is analytic (can be expanded as a convergent power series), then the periodic orbits in $\Omega$ are limit cycles. Since $\sigma$ is not analytic, our attention is devoted to the existence of periodic orbits, instead of making more endeavors to assert that of limit cycles. Essentially, they do not make a significant difference to the oscillatory behavior, though they are different in geometrical meaning.

**Lemma 14.2.** *The two-cell network (14.1) has only one equilibrium in the interior of every periodic orbit.*

*Proof.* It is known that there must be an equilibrium in the bounded domain enclosed by a closed orbit in a continuous planar vector field (Zhang, 1992). Now suppose the network has two or more equilibria, then there exists a straight line of equilibria, i.e., an invariant line exists, which does not allow any closed orbit.

Let $L_1$ and $L_2$ be the two lines that satisfy $\dot{x}_1 = 0$ and $\dot{x}_2 = 0$, respectively, i.e.,

$$\begin{cases} L_1 : (w_{11} - 1)x_1 + w_{12}x_2 + h_1 = 0, \\ L_2 : w_{21}x_1 + (w_{22} - 1)x_2 + h_2 = 0. \end{cases}$$

When $w_{12} < 0, w_{11} > 1, w_{22} < 1, w_{21} > 0$, (14.26) describes a vector field as shown in Figure 14.2. The trajectories that evolve in $D_1$ and $D_2$ are also intuitively illustrated. $\Gamma_R$ is the right trajectory in $D_1$ that starts from the point $p$ and ends at the point $q(0,q)$. $\Gamma_L$ is the left trajectory in $D_2$ that starts from the point $q$ and ends at the point $r(0,r)$ where it exists. The arrows indicate the directions of the vector fields. It can be seen that there exists a rotational vector field around the equilibrium.

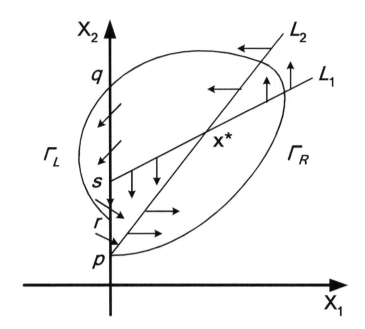

**Fig. 14.2.** The vector fields described by the oscillation prerequisites. The trajectories in $D_1$ and $D_2$ ($\Gamma_R$ and $\Gamma_L$, respectively) forced by the vector fields are intuitively illustrated.

**Theorem 14.3.** *The two-cell LT network (14.1) must have one periodic orbit, if it satisfies $w_{12} < 0, w_{22} < 1$ and (14.26).*

*Proof.* The equilibrium in $D_2$ is computed by

$$x^* = \frac{1}{1-w_{22}}\begin{pmatrix} w_{12}h_2 + (1-w_{22})h_1 \\ h_2 \end{pmatrix}.$$

It is noted that $x_1^* > 0, x_2^* > 0$, hence it is a virtual equilibrium lying in $D_1$. Although it does not practically exist, it still has a trend that drives the trajectory to approach it or diverge from it along its invariant eigenspace.

When $w_{22} \neq 0$, the Jacobian matrix in $D_2$ has two distinct eigenvalues, $\lambda_1 = -1$, $\lambda_2 = w_{22} - 1$ as well as two linear independent eigenvectors, $e_1 = (1,0)^T$, $e_2 = (\frac{w_{12}}{w_{22}}, 1)^T$. Firstly, consider the two cases for $w_{22} < 1$, where the subspaces spanned by $e_1$ and $e_2$ are both stable.

(i) $0 < w_{22} < 1$. In this case $\lambda_1 < \lambda_2$ and let $E^{ss} = \text{Span}\{e_1\}$ and $E^{ws} = \text{Span}\{e_1\}$ denote the strong stable subspace and weakly stable subspace, respectively. Figure 14.3 shows the phase portrait. Each trajectory except the invariant line $E^{ss}$ approaches the equilibrium along the well-defined tangent line, $E^{ws}$. Since $(0,p)$ lies in $E^{ss}$, each trajectory must intersect $x_2$-axis at some point $(0,r)$ where $r > p$. Therefore, a simple closed curve $\Gamma$ can be constructed such that $\Gamma = \Gamma_R \cup \Gamma_L \cup \overline{rp}$, which can be used as an outer boundary of an annular region.

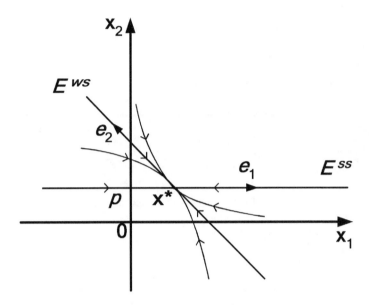

**Fig. 14.3.** Phase portrait for $0 < w_{22} < 1$.

(ii) $w_{22} < 0$. In this case $\lambda_2 < \lambda_1$ and let $E^{ss} = \mathrm{Span}\{e_2\}$ and $E^{ws} = \mathrm{Span}\{e_1\}$ denote the strong stable subspace and weakly stable subspace, respectively. The phase portrait is shown in Figure 14.4. Each trajectory except the invariant line $E^{ss}$ approaches the equilibrium along the well-defined tangent line, $E^{ws}$. Since $(0,p)$ lies in $E^{ss}$, each trajectory that starts from $(0,q)$ must intersect $x_2$-axis at some point $(0,r)$ where $r > p$. Thus, a simple closed curve $\Gamma$ can be constructed such that $\Gamma = \Gamma_R \cup \Gamma_L \cup \overline{rp}$.

(iii) $w_{22} = 0$. In this case there is only a single eigenvector $e_1$ which corresponds to the negative eigenvalue of $-1$. Figure 14.5 shows there is only a single invariant subspace. Each trajectory approaches the equilibrium along a well-defined tangent line determined by $e_1$. Thus the left trajectory intersects $x_2$-axis at a point which is above $(0,p)$.

To summarize the results of above cases, an outer boundary $\Gamma$ of an annular region can be constructed such that $\Gamma = \Gamma_R \cup \Gamma_L \cup \overline{rp}$. According to *Poincaré-Bendixson Theorem*, there must be a periodic orbit in the interior of the domain enclosed by $\Gamma$. This completes the proof of the theorem.

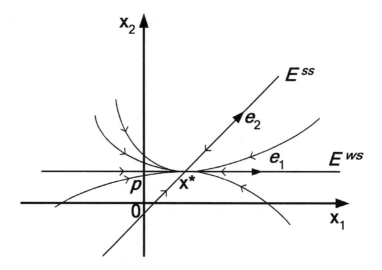

**Fig. 14.4.** Phase portrait for $w_{22} < 0$.

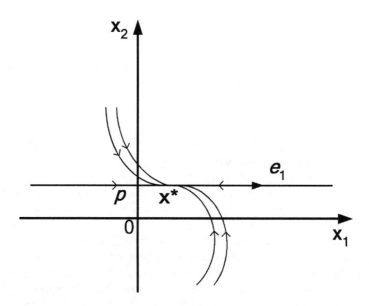

**Fig. 14.5.** Phase portrait for $w_{22} = 0$.

**Theorem 14.4.** *The two-cell network (14.1) must have no limit cycle, if $w_{22} > 1$ and $w_{12} < 0$.*

*Proof.* As analyzed in the previous theorem, $x_1^* < 0, x_2^* < 0$, the equilibrium is a virtual equilibrium lying in $D_3$. Notice that the system has two linear independent eigenvectors, $e_1 = (1,0)^T$, $e_2 = (\frac{w_{12}}{w_{22}}, 1)^T$, thus it has a stable subspace $E^s = \text{Span}\{e_1\}$ and an unstable subspace $E^u = \text{Span}\{e_2\}$. All the trajectories in $D_2$ move away from $x^*$ along the invariant space $E^u$. Therefore, in this case no limit cycle exists. See the phase portrait in Figure 14.6.

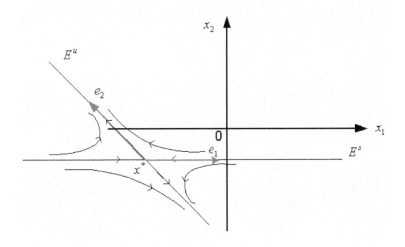

**Fig. 14.6.** Phase portrait for $w_{22} > 1$.

The periodic orbits resulting from a center-type equilibrium is investigated below.

**Theorem 14.5.** *If $\tau = 0, w_{12} < 0, w_{22} < 1$ and (14.26) holds for network (14.1), then the network produces periodic orbits of center type. The amplitude of its outermost periodic orbit is confined in the closed curve $\Gamma = \Gamma_R \cup \Gamma_L \cup \overline{rp}$, as shown in Figure 14.2, where*

$$p = \frac{h_2}{1 - w_{22}}, \quad \text{and}$$

$$q = -w_{22}\frac{w_{12}h_2 + (1 - w_{22})h_1}{(1 - w_{22})w_{12}} + \frac{(w_{11} - 1)h_2 - w_{21}h_1}{w_{12}w_{21} - (w_{11} - 1)(w_{22} - 1)}.$$

*Proof.* Based on the vector field analysis for $D_1$ and $D_2$ in the above discussions, it is known that the trajectory starting from a point $(0, q)$ with $q > s$ will intersect the $x_2$-axis again at a point $(0, r)$ above $(0, p)$. The trajectory starting from $(0, r)$ must approach a periodic orbit at $t \to \infty$. Thus the right trajectory $\Gamma_R$ which starts from $(0, p)$ and ends at $(0, q)$ and its continued left trajectory $\Gamma_L$ prescribes an outermost boundary such that every periodic orbit lies in the interior of the closed curve $\Gamma$.

Define $\widetilde{x}_1(t) = x_1(t) - x_1^*, \widetilde{x}_2(t) = x_2(t) - x_2^*$ for $t \geq 0$. Applying (14.8) for the trajectory equation in $D_1$, it is obtained that

$$\frac{w_{12}^2}{\omega^2}\widetilde{x}_2(t)^2 + \frac{2(w_{11}-1)w_{12}}{\omega^2}\widetilde{x}_1(t)\widetilde{x}_2(t) + (1 + \frac{(w_{11}-1)^2}{\omega^2})\widetilde{x}_1(t)^2$$
$$-(\frac{w_{11}-1}{\omega}\widetilde{x}_1(0) + \frac{w_{12}}{\omega}\widetilde{x}_2(0))^2 - \widetilde{x}_1(0)^2 = 0. \quad (14.27)$$

Suppose the trajectory intersects the $x_2$-axis again at $(0, q)$ where it exists after time $t_R$. Let $\widetilde{x}_1(t_R) = \widetilde{x}_1(0) = 0 - x_1^* = -x_1^*$, then

$$\frac{w_{12}^2}{\omega^2}\widetilde{x}_2(t_R)^2 + \frac{2(w_{11}-1)w_{12}}{\omega^2}\widetilde{x}_1(0)\widetilde{x}_2(t_R) + (1 + \frac{(w_{11}-1)^2}{\omega^2})\widetilde{x}_1(0)^2$$
$$-\frac{(w_{11}-1)^2}{\omega^2}\widetilde{x}_1(0)^2 - \frac{w_{12}^2}{\omega^2}\widetilde{x}_2(0)^2 - \frac{2w_{12}(w_{11}-1)}{\omega^2}\widetilde{x}_1(0)\widetilde{x}_2(0) - \widetilde{x}_1(0)^2 = 0,$$

that is

$$\frac{w_{12}^2}{\omega^2}\widetilde{x}_2(t_R)^2 + \frac{2(w_{11}-1)w_{12}}{\omega^2}\widetilde{x}_1(0)\widetilde{x}_2(t_R) - \frac{w_{12}^2}{\omega^2}\widetilde{x}_2(0)^2 \quad (14.28)$$
$$-\frac{2w_{12}(w_{11}-1)}{\omega^2}\widetilde{x}_1(0)\widetilde{x}_2(0) = 0. \quad (14.29)$$

It holds that

$$w_{12}^2\widetilde{x}_2(t_R)^2 + 2(w_{11}-1)w_{12}\widetilde{x}_1(0)\widetilde{x}_2(t_R) - w_{12}^2\widetilde{x}_2(0)^2$$
$$-2w_{12}(w_{11}-1)\widetilde{x}_1(0)\widetilde{x}_2(0) = 0. \quad (14.30)$$

To solve the above equation, let

$$\phi = 4(w_{11}-1)^2 w_{12}^2 \widetilde{x}_1(0)^2 + 4w_{12}^2(w_{12}^2\widetilde{x}_2(0)^2 + 2w_{12}(w_{11}-1)\widetilde{x}_1(0)\widetilde{x}_2(0))$$
$$= 4w_{12}^2((w_{11}-1)\widetilde{x}_1(0) + w_{12}\widetilde{x}_2(0))^2,$$

thus it is obtained that

$$\widetilde{x}_2(t_R) = \frac{-2(w_{11}-1)w_{12}\widetilde{x}_1(0) \pm \sqrt{\phi}}{2w_{12}^2}$$
$$= \frac{-(w_{11}-1)\widetilde{x}_1(0) \mp |(w_{11}-1)\widetilde{x}_1(0) + w_{12}\widetilde{x}_2(0)|}{w_{12}}. \quad (14.31)$$

Since

$$(w_{11}-1)\widetilde{x}_1(0) + w_{12}\widetilde{x}_2(0)$$
$$= (w_{11}-1)(x_1(0) - x_1^*) + w_{12}(x_2(0) - x_2^*)$$
$$= (w_{11}-1)x_1(0) + w_{12}x_2(0) - ((w_{11}-1)x_1^* + w_{12}x_2^*),$$

recall that $x_1(0) = 0, x_2(0) = p = -\frac{h_2}{w_{22}-1}$, and

$$(w_{11} - 1)x_1^* + w_{12}x_2^* + h_1 = 0,$$

thus

$$(w_{11} - 1)\tilde{x}_1(0) + w_{12}\tilde{x}_2(0) = 0 + w_{12}(-\frac{h_2}{w_{22} - 1}) - (-h_1)$$

$$= \frac{w_{12}h_2}{1 - w_{22}} + h_1$$

$$= \frac{w_{12}h_2 + (1 - w_{22})h_1}{1 - w_{22}}$$

$$> 0. \tag{14.32}$$

Therefore,

$$\tilde{x}_2(t_R) = \frac{-(w_{11} - 1)\tilde{x}_1(0) \mp ((w_{11} - 1)\tilde{x}_1(0) + w_{12}\tilde{x}_2(0))}{w_{12}}, \tag{14.33}$$

it is noted that one of the roots is $\tilde{x}_2(t_R) = \tilde{x}_2(0)$ when $t_R = 0$, and as $t_R > 0$,

$$\tilde{x}_2(t_R) = \frac{-(w_{11} - 1)\tilde{x}_1(0) - ((w_{11} - 1)\tilde{x}_1(0) + w_{12}\tilde{x}_2(0))}{w_{12}}$$

$$= \frac{-(w_{11} - 1)(0 - x_1^*) - \frac{w_{12}h_2 + (1-w_{22})h_1}{1-w_{22}}}{w_{12}}$$

$$= \frac{(w_{11} - 1)\frac{w_{12}h_2+(1-w_{22})h_1}{1-w_{22}} - \frac{w_{12}h_2+(1-w_{22})h_1}{1-w_{22}}}{w_{12}}$$

$$= \frac{(w_{11} - 1)w_{12}h_2 + (w_{11} - 1)(1 - w_{22})h_1 - w_{12}h_2 - (1 - w_{22})h_1}{w_{12}(1 - w_{22})}$$

$$= -\frac{w_{22}h_2}{1 - w_{22}} - \frac{w_{22}h_1}{w_{12}}$$

$$= -w_{22}\frac{w_{12}h_2 + (1 - w_{22})h_1}{(1 - w_{22})w_{12}}. \tag{14.34}$$

Hence

$$q = \tilde{x}_2(t_R) + x_2^* = -w_{22}\frac{w_{12}h_2 + (1 - w_{22})h_1}{(1 - w_{22})w_{12}} + \frac{(w_{11} - 1)h_2 - w_{21}h_1}{w_{12}w_{21} - (w_{11} - 1)(w_{22} - 1)}. \tag{14.35}$$

This completes the proof.

From the geometrical view as discussed, it reveals that the trajectory can only switch between a stable mode and an unstable mode, i.e., switching between two unstable modes or two stable modes is not permitted. It is difficult to calculate the amplitude and period of a periodic orbit. Fortunately, by virtue of the analytical results for center-type periodic orbits in Theorem 14.5

and the trajectory equations in temporal domain (referred to the Appendix), it is possible to estimate their variation trends, i.e., how they change with respect to $\theta_i, \tau$ and $h_i$. An approximate relationship describing the amplitude, period and external inputs can be stated as: (a) $q \propto h_i, q \propto \tau^2$, $q$ indicating the amplitude; (b) the period increases as $\tau$ increases, while analysis procedures are omitted.

## 14.7 Winner-take-all Network

A winner-take-all network is an interesting and meaningful application of LT network. The study of its dynamics, fixed points and stability is of great benefit. A biologically motivated WTA model with local excitation and global inhibition was established and studied by (Hahnloser, 1998). A generalized model of such a WTA network is formulated as the following differential equations:

$$\dot{x}_i(t) = -x_i(t) + \theta_i \sigma(x_i(t)) - L + h_i, \tag{14.36a}$$

$$\tau \dot{L}(t) = -L(t) + \sum_{j=1}^{n} \theta_j \sigma(x_j(t)), \tag{14.36b}$$

where $\theta_i > 0$ denotes local excitation, $i = 1, ..., n$, and $L$ is a global inhibitory neuron, $\tau$ a time constant reflecting the delay of inhibition. It is noted that the dynamics of $L$ has a property of nonnegativity: if $L(0) \geq 0$ then $L(t) \geq 0$ and if $L(0) < 0$ then $L(t_1) \geq 0$ after some transient time $t_1 > 0$. In an explicit and compact form for $x = (x_1, ..., x_n)^T$, the WTA model can be re-written as

$$\begin{bmatrix} \dot{x}(t) \\ \dot{L}(t) \end{bmatrix} = -\begin{bmatrix} x(t) \\ L(t) \end{bmatrix} + \begin{bmatrix} \Theta & -\mathbf{1} \\ \mathbf{v} & 1-\frac{1}{\tau} \end{bmatrix} \begin{bmatrix} \sigma(x(t)) \\ L(t) \end{bmatrix} + \begin{bmatrix} h \\ 0 \end{bmatrix}, \tag{14.37}$$

where $\Theta = diag(\theta_1, ..., \theta_n)$, $\mathbf{v} = \frac{1}{\tau}(\theta_1, ..., \theta_n)$, and $\mathbf{1}$ a columnar vector of ones.

Unlike in (Hahnloser, 1998), the generalized WTA model (14.36) allows each neuron to have an individual self-excitation $\theta_i$, which can be different from each other. Hence, it is believed to be more biologically plausible than requiring the same excitation for all neurons. By generalizing the analysis (Hahnloser, 1998), it can be concluded that if $\theta_i > 1$ and $\tau < \frac{1}{\theta_i - 1}$ for all $i$, the network performs the WTA computation:

$$x_k = h_k, \tag{14.38a}$$
$$x_i = h_i - \theta_k h_k, \ i \neq k, \tag{14.38b}$$
$$L = \theta_k h_k, \tag{14.38c}$$

where $k$ indicates the winning neuron, nevertheless, it may not be the neuron having the largest external input. And all $\theta_i = 1$ will produce an absolute WTA network, in the sense that the winner $x_k$ is the neuron which receives the largest external input $h_k$.

It was claimed (Hahnloser, 1998) that a periodic orbit may occur by slowing down the global inhibition (but not excessively). However, its occurrence is not known *a priori*. Consider a single excitatory-inhibitory pair of the dynamic system (14.36), that is

$$\begin{bmatrix} \dot{x}(t) \\ \dot{L}(t) \end{bmatrix} = -\begin{bmatrix} x(t) \\ L(t) \end{bmatrix} + \begin{bmatrix} \theta & -1 \\ \frac{\theta}{\tau} & 1-\frac{1}{\tau} \end{bmatrix} \begin{bmatrix} \sigma(x(t)) \\ L(t) \end{bmatrix} + \begin{bmatrix} h \\ 0 \end{bmatrix}. \qquad (14.39)$$

Since each neuron is only coupled with a global inhibitory neuron (though such globally inhibitory neurons reflect collective activities of all the excitatory neurons), it does not interact directly with other neurons. Therefore, the dynamics of $n$-dimensional WTA network is possible to be clarified by studying a two-cell network with a single excitatory neuron. The established results for any two-dimensional network in Theorem 14.3 allow us to put forward the theorem on the existence of periodic orbits for the two-cell WTA model.

**Theorem 14.6.** *The single excitatory-inhibitory WTA network (14.39) must have a periodic orbit if its local excitation and global inhibition satisfy $\theta > 1$ and $\frac{1}{\theta-1} < \tau < \frac{(\sqrt{\theta}+1)^2}{(\theta-1)^2}$.*

**Proof.** The connection matrix and external inputs are respectively,

$$W = \begin{bmatrix} \theta & -1 \\ \frac{\theta}{\tau} & 1-\frac{1}{\tau} \end{bmatrix}, \text{ and } \hat{h} = \begin{bmatrix} h \\ 0 \end{bmatrix}.$$

Obviously, the first two conditions of Theorem 14.3 are met since $-1 < 0$, $1 - \frac{1}{\tau} < 1$. Now it remains to prove that condition (14.26) is also met by synaptic weights and external inputs. From $\frac{1}{\theta-1} < \tau$, it is easy to show that $\theta + 1 - \frac{1}{\tau} > 2$, which verifies the last inequality of (14.26). On the other hand, its first inequality is equivalent to

$$-\frac{\theta}{\tau} < -\frac{1}{4}(\theta-1+\frac{1}{\tau})^2, \qquad (14.40)$$

then it yields immediately that

$$\frac{(\sqrt{\theta}-1)^2}{(\theta-1)^2} < \tau < \frac{(\sqrt{\theta}+1)^2}{(\theta-1)^2}. \qquad (14.41)$$

Apparently, $\frac{(\sqrt{\theta}-1)^2}{(\theta-1)^2} < \frac{1}{\theta-1}$. Recall $\tau > \frac{1}{\theta-1}$, it is obtained that

$$\frac{1}{\theta-1} < \tau < \frac{(\sqrt{\theta}+1)^2}{(\theta-1)^2}. \qquad (14.42)$$

The next two inequalities of (14.26) are easily verified by considering that $\hat{h}_1 > 0, \hat{h}_2 = 0$. Therefore, according to Theorem 14.3, the WTA network must have a periodic orbit. This completes the proof.

The above analysis indicates that the parameters ($\theta$ and $\tau$), i.e., the synaptic connections between excitatory and inhibitory neurons are the dominating factors on the stability and oscillatory behavior of the network. On the other hand, introducing a new excitatory neuron to (14.39) only increases the strength of the global inhibition $L$, which would resemble adding a new excitatory-inhibitory pair $(x_2, L)$ to the existing one $(x_1, L)$. As a consequence, the stability of the existing pair would not change if the new pair is stable, hence the whole network $(x_1, x_2, L)$ retains stability. Otherwise, the existing pair would lose stability if the new pair is oscillating, and hence the whole network will begin to oscillate.

Based on the heuristic discussion above, we conjecture the following conditions that lead to oscillation behaviors of $n$-dimensional WTA networks (14.36). Rigorous proof of them is difficult. Nevertheless, extensive simulation studies support our observations.

**Proposition 14.7.** *The WTA network (14.36) loses stability and the periodical oscillation occurs if its local excitation and global inhibition satisfy $\theta_i > 1$ and $\frac{1}{\theta_i - 1} < \tau < \frac{(\sqrt{\theta_i}+1)^2}{(\theta_i-1)^2}$ for any $i = 1, ..., n$.*

As aforementioned, the strengths of local excitation and global inhibition ($\theta_i$ and $\tau$) play a determining role on the dynamical properties of the recurrent model. A complete description of the dynamics of the WTA model is thus summarized as follows: $i = 1, ..., n$,

i. $\theta_i \leq 1$ for all $i$ results in a global stability;
ii. $\theta_i > 1$ and $\tau < \frac{1}{\theta_i - 1}$ for all $i$ give rise to a WTA computation; it is absolute if all $\theta_i = 1$;
iii. $\theta_i > 1$ and $\frac{1}{\theta_i - 1} < \tau < \frac{(\sqrt{\theta_i}+1)^2}{(\theta_i-1)^2}$ for any $i$ lead to periodic oscillations;
iv. $\theta_i > 1$ and $\tau > \frac{(\sqrt{\theta_i}+1)^2}{(\theta_i-1)^2}$ for any $i$, implying too slow a global inhibition, would result in unstable dynamics due to a very strong unstable mode (in two dimensions, it is exactly an unstable focus) that forces the activities to diverge.

The study on cyclic dynamics of winner-take-all networks is of special interest particularly in cortical processing and short term memory and significant efforts have been made thus far (Ellias and Grossberg, 1975; Ermentrout, 1992). A different WTA model using global inhibition was analyzed (Ermentrout, 1992) and conditions on the existence of periodic orbits were proven for a single excitatory-inhibitory pair by virtue of the famous *Poincaré-Bendixson Theorem*. This is unlike the model (14.39), where the neuronal activities were saturated, thus the outer boundary of an annular region is relatively straightforward to be found. In contrast, it is nontrivial to construct such an outer boundary for the network (14.39) since nonsaturating transfer functions may give rise to unbounded activities. Different treatment of applying *Poincaré-Bendixson Theorem* is also required by the fact that the WTA network of LT neurons differs from various models studied in (Ermentrout, 1992) due to the non-differentiability of LT transfer functions, which imposes another difficulty on the analysis of cyclic dynamics. More interestingly, however, similar observations to our analysis were made: the topology of the connections is the main determining factor for such WTA dynamics and oscillatory behavior begins when the inhibition slows down.

## 14.8 Examples and Discussions

In this section, examples of computer simulations are provided to illustrate and verify the theories developed.

### 14.8.1 Nondivergence Arising from A Limit Cycle.

Consider the two-cell LT network,

$$\dot{x}(t) = -x(t) + \begin{pmatrix} 2 & -1 \\ 1 & 0.5 \end{pmatrix} \begin{pmatrix} \sigma(x_1(t)) \\ \sigma(x_2(t)) \end{pmatrix} + \begin{pmatrix} 3 \\ 1 \end{pmatrix}. \tag{14.43}$$

The network has an unstable focus in $D_1$, thus any trajectory starting from its neighborhood must move away from it. However, it does not result in the divergence of its dynamics. It is easy to check that the connection matrix $W$ and external inputs $h$ satisfy the conditions in the above theorem. Therefore, it is claimed that there must exist one limit cycle which encircles the unstable focus. Fig. 14.7 shows the limit cycle in phase plane and the neural states $x_1$ and $x_2$ in temporal domain. As expected, the neural states switches endlessly between the unstable $D_1$ and $D_2$.

### 14.8.2 An Example of WTA Network

Consider an example of WTA network (14.36) with 6 neurons. When $\tau = 2, \theta_i = 2$ for all $i$, the network becomes periodically oscillating (Figures 14.8 and 14.9). It is verified by Proposition 14.7, since $\tau$ is in the region $1 < \tau < (\sqrt{2}+1)^2$ where periodic orbit (undamped oscillation) occurs. As $\tau < 1$, the network is asymptotically stable and performs the winner-take-all computation, whereas it becomes unstable when $\tau > (\sqrt{2}+1)^2$.

### 14.8.3 Periodic Orbits of Center Type

To show the periodic orbits of center type, a simple network is taken as an example:

$$\frac{dx}{dt} = -x + \begin{pmatrix} 2 & -1 \\ 2 & 0 \end{pmatrix} \begin{pmatrix} \sigma(x_1(t)) \\ \sigma(x_2(t)) \end{pmatrix} + \begin{pmatrix} 2 \\ 1 \end{pmatrix}.$$

Figure 14.10 gives an illustration of the trajectories starting from three different points, $(0, 0.5), (0, 1)$ and $(0, 1.2)$. As can be seen, each approaches a periodic orbit finally, where the inner closed curve shows a center. It is verified by Theorem 14.5, as a computer illustration of Figure 14.2, where $p = 1$ and $q = 3$. The value of $q$ can also be determined by solving the temporal equations of the neural states. Further evaluation of the temporal equations (14.21–14.23) shows exactly $1 < r < 2$, as given below.

Its equilibrium in $D_1$ is $x^* = (1, 3)$ which is a topological center. Define $\tilde{x} = (\tilde{x}_1, \tilde{x}_2) = x - x^*$, applying equation (14.21), we get

$$\tilde{x}(t) = \begin{pmatrix} \cos t + \sin t & -\sin t \\ 2\sin t & \cos t - \sin t \end{pmatrix} \tilde{x}(0). \tag{14.44}$$

Since in $D_1$ the orbit starts from the point $(0, p)$ and ends at the point $(0, q)$, where $p, q > 0$ exists. Let $p = 1$ and let $\Gamma_R$ denote the orbit starting from $(0, p)$ and ending at $(0, q)$. Thus $\tilde{x}_1(0) = -1, \tilde{x}_2(0) = -2$, we get

$$\tilde{x}_1(t) = \tilde{x}_1(0)(\cos t + \sin t) - \tilde{x}_2(0)\sin t,$$
$$\tilde{x}_2(t) = 2\tilde{x}_1(0)\sin t + \tilde{x}_2(0)(\cos t - \sin t).$$

Let $\tilde{x}_1(t) = \tilde{x}_1(0)$, then it is obtained $t_R = \frac{3}{2}\pi$. Thus we get $\tilde{x}_2(\frac{3}{2}\pi) = 0$, then $q = 3$. To avoid calculating time $t_R$, we can adopt equation (14.35) which also gives that

$$q = \frac{(w_{11} - 1)h_2 - w_{21}h_1}{w_{12}w_{21} - (w_{11} - 1)(w_{22} - 1)} = 3.$$

Subsequently, we compute the orbit starting from the point $(0, 3)$ and ending at the point $(0, r)$ in $D_2$, where $x^* = (1, 1)$. Again, taking $\tilde{x} = x - x^*$, it is derived from (14.23) that

$$\tilde{x}(t) = \begin{pmatrix} \exp(-t) & -t\exp(-t) \\ 0 & \exp(-t) \end{pmatrix} \tilde{x}(0). \tag{14.45}$$

By solving the above equations, it is obtained

$$r = \exp(-t_L)(3 - 1) + 1 = 2\exp(-t_L) + 1 = \frac{2}{1 + 2t_L}.$$

It is easy to show that $t_L > 1$. for all $t > 0$, hence $1 < r < 2$. It is consistent with the result obtained from the analysis of vector fields.

Therefore, an outer boundary $\Gamma = \Gamma_1 \cup \Gamma_2 \cup \overline{rp}$ has been constructed. All the periodic orbits lie in its interior.

The established analytical results imply that slowing down the global inhibition in a multistable WTA network (Hahnloser, 1998) is equivalent to giving rise to an unstable focus, which would result in occurrence of periodic orbits. Another important issue for LT networks is on continuous attractors, for example, line attractors as discussed in (Seung, 1998; Hahnloser et al., 2003). On the other hand, as shown in this letter, a stable periodic orbit is another type of continuous attractor, which encircles an unstable stationary state.

## 14.9 Conclusion

By concentrating on the analysis of a general parameterized two-cell network, this chapter studied the geometrical properties of equilibria and oscillatory behaviors of LT networks. The conditions for existence of periodic orbits were established, and the factors affecting its amplitude and period also revealed. Generally, the theoretical results are only applicable to a two-cell network; however, they can be extended to a special class of biologically motivated networks of more than two neurons, i.e., a WTA network using local excitation and global inhibition. The theory for the cyclic dynamics of such a WTA network was presented. Lastly, simulation results verified the theory that was developed.

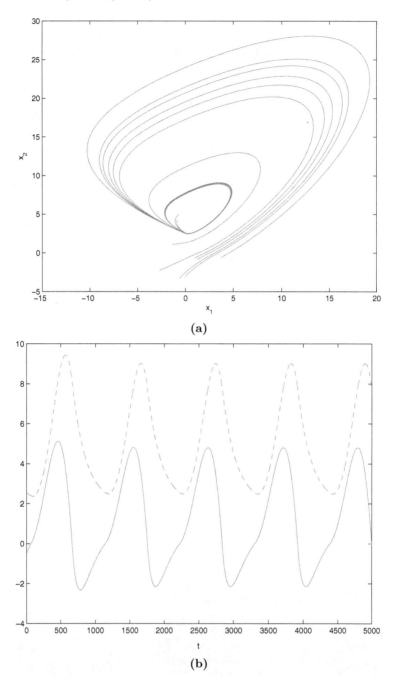

**Fig. 14.7.** Nondivergence arising from a stable limit cycle. (a) The limit cycle in the phase plane. Each trajectory from its interior or its outside approaches it. It is obviously an attractor. (b) The oscillations of the neural states. One of the neural states endlessly switches between stable mode and unstable mode. An Euler algorithm is employed for the simulations with time step 0.01.

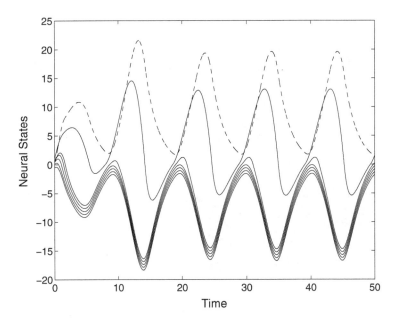

**Fig. 14.8.** Cyclic dynamics of the WTA network with 6 excitatory neurons. $\tau = 2, \theta_i = 2$ for all $i$, $h = (1, 1.5, 2, 2.5, 3, 3.5)^T$. The trajectory of each neural state is simulated for 50 seconds. The dashed curve shows the state of the global inhibitory neuron $L$.

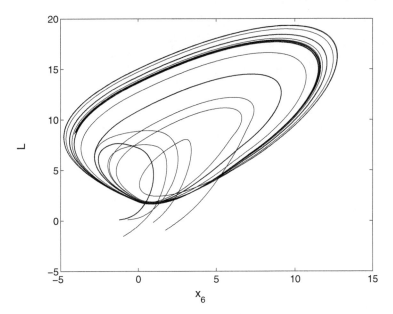

**Fig. 14.9.** A periodic orbit constructed in $x_6 - L$ plane. It shows the trajectories starting from five random points approaching the periodic orbit eventually.

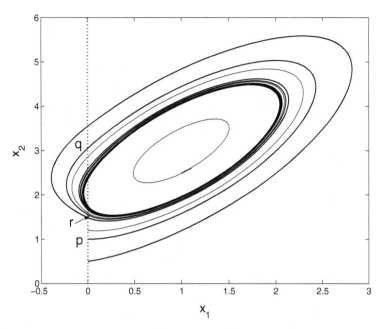

**Fig. 14.10.** Periodic orbits of center type. The trajectories with three different initial points $(0, 0.5), (0, 1)$ and $(0, 1.2)$ approach a periodic orbit that crosses the boundary of $D_1$. The trajectory starting from $(0, 1)$ constructs an outer boundary $\Gamma = \widehat{pqr} \cup \overline{rp}$ such that all periodic orbits lie in its interior.

# 15
# LT Network Dynamics and Analog Associative Memory

## 15.1 Introduction

[1]Although recurrent networks with linear threshold (LT) neurons were first examined in 1958 in a study of the limulus in the anatomical make-up of the eye (Hartline and Ratliff, 1958), only in recent years has interest in the biologically based models been growing (Douglas et al., 1995; Ben-Yishai et al., 1995; Salinas and Abbott, 1996; Bauer et al., 1999). The lack of an upper saturating state in neurons has been supported by experiments of cortical neurons that rarely operate close to saturation (Douglas et al., 1995), which suggests that the upper saturation may not be involved in actual computations of the recurrent network with LT neurons. However, this non-saturating characteristic of a neuron increases the likelihood that the network dynamics may be unbounded, as well as the possibility that no equilibrium point might exist (Forti and Tesi, 1995). Therefore, it is worthy of investigating and establishing conditions of boundedness and convergence of LT dynamics to design or exploit the recurrent networks.

Hahnloser (1998) analyzed the computational abilities and dynamics of linear threshold networks, particularly in explaining winner-take-all (WTA) multistability of the network with self-excitation and global inhibition, as well as oscillating behavior of the dynamics when the inhibition delay of the network is increased in an asymmetrical network. Of more recent interest, a functional silicon-based circuit design for a linear threshold network (Hahnloser et al., 2000) has demonstrated to be able to verify the theory of digital selection (groups of active and inactive neurons) and analog amplification (active neurons are amplified in magnitude, as represented by the gain in neuronal activity). A convergence condition was given by constructing a Lyapunov function for asynchronous neural networks with non-differentiable input-output

---

[1] Reuse of the materials of "Dynamics analysis and analog associative memory of networks with LT neurons", 17(2), 2006, 409–418, IEEE Trans. on Neural Networks, with permission from IEEE.

characteristics (Feng, 1997). Conditions for generalized networks of linear threshold neurons that ensure boundedness and complete convergence, for symmetrical and asymmetrical weight matrices were established (Wersing et al., 2001b; Yi et al., 2003). Meanwhile, the studies (Xie et al., 2002; Hahnloser, 1998) put forward the underlying framework for analyzing networks of linear threshold neurons by examining the system eigenvalues and classifying neurons according to permitted and forbidden sets, where subsets of permitted groups of neurons are likewise permitted, and supersets of forbidden groups of neurons are likewise forbidden. This is a consequence of the digital selection abilities of the network. This chapter places an emphasis on the dynamics analysis for the LT networks, to extend existing results by providing some new and milder conditions for boundedness and asymptotical stability.

It has been shown that complex perceptual grouping tasks, such as feature binding and sensory segmentation, can be accomplished in an LT network composed of competitive layers (Wersing et al., 2001a). Yet the computational ability of associative memory (also called content-addressable memory) has not been made clear. Recurrent neural network architectures have long been the focus of dynamical associative memory studies since Hopfield's seminal work (Hopfield, 1982). However, while the majority of efforts made since then have centered about the use of binary associative memory with discrete-state and saturating neural activities, few in the literature have generalized the original idea to the non-binary case, and even less so to the continuous case. Another focus of this chapter is on the associative memory (especially analog, real-valued auto-associative memory of gray-level images). A design method for the LT network's analog associative memory is proposed and discussed extensively.

## 15.2 Linear Threshold Neurons

The analog of the firing-rates of actual biological neurons in an artificial neuron is represented by an activation function. Various types of activation functions have been used in modeling the behavior of artificial neurons, and it is from these (nonlinear) activation functions that a network of neurons is able to derive its computational abilities. In recent years, a class of neurons which behaves in a linear manner above a threshold, known as linear threshold (LT) or threshold linear neurons, have received increased attention in the realm of neural computation. Previously, saturation points were modeled in activation functions to account for the refractory period of biological neurons when it 'recharges', and hence cannot be activated.

Computational neuroscientists believe that artificial neurons based on an LT activation function are more appropriate for modeling actual biological neurons. This is quite a radical departure from the traditional line of thought which assumed that biological neurons operate close to their saturating points, as conventional approaches to neural networks were usually based on saturating neurons with upper and lower bounded activities. Studies have demonstrated that cortical neurons rarely operate close to their saturating points, even in the presence of strong recurrent excitation (Douglas et al., 1995). This suggests that although saturation is present, most (stable) neural activities will seldom reach these levels.

It is believed that there are three generations of artificial neuron models. First-generation artificial neurons were discrete and binary-valued (*digital representation*), such as the classical binary threshold *McCulloch-Pitts* neuron. Second-generation models were based on neurons with continuous activation functions (*analog representation*), with sigmoidal neurons being archetypical. A third-generation neuron model is the *spiking neuron model*, which is represented by an LT neuron – a non-differentiable, unbounded activation function that behaves linearly above a certain threshold, similar in functionality to (half-wave) rectification in circuit theory. The general form of the LT activation function is defined as

$$\sigma(x) = [x]^+ = k \times \max(\theta, x), \qquad (15.1)$$

where $k$ and $\theta$ respectively denotes the *gain* ( usually 1) and *threshold* (usually 0), of the activation function (Figure 15.1). This form of nonlinearity is also known as threshold, or rectification nonlinearity.

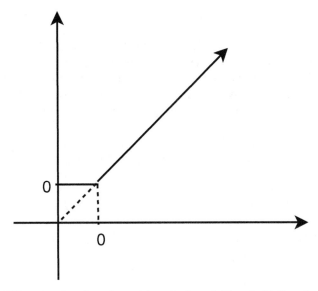

**Fig. 15.1.** LT activation function with gain $k = 1$, threshold $\theta = 0$, relating the neural activity output to the induced local field.

## 15.3 LT Network Dynamics (Revisited)

The network with $n$ linear threshold neurons is described by the additive recurrent model,

$$\dot{x}(t) = -x(t) + W\sigma(x(t)) + h, \qquad (15.2)$$

where $x, h \in \Re^n$ are the neural states and external inputs, $W = (w_{ij})_{n \times n}$ the synaptic connection strengths. $\sigma(x) = (\sigma(x_i))$ where $\sigma(x_i) = \max\{0, x_i\}$, $i = 1, \cdots, n$.

There is another equivalent model which is described by

$$\dot{y}(t) = -y(t) + \sigma(Wy(t) + h). \qquad (15.3)$$

The former one is more dynamically plausible than the latter one as suggested in (Yi et al., 2003), because system (15.3) can be transformed to system (15.2) by a simple linear transformation while it is not this case when transforming (15.2) to (15.3).

The nonlinearity of an LT network is a consequence of (i) the rectification nonlinearity of the LT neurons on an individual level, and (ii) the switching between active and inactive partitions of neurons on a collective level. Although (i) is apparent from the structure of the activation function, (ii) is a more subtle effect, yet contributes significantly to the computational abilities of the network. Specifically, the latter arises from the fact that *an inactive partition of LT neurons do not contribute to the feedback within the recurrent neural network.*

Such a network is said to exhibit hybrid analog-digital properties; while most digital circuits display some form of multistability (recall that feedback in digital flip-flops promotes some form of inherent differential instability), analog circuits in contrast, demonstrates linear amplification. Together, both characteristics co-exist in an LT network, and have been shown to manifest in an actual silicon circuit (Hahnloser et al., 2000), and is suggested as an underlying mechanism of the neocortex in response to sensory stimulation.

Since at any time $t$, each LT neuron is either firing (active) or silent (inactive), the whole ensemble of the LT neurons can be divided into a partition $P^+$ of neurons with positive states $x_i(t) \geq 0$ for $i \in P^+$, and a partition $P^-$ of neurons with negative states $x_i(t) < 0$ for $i \in P^-$. Clearly $P^+ \cup P^- = \{1, \cdots, n\}$.

Define a matrix $\Phi = \text{diag}(\phi_1, \cdots, \phi_n)$, where $\phi_i = 1$ for $i \in P^+$ and $\phi_i = 0$ for $i \in P^-$. The matrix $\Phi$ carries the digital information about which neurons are active. It is obviously

$$W\sigma(x) = W\Phi x. \qquad (15.4)$$

Let $W^e = W\Phi$, called the *effective recurrence matrix* (Hahnloser, 1998), which describes the connectivity of the operating network, where only active neurons are dynamically relevant.

**Theorem 15.1.** *Let $\Phi$ define a digital carry matrix. If there exists a matrix $\hat{W} \geq W$ satisfying one of the conditions*

## 15.3 LT Network Dynamics (Revisited)

$$w_{ii} + \sum_{j} |\hat{w}_{ij}|\phi_j - 1 < 0, \tag{15.5a}$$

$$Re\{\lambda_{max}\{\hat{W}\Phi - I\}\} < 0, \tag{15.5b}$$

$$Re\{\lambda_{max}\{\hat{W}\Phi\}\} < 1. \tag{15.5c}$$

*for all $i$, then the LT dynamics (15.2) is bounded in the partitions defined by $\Phi$.*

*Proof.* It holds that

$$\begin{aligned}\dot{x} &= -x + W\sigma(x(s)) + h \\ &\leq -x + \hat{W}\sigma(x(s)) + h \\ &= -(I - \hat{W}\Phi)x(s) + h.\end{aligned} \tag{15.6}$$

Let $y(t) = x(t) - h$, then

$$\dot{y} \leq -(I - \hat{W}\Phi)y(t). \tag{15.7}$$

Define $\dot{z}(t) = -(I - \hat{W}\Phi)z(t)$. Let $\lambda_i$ and $\mathbf{v}_i, i = 1, ..., n$ be the eigenvalues and $n$ linearly independent eigenvectors of $-(I - \hat{W}\Phi)$, respectively. According to the linear system theory (Bay, 1999), then $z(t)$ can be decomposed as $z(t) = \sum_i \xi_i(t)\mathbf{v}_i$, where $\xi_i(t), i = 1, ...n$ are functions of time. Therefore

$$z(t) = \sum_{i=1}^{n} e^{\lambda_i(t-t_0)}\mathbf{v}_i\xi_i(0). \tag{15.8}$$

Obviously $z(t)$ has an upper bound $m > 0$ if $\lambda_i < 0$. According to the comparison principle, $y(t) \leq z(t) \leq m$. Thus

$$x(t) \leq m + h$$

for all $t \geq 0$. This shows that $x_i(t)$ is upper bounded. Furthermore, the trajectory $x(t)$ of network (15.2) satisfies

$$x_i(t) = x_i(t_0)e^{-(t-t_0)} + \int_{t_0}^{t} e^{-(t-s)}\left(\sum_{j=1}^{n} w_{ij}\sigma(x_j(s)) + h_i\right) ds, \tag{15.9}$$

for all $t \geq t_0$. Then

$$|x_i(t)| \leq |x_i(t_0)|e^{-(t-t_0)} + (m+h)\sum_{j=1}^{n}|w_{ij}| + |h_i|, \quad i = 1, ..., n. \tag{15.10}$$

This proves that the LT dynamics is bounded in the partition defined by $\Phi$ if (15.5b) holds. It is obviously to derive (15.5c). Furthermore, by applying Gerschgorin's theorem (Goldberg, 1992), (15.5a) is easily obtained from (15.5b). The proof is completed.

A set of active neurons is called a *permitted set* if they can be coactivated at a stable steady state by some input; otherwise, it is termed a *forbidden set*. Permitted set and forbidden set were firstly studied in (Hahnloser et al., 2000), then further discussed in (Xie et al., 2002) and (Hahnloser et al., 2003). The assertion given for the superset of a forbidden set and the subset of a permitted set, nevertheless, was addressed in a more likely intuitive manner. Herein, from the above established theorem, such results is proven in a mathematical manner in the next.

**Corollary 15.2.** *Any subset of a permitted set is also permitted; any superset of a forbidden set is also forbidden.*

*Proof.* Let $\Gamma = \{\gamma_i\}$ denote a set of $m$ active neurons, $\gamma_i \in \{1, ..., n\}, i = 1, ..., m$. Firstly, suppose $\Gamma$ is a forbidden set. Since $\phi_j = 1$ if $j \in \Gamma$ and $\phi_j = 0$ otherwise, according to Theorem 1, the following condition is met,

$$w_{ii} + \sum_{j \in \Gamma} |\hat{w}_{ij}| \phi_j > 1.$$

Then its superset $\hat{\Gamma}$, with more active neurons, obviously results in

$$w_{ii} + \sum_{j \in \hat{\Gamma}} |\hat{w}_{ij}| \phi_j > w_{ii} + \sum_{j \in \Gamma} |\hat{w}_{ij}| \phi_j > 1.$$

Thus $\hat{\Gamma}$ is also forbidden. It is similarly to prove for permitted set.

Next, the conditions which ensure that the LT dynamics is bounded regardless of partitions are presented. The following lemma is firstly introduced.

**Lemma 15.3.** *Let $y(t)$ denote the solution of the following equation*

$$\dot{y}(t) = -y(t) + \sigma \left( Wy(t) + h + (x(0) - Wy(0) - h)e^{-t} \right).$$

*Then for any $x(0) \in \Re^n$, the solution $x(t)$ of network (15.2) starting from $x(0)$ can be represented by*

$$x(t) = Wy(t) + h + (x(0) - Wy(0) - h) e^{-t}.$$

*This proof can be referred to (Yi et al., 2003).*

**Theorem 15.4.** *If there exists a matrix $\hat{W} \geq W$ satisfying one of the following:*

$$w_{ii} + \sum_j |\hat{w}_{ij}| - 1 < 0, \qquad (15.11a)$$

$$Re\{\lambda_{max}\{\hat{W} - I\}\} < 0, \qquad (15.11b)$$

$$Re\{\lambda_{max}\{\hat{W}\}\} < 1, \qquad (15.11c)$$

*for all $i$, then the LT dynamics (15.2) is bounded.*

## 15.3 LT Network Dynamics (Revisited)

*Proof.* Consider the following dynamical system

$$\dot{y}(t) = -y(t) + \sigma(Wy(t) + h), \ t \geq 0, \tag{15.12}$$

of which the solution is calculated as

$$y(t) = e^{-t}y(0) + \int_0^t e^{-(t-s)}\sigma(Wy(s) + h)ds. \tag{15.13}$$

By taking $x(0) = Wy(0) + h$, recall Lemma 1, the dynamics (15.2) can be associated with (15.12) as

$$x(t) = Wy(t) + h. \tag{15.14}$$

It means that any trajectory $y(t)$ starting from $y(0)$ can be transformed to the trajectory $x(t)$ starting from $Wy(0) + h$. Consider the trajectory formulation (15.13), the dynamics (15.12) possess a property of nonnegativity: if $y(0) \geq 0$ then $y(t) \geq 0$ and if $y(0) < 0$ then $y(t_1) \geq 0$ after some transient time $t_1 > 0$. Consequently, in (15.12) it is assumed that $y_i(t) \geq 0$ for all $i$. Now it holds either $\dot{y}(t) = -y(t) \leq 0$ or

$$\begin{aligned}\dot{y}(t) &= -y + Wy + h \\ &\leq -y + \hat{W}y + h \\ &= -(I - \hat{W})y + h.\end{aligned} \tag{15.15}$$

Thus if $Re\{\lambda_{max}\{I - \hat{W}\}\} > 0$, following the method in the proof of Theorem 15.1 $y(t)$ is bounded for all $t \geq 0$. Therefore, in light of (15.14), $x(t)$ is also bounded. The rest conditions are easily derived.

In contrast to that of Theorem 15.1, the condition of Theorem 15.4 implies that the dynamics (15.2) is bounded in any of partitions, thus it results in global boundedness.

Note that the components of $\hat{W}$ need not be positive. Define a matrix $W^+ = (w_{ij}^+)$, where

$$w_{ij} = \begin{cases} w_{ii}, & i = j, \\ \max(0, w_{ij}), & i \neq j. \end{cases}$$

Clearly, $W^+$ provides a straightforward choice of $\hat{W}$ with nonnegative off-diagonal elements. As a consequence, all the analytical results for $\hat{W}$ are applicable to $W^+$, which give rise to equivalent conditions as those of (Wersing et al., 2001b) for nonsymmetric networks. Unlike their argument that required $W$ and $\hat{W}$ to be symmetric, furthermore, note that Theorem 2 where $\hat{W}$ need not be symmetric is applicable to both symmetric and nonsymmetric networks. Hence it gives a wider rule.

**Lemma 15.5.** *For any $x, y \in \Re^n$, the linear threshold transfer function $\sigma(\cdot)$ has the following properties:*

i). $\|\sigma(x) - \sigma(y)\|^2 \leq (x-y)^T[\sigma(x) - \sigma(y)]$,

ii). $[\sigma(y) - y]^T[\sigma(y) - u] \leq 0$, $u_i \geq 0 (i = 1,...n)$,

iii). $x^T[\sigma(u-x) - u] \leq -\|\sigma(u-x) - u\|^2$, $u_i \geq 0 (i = 1,...n)$.

*Proof.* (i) can be concluded from Lemma 2 of (Yi et al., 2003). To prove (ii), it is derived

$$\begin{aligned}[\sigma(x) - x]^T[\sigma(x) - u] &= [\sigma(x) - x]^T[\sigma(x) - \sigma(u)] \\ &= [\sigma(x) - \sigma(u)]^T[\sigma(x) - \sigma(u)] + [\sigma(u) - x]^T[\sigma(x) - \sigma(u)] \\ &= \|\sigma(x) - \sigma(u)\|^2 - [x - u]^T[\sigma(x) - \sigma(u)] \\ &\leq 0.\end{aligned}$$

The last inequality is resulted from (i). Let $y = u - x$ and substitute into (ii), it follows

$$[\sigma(u-x) - u + x]^T[\sigma(u-x) - u] \leq 0,$$

which is equivalent to

$$[\sigma(u-x) - u]^T[\sigma(u-x) - u] + x^T[\sigma(u-x) - u] \leq 0.$$

Thus

$$x^T[\sigma(u-x) - u] \leq -\|\sigma(u-x) - u\|^2.$$

This completes the proof.

**Theorem 15.6.** *If $W$ is symmetric and there exists a $\hat{W} \geq W$ satisfying one of conditions (15.11), then the LT dynamics (15.2) is Lyapunov asymptotically stable.*

*Proof.* The LT dynamics (15.3) admits an energy-like function

$$E(t) = \frac{1}{2}y^T(t)(I - W)y(t) - h^Ty(t) \tag{15.16}$$

for all $t \geq 0$. Obviously, $E(0) = 0$. According to Theorem 15.4, $y(t)$ is bounded. It is easy to show that $E(t)$ is also bounded. It is derived from Lemma 1 that

$$\begin{aligned}\dot{E}(t) &= [y(t)(I - W) - h]^T\dot{y}(t) \\ &= [y(t) - (Wy(t) + h)]^T(-y(t) + \sigma(Wy(t) + h)) \\ &= -[y(t) - (Wy(t) + h)]^T[\sigma(y(t)) - \sigma(Wy(t) + h)] \\ &\leq -\|y(t) - \sigma(Wy(t) + h)\|^2 \\ &= -\|\dot{y}(t)\|^2. \end{aligned} \tag{15.17}$$

Hence $E(t)$ is monodecreasing. Since it is bounded below, $E(t)$ must approach a limit as $t \to \infty$ and thus $\dot{E}(t) = 0$ as $t \to \infty$. So $\dot{y}(t) \to 0$ as $t \to \infty$. It follows that

$$\lim_{t\to\infty} \dot{x}(t) = W \lim_{t\to\infty} \dot{y}(t) = 0.$$

Eventually, based on an analogous analysis as (Yi et al., 2003), it is concluded that any trajectory of the network approaches one of its equilibria, that is to say the dynamics is Lyapunov asymptotically stable.

In contrast to global asymptotical stability that admits exactly one equilibrium, the above theorem does not obey such restriction and thus allows for *multistable* dynamics stated in (Hahnloser, 1998; Wersing et al., 2001b).

While the synaptic weight matrix $W$ determines the stability of the dynamics, the input vector $h$ determines the analog responses of the active neurons at steady state. By replacing the rectification function $\sigma(x)$ with $\Phi$ (only when the identities of the active neurons are known *a priori*), the steady state is computed by

$$(I - W\Phi)x = h.$$

Thus $x = (I - W\Phi)^{-1}h$. Using a power series expansion, it gives

$$(I - W\Phi)^{-1}h = (I + W\Phi + (W\Phi)^2 + ...)h. \qquad (15.18)$$

This can be interpreted as the signal flowing through the partition of active neurons in a first synapse pass, a second synapse pass, and so on. Mathematically, the power series expansion allows the determination of the analog responses of the active neurons without explicitly computing the inverse of (I-W), which may not be possible in large networks.

Together, $W$ and $h$ determines the identity and response of the active neurons at steady state, thus the LT network can perform computations involving both digital selection and analog amplification, as realized in a silicon circuit (Hahnloser et al., 2000).

## 15.4 Analog Associative Memory

### 15.4.1 Methodology

From a physiological or biological viewpoint, memory can be divided into two main categories, short-term and long-term memory. Short-term memory, it is believed, arises from persistent activity of neurons, where the timeframe involved is in the order of seconds to minutes. On the other hand, long-term memory involves storage of memory patterns through actual physical/neuronal modifications of synaptic strengths. Such changes, which usually persist for period from tens of minutes or longer (Dayan and Abbott, 2001) are generally known as long-term potentiation (LTP) and long-term depression (LTD), and of these types, the longest lasting forms of LTP and LTD require some kind of protein synthesis to modify the synaptic connections.

Another categorical measure of associative memory is the input stimulus used (Kohonen, 1989). *Auto-associative memory* refers to the presentation of a partial or noisy version of the desired memory state which is to be recalled. An example would be presentation of a noisy version of $A$, say $A'$ to recall $A$. *Hetero-associative memory*, in contrast, is said to occur when an input stimulus that is different from the memory to be recalled is presented to the network. An example would be presentation of an input stimulus $B$ to recall $A$. The focus of this work is centered on long-term auto-associative memory. Various encoding methods for associative memory were proposed (Costantini et al., 2003; Zurada et al., 1996; Jankowski et al., 1996; Müezzinoğlu et al., 2003; Wang and Liu, 2002).

The dynamics of early associative memory models involved the use of input patterns (in a partial or approximate form) $h$ as the starting conditions for the network dynamics. The most recognizable type of network based upon such a principle is the Hopfield network (Hopfield, 1982). In such networks, the stored patterns (the desired or intended patterns to be recalled) are encoded as fixed-points of the network with the 'noisy' or corrupted versions as the basin of attraction (or also known as the probe) - these initial conditions that are started of with then 'flow' into a state-space trajectory that will (hopefully) converge to their equilibrium points. Error correction or pattern retrieval is achieved by these initial states or probes being attracted to the stored fixed-points.

However, to which of these fixed equilibrium points are arrived at are largely dependent on where these initial conditions begin: only if these starting points lie within the basin of attraction of the desired pattern (hence the use of the Hamming distance measure of the difference between the initial and the desired pattern) will these starting points converge to an intended pattern. Useful pattern matching will in turn require each fixed-point to have a sufficiently large basin of attraction.

### 15.4.2 Design Method

Essentially, the key proposition of the Hebbian rule is that, during the learning phase, pre-synaptic signals and post-synaptic activity that fire at the same time or within a short time period of each other are more closely correlated, which is reflected in the synaptic weight connection between these two neurons. There should then be a greater (excitatory) connection between neurons $i$ and $j$. Such adaptation method is described as follows ($\eta > 0$ denotes the learning rate)

$$\Delta w_{ij} \propto \xi_i \xi_j \qquad (15.19a)$$
$$\Delta w_{ij} = \eta \xi_i \xi_j \qquad (15.19b)$$

During the training phase, the network learns by potentiating the synapse between two neurons if both neurons are active (or inactive), and depressing the synapse if one neuron is active (inactive) and the other is inactive (active). The inclusion of depressive action in reducing the synaptic strength is a more

## 15.4 Analog Associative Memory

general form of the Hebb's original postulate. Other, more general form of this rule is that synapses should change according to, and proportionally to the degree of correlation or covariance of the activities of both the pre-synaptic and post-synaptic neurons (Dayan and Abbott, 2001).

For gray-level image storage of an $N$-dimensional pattern/memory vector, the synaptic weight matrix is set-up using the following design approach (which is not unlike a generalized Hebbian rule for learning)

$$\mathbf{W} = \{\alpha\{\frac{\delta}{m}\sum_{\mu=1}^{m}\{\nu(\mu)\sqrt{\xi^\mu\xi^{\mu T}} - \kappa\mathbf{I}\}\} - \beta\mathbf{11}^T\} - \omega\mathbf{I} \quad (15.20a)$$

$$\mathbf{W} = \{\alpha\mathbf{Wc} - \beta\mathbf{11}^T\} - \omega\mathbf{I} \quad (15.20b)$$

The proposed approach uses small, positive (excitatory) synaptic weights mediated by inhibition for non-divergence of the linear threshold network dynamics. Inhibitive self-coupling further allows the fine-tuning of the stability margin of the system. Because self-couplings are denoted as the diagonal entries in the synaptic weight matrix $\mathbf{W}$, control of the eigenvalues can be done directly (however, without knowledge of the actual values). Suppose $\lambda_i, i = 1, ..., n$, are eigenvalues of $\mathbf{W}$. By introducing an exogenous variable $\omega$ that only shifts the diagonals of $\mathbf{W}$, the eigenvalues of the system are shifted uniformly by $(\lambda_i + \omega)$. In a further step, after scaling by a factor $\alpha$, i.e., applying a transformation to $\mathbf{W}$ as $\alpha\mathbf{W} + \omega\mathbf{I}$, the eigenvalues are now changed to $\alpha\lambda_i + \omega$ for all $i$, where $\omega$ can be either positive or negative.

The three parameters (termed as 'external' parameters, as opposed to the 'internal' parameters as those found within the correlation matrix $\mathbf{Wc}$) are $\alpha$, $\beta$ and $\omega$. These three parameters are considered to be 'external' by virtue that they are not dependent on the input patterns (images).

Table 15.1. Nomenclature

| | |
|---|---|
| $m$ | no. of patterns |
| $\delta$ | normalization term |
| $\nu(\mu)$ | contribution of each pattern, $\mu = 1, ..., m$. |
| $\kappa$ | removal of self-correlation term |
| $\xi^\mu$ | $\mu$th pattern vector $\xi$ |
| $\mathbf{I}$ | identity matrix |
| $\mathbf{11}^T$ | matrix of ones |
| $\alpha$ | $> 1$ :global excitation; $< 1$ : divisive inhibition |
| $\beta$ | global inhibition |
| $\omega$ | self-coupling for eigenvalue 'placement' |
| $\mathbf{Wc}$ | analog correlation matrix |

The nomenclature for the terms used in the proposed design approach is listed in Table 15.1. The primary difficulties with using the Hebbian rule for storing analog patterns, are (i) the possibility of instability, and (ii) excessive, unbalanced inhibition. (i) arises because the possibility of instability is caused by the method used to set up the synaptic weights (as correlations between pixel gray levels). If saturating neurons were used instead of LT neurons, neural activities would be driven to their saturation values. On the other hand, (ii) occurs because of the large range of possible gray level values: leaving darker pixels inhibited and brighter pixels excited, in effect, a partitioning occurs between active (very large activities) neurons and silent neurons (at threshold).

To further elucidate, consider the following: for gray scale images, possible gray-level values reside in the range [0, 255]. The idea that is being put forward here is that neurons corresponding to higher gray-level values should have, accordingly greater excitatory connections with other neurons. However, the problem with the preceding approach is that neurons corresponding to lower gray level values have less excitatory connections with other neurons. Thus some of these neurons are inhibitive (negative connective weights) after introducing inhibition. This in turn leads to the (perpetual) state at which the group of neurons with lower gray level values will always be inhibited by those more active neurons (because of the inhibitive connections), and hence are subsequently silent at steady state.

In view of the aforementioned problem, the key idea in the proposed design method to be used in a network of LT neurons for continuous, real-valued associative memory is to limit the degree of inhibition introduced, such that most synaptic connections are excitatory ($w_{ij} > 0$) only to a small extent. Divisive inhibition is preferably used ($\alpha < 1$), as opposed to subtractive inhibition ($\beta$). However, the self-couplings of the neurons in the network are usually made inhibitive to prevent runaway excitation in the dynamics of the network (in simpler terms, this means that the diagonals of the network matrix are negative), though this condition is not always necessary; small, finitely positive values can also be used for self-excitation, and yet retain stability. Moreover, the off-diagonals are also small but finite values, which can either be excitatory or inhibitory.

A modulating term ($\delta$) that is introduced to the weight matrix **W** assists in normalizing the synaptic weights that connects a neuron $i$ to all other neurons and vice-versa. It is a global parameter, as it is applied to all synaptic connections in the weight matrix **W**. Suppose there be $N$ neurons and the maximum gray level value in the stored image/pattern be $I_{\max}$. In the worst case scenario, all $N$ neurons would consist of the maximum gray level value ($I_{\max}$). To approximate the value of $\delta$ to be used,

$$\delta N I_{\max} < 1 \Rightarrow \delta < \frac{1}{N I_{\max}} \qquad (15.21)$$

### 15.4.3 Strategies of Measures and Interpretation

A measure of the difference between two analog patterns, can be quantified by comparing the absolute differences in the original pattern and the distorted pattern, to the difference in the original pattern and the corrected/retrieved pattern, namely, the signal-to-noise ratio (SNR) computed as

$$SNR = \frac{\sum_i^N \sum_j^N (I_{original}(i,j) - I_{noise}(i,j))^2}{\sum_i^N \sum_j^N (I_{original}(i,j) - I_{retrieval}(i,j))^2}. \tag{15.22}$$

The greater the $SNR$ value, the better the correction (retrieval) process is, as performed by the associative memory network. An $SNR$ value larger than unity indicates that error-correction has occurred - however, the tasks that are performed by an image filter and that of an associative memory system are, at a conceptual level, quite different. While a filter is expected to remove noise from any signal, the associative memory system is designed to only filter the noise on prototype/probe vectors only (Müezzinoğlu et al., 2003).

The $SNR$ value measures the absolute differences in gray-level values by quantifying the degree to which error correction has been made. However, this measure may not be particularly good for measuring the difference in content, or structure of the retrieved image with the original stored image (where relative difference is more important).

Interpretation of the neural activities at steady-state is not possible unless some form of quantization is imposed on the final neural activities. The final image that is obtained from simulating the network dynamics can be interpreted by linearly scaling these neural activities according to the minimum and maximum gray level value ($I_{\min}$ and $I_{\max}$ respectively)

$$\bar{x} = I_{\min} + x \times \frac{I_{\max} - I_{\min}}{x_{\max} - x_{\min}} \tag{15.23}$$

This process if less of a biological mechanism, but is useful in simulations to better visualize the readout process of the final neural activities because of the large range of their values.

## 15.5 Simulation Results

The dynamics of an LT network is capable of amplifying arbitrarily small differences in the inputs – the explicit presence of the input ($h_i$) in the state update equation also introduces some interesting behavior into the dynamics of the network, and as (Wersing et al., 2001b) has illustrated, multistability allows a system to exhibit symmetry-breaking, where the dynamics allows for nonlinear amplification of small input differences. Moreover, assuming that the state ($x^*$) is a fixed-point in the network (or a memory state), ($\rho x^*$) is also a fixed-point (memory state) of the network, so that uniform scaling of active neurons is possible while leaving all inactive neurons unchanged. Care, however, has to be taken in ensuring that the input vector $h_i$ is approximately of the same order as $\sum_j^N w_{ij}x_j$ to prevent either term from dominating.

The update of neuron states can be achieved either synchronously (all neurons in parallel at once), or asynchronously (one neuron at a time in a serial manner and at any time step, each neuron has a uniform probability of being selected for updating). From the previous analysis for the dynamics, the energy function of an LT network assumes the form

$$E(x) = \frac{1}{2}x(\mathbf{I} - \mathbf{W})x^T - h^T x \qquad (15.24)$$

A system which is stable and convergent would tend to a certain energy limit after some transient time or a number of iterations. In the case of an unstable system, the energy would increase exponentially as the individual neural activities of neurons experience runaway excitation. An oscillatory system is characterized by fluctuations (which may be upper and lower bounded) without tending to a fixed energy value. The trend of the energy function is more important than the final value, unless the problem is that of optimization, such as quadratic optimization (Tan et al., 2004).

### 15.5.1 Small-Scale Example

In this example, a smaller network of $N=9$ neurons was used. Smaller networks are easier to analyze in terms of its eigen-structure. A 2-dimensional analog pattern was stored in the network, and subsequently recalled using a noisy input. The network parameters are: $\alpha = 4.5, \beta = 0.6, \omega = -0.2$ and $\max(w_{ii} + \sum_{j \neq i}^N w_{ij}^+) = 0.9365$. The $N$ eigenvalues of the system matrix $W$ are

$$-3.8805, -1.7750, -1.7750, -1.7750, -1.7750, -1.7750, -1.7750, -1.7750, 1.2638,$$

of which the largest eigenvalue is above unity. Since $\max(w_{ii} + \sum_{j \neq i}^N w_{ij}^+) < 1$ and $W$ is symmetric, according to Theorem 3, the network is asymptotically stable. The original pattern and the noisy input (probe vector) are respectively,

$$\begin{bmatrix} 7 & 7 & 3 \\ 1 & 5 & 9 \\ 0 & 4 & 2 \end{bmatrix} \text{ and } \begin{bmatrix} 6.9158 & 7.9138 & 3.8590 \\ 0.0871 & 5.2176 & 8.8599 \\ -0.8009 & 3.4935 & 1.0980 \end{bmatrix}.$$

After convergence, the final neural activities $x$ and the retrieved pattern vector (after linear scaling) are respectively,

$$\begin{bmatrix} 27.9295 & 27.9600 & 7.7582 \\ 0 & 18.5496 & 35.5272 \\ 0 & 13.7720 & 1.1145 \end{bmatrix} \text{ and } \begin{bmatrix} 8 & 8 & 2 \\ 1 & 5 & 9 \\ 0 & 4 & 1 \end{bmatrix}.$$

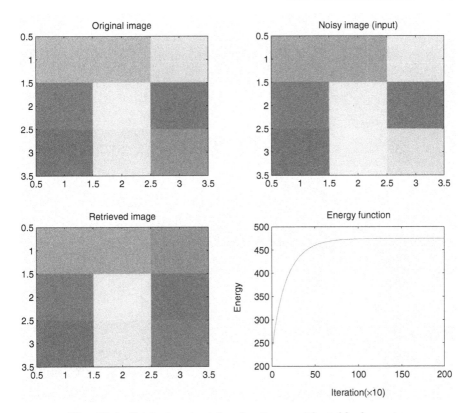

**Fig. 15.2.** Original and retrieved patterns with stable dynamics.

Fig. 15.2 shows the retrieval process and the convergence of energy function (for the sake of illustration, the energy function takes its reverse sign herefrom). Fig. 15.3 depicts the convergence of individual neuron activity. Note that if $\omega = -0.1$ was to be shifted to $\omega = 0.1$, the eigenvalues would all be shifted by $+0.2$, making the largest eigenvalue of $W^+$ to be larger than unity (also $\max(w_{ii} + \sum_{j \neq i}^{N} w_{ij}^+) = 1.2365 > 1$) and hence create an unstable system. On the contrary, if $\omega$ to be shifted to a more negative value, it would result in a faster convergence. *Self-coupling can therefore be used to tune the stability margins of the system.*

### 15.5.2 Single Stored Images

Four different gray scale images of resolution $32 \times 32$ with 256 gray-levels were stored in the network independently of each other (Figure 15.4).

The scaling parameter $\alpha$ proves to be an interesting parameter for consideration because its role is two-fold: if $\alpha > 1$, it controls the excitation; if $\alpha < 1$, it controls the amount of (divisive) inhibition in the network. Because

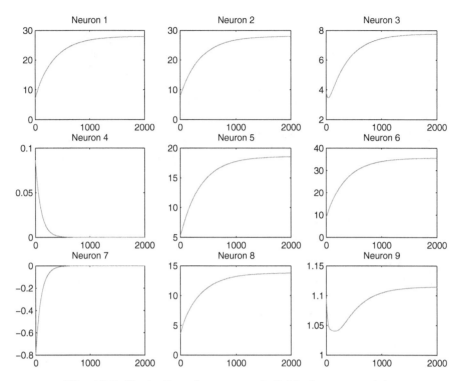

**Fig. 15.3.** Illustration of convergent individual neuron activity.

excitation is inversely related to divisive inhibition, increasing $\alpha$ is similar to increasing the instability (excitation) while decreasing the amount of (divisive) inhibition. The input (probe) vectors were noisy versions of the original gray scale images (Gaussian noise with mean 0 and variance of 10% or 50% salt-&-pepper noise). Because of the input dependency of the LT dynamics, certain types of noise will not result in reliable retrieval because neurons with zero inputs will remain inactive at steady-state. Unless otherwise specified, the network parameters used in the simulations are, $\nu = 1$, $\kappa = 0.35$; $m$ and $\sigma$ are pattern-dependent. As to the other parameters, $\alpha$ was varied such that the excitation measured by $w_{ii} + \sum_{j \neq i}^{N} w_{ij}^{+}, i = 1, ..., N$ was increased; $\beta$ and $\omega$ were kept at zero and the resulting $SNR$ value calculated.

Figures 15.5–15.6 illustrate the variation in the quality of recall measured by the $SNR$ value in the retrieval process of the four gray scale images as the excitation is increased (correspondingly, divisive inhibition is decreased), where the excitation strength, denoted by $\text{MaxW}^{+}$, is calculated as $\max_i \{w_{ii} + \sum_{j \neq i}^{N} w_{ij}^{+}\}$.

From the resulting plots, it is interesting to note the range of values of $\alpha$ that would result in the best recalled image, as measured by the $SNR$. These figures gives a general trend of how the $SNR$ measure varies: as $\alpha$ increases up

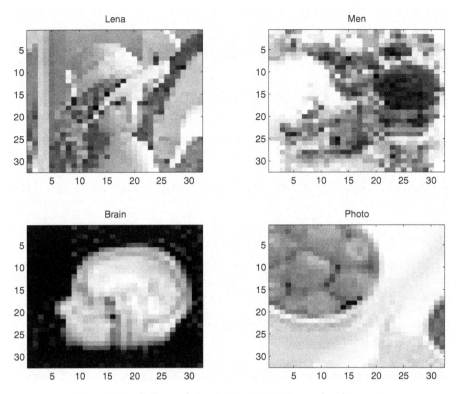

**Fig. 15.4.** Collage of the 4, 32 × 32, 256 gray-level images.

to a certain point (competition/instability likewise becomes greater), the $SNR$ will correspondingly increase. After this peak, greater competition will result in overall instability in the network; consequently, divergence of individual neural activity will occur resulting in a marked decrease in the $SNR$. The optimal $\alpha$ value is image-dependent, closely related to the overall composition and contrast of the image. More precise tuning of the network parameters ($\alpha$, $\beta$, and $\omega$) will result in better quality memory retrieval as shown in Figures 15.7–15.9.

### 15.5.3 Multiple Stored Images

Now the four images (Fig. 15.4) were stored together in the same network (not independently, as in the previous approach). Expectedly, the quality of recall decreased, yet the retrieved image was still quite apparent (see Fig. 15.10). 50% Salt-&-Pepper noise was used (Gaussian noise was also used with similar results). The network parameters were set as $\alpha = 0.24, \beta = 0.0045, \omega = 0.6$, with $SNR = 1.8689$.

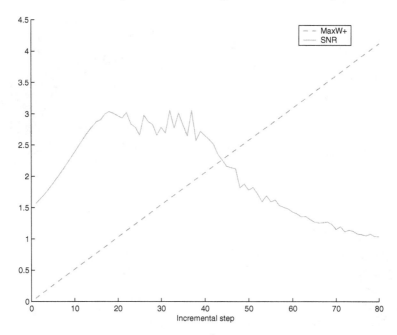

**Fig. 15.5.** Lena: $SNR$ and MaxW$^+$ with $\alpha$ in increment of 0.0025.

## 15.6 Discussion

### 15.6.1 Performance Metrics

Typical performance metrics of associate memory includes (i) the ability of the network to recall *noisy* or *partial* memory vectors presented at the inputs. This ability is in turn, dependent on the sparsity of the stored memory vectors (Hopfield, 1982). This requirement of sparsity places restrictive constraints on the types of images that are stored in an associative memory network, which is even more severe when dealing with gray scale images because large regions of gray scale images have positive gray-levels. Likewise, the *information capacity per neuron* for an LT neuron is difficult to quantify because of its unbounded, continuous characteristic (unlike discrete-valued binary neurons).

### 15.6.2 Competition and Stability

Competition is a corollary of instability, which promotes separation of neural activities in a recurrent neural network. It can be observed from Figs. 15.5–15.6 that the $SNR$ increases as $w_{ii} + \sum_{j \neq i}^{N} w_{ij}^+$ (regarded as a measure of competition/instability) increases, until it results in divergent neural activities and hence a decrease in the $SNR$. The use of a saturating neuron in competitive dynamics would result in activities that are close to the saturation points of the activation function, the LT neuron, in a stable LT network,

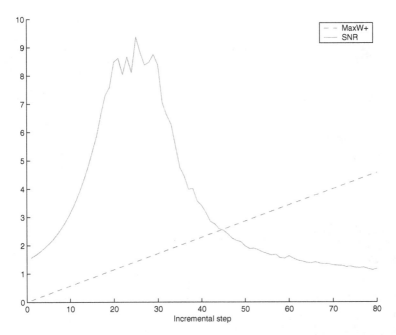

**Fig. 15.6.** Brain: $SNR$ (solid line) and MaxW$^+$ with $\alpha$ in increment of 0.005.

however, allows its neural activities to run to arbitrarily high value, until convergence.

Multistability is a manifestation of instability in an LT network that arises from the existence of unstable differential-mode eigenvalues (Hahnloser et al., 2000). The LT dynamics results in nonlinear switching between active and inactive partitions of neurons until neurons corresponding to unstable eigenvalues are eventually inactivated. Convergence then occurs after the nonlinear switching selects a permitted set of neurons and the linear dynamics of the LT network then takes over to provide for the analog responses of these active sets of neurons.

### 15.6.3 Sparsity and Nonlinear Dynamics

The computational abilities of any neural network is a result of the degree of nonlinearity present in the network. Because the nonlinear dynamics of the LT network is a consequence of the switching between active and inactive partitions of neurons (inactive neurons do not contribute to the strength of feedback in the network). This form of *dynamic feedback* depends on the ratio of inactive neurons to active neurons, which is a measure of the *sparsity* of the memory vectors that are stored. Recall that an LT neuron behaves linearly above threshold, hence if the pattern vector that is stored in the associative memory is dense (less sparse), the network can be expected to behave more

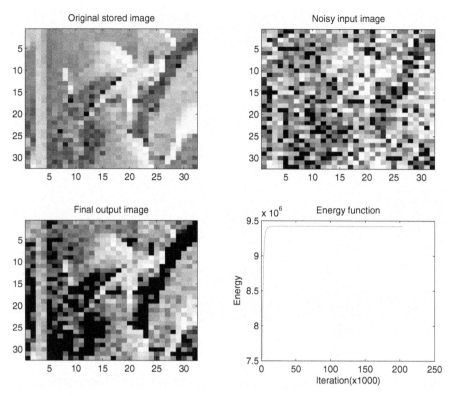

**Fig. 15.7.** Lena: $\alpha = 0.32, \beta = 0.0045, \omega = -0.6, SNR = 5.6306$; zero mean Gaussian noise with 10% variance.

like a linear system, and in a way, lose its computational abilities derived from its nonlinear switching activities.

A pioneer work about optimal sparsity for random patterns was made (Xie et al., 2002). In determining the maximum storage capacity of an associative memory network, it is necessary to know the optimal sparsity of the memory vectors to be stored. The reason for this is mostly related to information theory; the network has a upper limit on the amount of information that it will be able to store. If each pattern was to contain more information (low sparsity), the number of patterns that can be stored and retrieved reliably decreases. In effect, a network would be able to store more sparse patterns than dense patterns. However, the difficulty in analyzing the storage capacity using conventional methods is the continuous and unbounded nature of an LT neuron. It might be interesting to derive the optimal sparsity of pattern vectors for an associative memory network of LT neurons, much like the classical Hopfield network, where it was calculated that $p = 0.138N$ (Hertz et al., 1991). Evidently, such a task is quite challenging because of the continuous and unbounded nature of LT neurons.

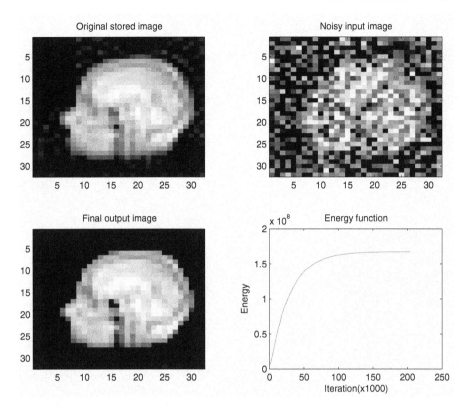

**Fig. 15.8.** Brain: $\alpha = 0.43, \beta = 0.0045, \omega = -0.6, SNR = 113.8802$; zero mean Gaussian noise with 10% variance.

## 15.7 Conclusion

This chapter has extensively analyzed the LT dynamics, presenting new and milder conditions for its boundedness and stability, which extended the existing results in the literature such that symmetric and nonnegative conditions for a deterministic matrix are not required to assert the bounded dynamics. Based on a fully-connected single-layered recurrent network of linear threshold neurons, this chapter also proposed a design method for analog associative memory. In the design framework, with the established analytical results and the matrix theory that underly LT networks, fine-tuning of network stability margins was realized in a simple and explicit way. The simulation results illustrated that the LT network can be successful to retrieve both single stored and multiple stored real-valued images.

Measures to quantify the performance of associative memory were also suggested, and its relationship with network competition and instability was also discussed. Because the computational abilities of the LT network is a consequence of its switching between active and inactive partitions, dense patterns

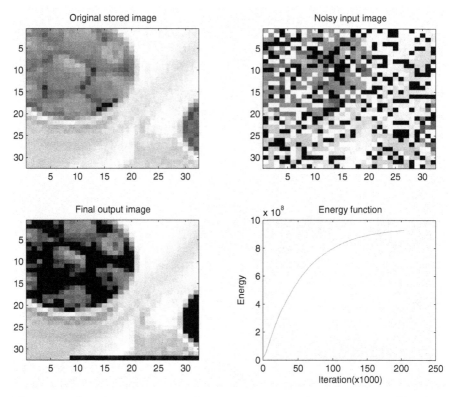

**Fig. 15.9.** Strawberry: $\alpha = 0.24, \beta = 0.0045, \omega = 0.6, SNR = 1.8689$; 50% Salt-&-Pepper noise.

result in low storage capacity. Although the performance of the associative memory capacity was largely limited to a few stored patterns of high density (low sparsity), LT networks have shown to exhibit interesting dynamics that can be further analyzed and addressed.

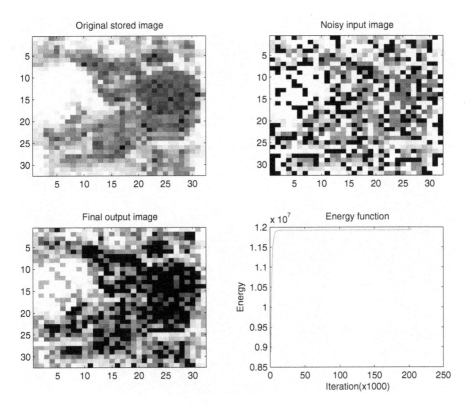

**Fig. 15.10.** Men: $\alpha = 0.24, \beta = 0.0045, \omega = 0.6, SNR = 1.8689$; 50% Salt-&-Pepper noise.

# 16

# Output Convergence Analysis for Delayed RNN with Time Varying Inputs

[1]This chapter studies the output convergence of a class of recurrent neural networks with time varying inputs. The model of the studied neural networks has a different dynamic structure from that of the well known Hopfield model, as it does not contain linear terms. Since different structures of differential equations usually result in quite different dynamic behaviors, the convergence of this model is quite different from that of Hopfield model. This class of neural networks has been found many successful applications in solving many optimization problems. Some sufficient conditions to guarantee output convergence of the networks are derived in this chapter.

## 16.1 Introduction

The model of neural networks we will study in this chapter is described by the following nonlinear differential equations

$$\frac{dx_i(t)}{dt} = \sum_{j=1}^{n}[a_{ij}g_j(x_j(t)) + b_{ij}g_j(x_j(t-\tau_{ij}(t)))] + I_i(t), (i=1,\cdots,n)$$

(16.1)

for $t \geq 0$, where each $x_i(t)$ denotes the state of neuron $i$, $a_{ij}, b_{ij}(i,j=1,\cdots,n)$ are connection weights, they are constants, the delays $\tau_{ij}(t) \geq 0 (i,j = 1,\cdots,n)$ are nonnegative continuous functions with $0 \leq \tau_{ij}(t) \leq \tau$ for $t \geq 0$, where $\tau \geq 0$ is a constant, the inputs $I_i(t)(i=1,\cdots,n)$ are some continuous functions defined on $[0,+\infty)$, $g_i(i=1,\cdots,n)$ are networks output functions. We call $x(t) = (x_1(t),\cdots,x_n(t))^T$ the state of the network at time $t$, and call $G(x(t)) = (g_1(x_1(t)),\cdots,g_n(x_n(t)))^T$ the output of the network at time $t$. Clearly, model (17.1) is a recurrent neural network (RNN).

---

[1] Reuse of the materials of "Output convergence analysis for a class of delayed recurrent neural networks with time varying inputs", 36(1), 2006, 87–95, IEEE Trans. on Systems, Man and Cybernetics-B, with permission from IEEE.

This model of neural networks is much more general than the networks proposed in (Wang, 1996; Wang, 1993; Wang, 1995; Wang, 1992; Wang, 1997a; Liu et al., 2004).This particular model has some important applications. In fact, some special cases of model (17.1) have been successfully used in (Wang, 1996; Wang, 1993; Wang, 1995; Wang, 1992; Wang, 1997a) to solve various optimization problems. It has been demonstrated that this class of neural networks could be easily implemented by electronic circuits (Wang, 1996; Wang, 1993; Wang, 1995; Wang, 1992; Wang, 1997a). This model has some interesting features. It is essentially different from the well known model of Hopfield neural networks (Hopfield, 1984) or Hopfield neural networks with delays (Yi et al., 2004). In fact, since (17.1) contains time varying inputs, it is a nonautonomous differential equation. More importantly, it does not contain linear terms as found in Hopfield networks. Hence, the dynamic structure of this class of neural networks is different from that of Hopfield networks. Since different structure of differential equations usually results in different dynamic behavior, the convergence of (17.1) would be quite different from that of Hopfield networks.

It is well known that if the connection matrix of a Hopfield network is symmetric, then each trajectory will converge to an equilibrium point (Hopfield, 1984). However, the network (17.1) does not possess this property. Let us consider a simple one dimensional RNN with time varying input and without delay

$$\frac{dx(t)}{dt} = -g[x(t)] - e^{-t}$$

for $t \geq 0$, where

$$g(x) = \frac{1}{1+e^{-x}}.$$

Clearly, the connection matrix is symmetric since the network is one dimensional. However, we have

$$\frac{d\left(e^{-x(t)} - x(t)\right)}{dt} = -\left(1 + e^{-x(t)}\right)\frac{dx(t)}{dt}$$
$$\geq 1$$

for all $t \geq 0$. Then, $e^{-x(t)} - x(t) \to +\infty$ as $t \to +\infty$, that is $x(t) \to -\infty$ as $t \to +\infty$. This shows that any state trajectory of this network will diverge. A natural problem is to consider is whether this class of networks (17.1) will converge?

It is well known that convergence analysis is one of the most important issues in studying recurrent neural networks. Understanding the convergence properties of recurrent neural networks is an initial and essential step towards their successful applications. There are two kinds of convergence for a given network: state convergence and output convergence, both of which are important and have practical applications. In some applications, such as content addressable memory, networks are required to have state convergence property, while in other applications such as optimization problems, output convergent is sufficient (Liu et al., 2004). State convergence analysis and output convergence analysis study the state behaviors and output behaviors of recurrent neural networks when time approaches infinity, respectively. Usually, state convergence implies output convergence, however, output convergence does not imply state convergence in general. Thus, conditions for guaranteeing state and output convergence would be different. Convergence of recurrent neural networks has been studied extensively in recent years; for examples of state convergence, see (Cao et al., 2003; J. Cao, 2003; Forti, 1994; Forti and Tesi, 1995; Liang and Si, 2001; Yi et al., 1999), while for examples of output convergence, see (Yi et al., 2001; Liu et al., 2004; Li et al., 2004).

The network model of (17.1) contains some time delays. Time delays are important parameters of recurrent neural networks. In fact, delays exist both in biological neural networks and artificial neural networks. For biological neural networks, delays are natural and essential. For artificial networks, in hardware implementation of recurrent neural networks, time delays occur due to the finite switching speed of the amplifiers. Some recurrent neural networks with delays have direct practical applications, for example, in processing of moving images, image compression (Venetianer and Roska, 1998), etc. On the other hand, delays may affect the dynamics of recurrent neural networks (Baldi and Atiya, 1994; Marcus and Westervelt, 1989; Civalleri et al., 1993). A stable network may become unstable by adding some delays. Thus, analysis of recurrent neural networks with delays is important both in theory and applications. Some reports on convergence of neural networks with delays can be found in (Cao et al., 2003; J. Cao, 2003; Gopalsamy and He, 1994; Li et al., 2004; Lu, 2001; Ye et al., 1994; Pakdaman et al., 1998; Driessche and Zou, 1998; Chen and Aihara, 1995; Chen and Amari, 2001; Liao and Wang, 2003; Hu and Wang, 2002; Yi et al., 2001; Yi and Tan, 2002).

This chapter analyzes the output convergence of the networks of (17.1). We will develop some sufficient conditions to guarantee the global output convergence of (17.1). The rest of this chapter is organized as follows. In Section 2, we will give some preliminaries. Convergence results and proofs will be given in Section 3. Examples and simulations will be given in Section 4. Finally, conclusions are presented in Section 5.

## 16.2 Preliminaries

We will use $D^+$ in this paper to denote the *upper right Dini derivative*. For any continuous function $f : R \to R$, the *upper right Dini derivative* of $f(t)$ is defined by

$$D^+ f(t) = \lim_{h \to 0^+} sup \frac{f(t+h) - f(t)}{h}.$$

Obviously, if $f(t)$ is locally Lipschitz, then $|D^+ f(t)| < +\infty$.

We assume that each output function $g_i$ is a continuous function which satisfies

$$0 \leq D^+ g_i(s) \leq \omega, (i = 1, \cdots, n) \quad (16.2)$$

for all $s \in R$, where $\omega > 0$ is a constant. Clearly, the requirement in (17.2) is equivalent to

$$0 \leq \frac{g_i(\xi) - g_i(\eta)}{\xi - \eta} \leq \omega \quad (16.3)$$

for all $\xi, \eta \in R$ and $\xi \neq \eta$. It should be noted here that the output functions under the above assumption may be not bounded. There are many frequently used output functions that satisfy the above assumption, say, $1/(1 + e^{-\theta})$, $(2/\pi)arctan(\theta)$, $max(0, \theta)$ and $(|\theta + 1| - |\theta - 1|)/2$, where $\theta \in R$.

Let $C = C([-\tau, 0], R^n)$ be the Banach space of continuous functions mapping the interval $[-\tau, 0]$ into $R^n$ with the topology of uniform convergence. For any $\phi = (\phi_1, \cdots, \phi_n)^T \in C = C([-\tau, 0], R^n)$, the initial condition in (17.1) is assumed to be

$$x_i(t) = \phi_i(t), -\tau \leq t \leq 0, (i = 1, \cdots, n).$$

Given any $\phi \in C$, under the assumptions in (17.2), there exists a unique trajectory of (17.1) starting from $\phi$ (Hale, 1977). Denote by $x(t, \phi)$ the trajectory of (17.1) starting from $\phi$.

Denote $I(t) = (I_1(t), \cdots, I_n(t))^T, G(x(t)) = (g_1(x_1(t)), \cdots, g_n(x_n(t)))^T$. We assume throughout this paper that

$$\lim_{t \to +\infty} I(t) = I, \quad (16.4)$$

where $I \in R^n$ is a constant vector, and the set

$$\Omega = \left\{ x^* \in R^n \big| (A+B)G(x^*) + I = 0 \right\}$$

is nonempty, i.e., $\Omega \neq \emptyset$. These assumptions are quite reasonable. The set $\Omega \neq \emptyset$ is crucial, in fact, if $\Omega$ is empty, then we will not know where a trajectory converges to. And, the trajectories may diverge. The example discussed in the last section is just this case.

Denote that

$$\begin{cases} \underline{g}_i = \inf_{-\infty < s < +\infty} (g_i(s)) \\ \overline{g}_i = \sup_{-\infty < s < +\infty} (g_i(s)), (i = 1, \cdots, n) \end{cases}$$

for all $s \in R$, where $\underline{g}_i$ and $\overline{g}_i$ may take the values $-\infty$ and $+\infty$, respectively. The following *Lemma* gives a method to check the set $\Omega$ to be nonempty.

**Lemma 16.1.** *Suppose* $(A+B)^{-1} = (c_{ij})_{n \times n}$ *exist, then* $\Omega \neq \emptyset$ *if and only if*

$$-\sum_{j=1}^{n} c_{ij} I_j \in \left( \underline{g}_i, \overline{g}_i \right), (i = 1, \cdots, n).$$

*Proof.* For each $i(1 \leq i \leq n)$, since

$$\underline{g}_i < -\sum_{j=1}^n c_{ij} I_j < \overline{g}_i$$

then by the continuity of $g_i$, there must exist a constant $x_i^*$ such that

$$g_i(x_i^*) = -\sum_{j=1}^n c_{ij} I_j.$$

And so $G(x^*) = -(A+B)^{-1} I$, i.e., $(A+B)G(x^*) + I = 0$. This completes the proof.

**Definition 16.2.** *The network (17.1) is said to be globally output convergent (GOC), if for any $\phi \in C$, there exists a $x^* \in \Omega$ such that*

$$\lim_{t \to +\infty} G(x(t,\phi)) = G(x^*).$$

We will use the following lemma which was verified in (Yi et al., 1999). It provides an interesting property of the output functions.

**Lemma 16.3.** *It holds that*

$$\int_\eta^\xi [g_i(s) - g_i(\eta)]\, ds \geq \frac{1}{2\omega}[g_i(\xi) - g_i(\eta)]^2$$

*for all $\xi, \eta \in R$ and $(i = 1, \cdots, n)$.*

*Proof.* Given any $\eta \in R$, define the continuous functions

$$E_i(\xi) = \int_\eta^\xi [g_i(s) - g_i(\eta)]\, ds - \frac{1}{2\omega}[g_i(\xi) - g_i(\eta)]^2$$

for all $\xi \in R$ and $(i = 1, \cdots, n)$. It follows that

$$D^+ E_i(\xi) = [g_i(\xi) - g_i(\eta)] \left[1 - \frac{D^+ g_i(\xi)}{\omega}\right].$$

From (17.2), $0 \leq D^+ g_i(s) \leq \omega$ for all $s \in R$, then

$$D^+ E_i(\xi) \begin{cases} \geq 0, \xi > \eta \\ \leq 0, \xi < \eta \\ = 0, \xi = \eta. \end{cases}$$

This shows that $\xi = \eta$ is a global minimum point of the function $E_i(\xi)$, i.e., $E_i(\xi) \geq E_i(\eta) = 0$ for all $\xi \in R$. Clearly,

$$\int_\eta^\xi [g_i(s) - g_i(\eta)]\, ds \geq \frac{1}{2\omega}[g_i(\xi) - g_i(\eta)]^2$$

for all $\xi, \eta \in R$ and $(i = 1, \cdots, n)$. The proof is completed.

A matrix $M = (m_{ij})_{n \times n}$ is called an *M-matrix* if $m_{ii} \geq 0 (i = 1, \cdots, n)$, $m_{ij} \leq 0 (i \neq j; i, j = 1, \cdots, n)$, and the real part of each eigenvalue of $M$ is nonnegative. $M$ is called a nonsingular *M-matrix* if $M$ is both an *M-matrix* and nonsingular. In (Berman and Plemmons, 1994), it gives that a matrix $M$ is a nonsingular $M - matrix$ if and only if there exist constants $\alpha_i > 0 (i = 1, \cdots, n)$ such that

$$\alpha_i m_{ii} + \sum_{j=1, j \neq i}^{n} \alpha_j m_{ij} > 0, (i = 1, \cdots, n).$$

or

$$\alpha_j m_{jj} + \sum_{i=1, i \neq j}^{n} \alpha_i m_{ij} > 0, (j = 1, \cdots, n).$$

## 16.3 Convergence Analysis

In this section, we will give some conditions to guarantee the GOC of the network (17.1). Two theorems will be proved for output convergence. We are interested to derive checkable and simple conditions of convergence, such conditions could be easy for designing networks in practical applications.

**Theorem 16.4.** *Define a matrix* $M = (m_{ij})_{n \times n}$ *by*

$$m_{ij} = \begin{cases} -a_{ii} - |b_{ii}|, & \text{if } i = j \\ -|a_{ij}| - |b_{ij}|, & \text{if } i \neq j. \end{cases}$$

*If $M$ is a nonsingular M-matrix, then the network (17.1) is GOC.*

*Proof.* Since $\Omega \neq \emptyset$, then a constant vector $x^* = (x_1^*, \cdots, x_n^*)^T \in R^n$ exists, such that

$$\sum_{j=1}^{n} (a_{ij} + b_{ij}) g_j(x_j^*) + I_i = 0, (i = 1, \cdots, n).$$

For any $\phi \in \mathcal{C}$, let $x_i(t) (i = 1, \cdots, n)$ be the trajectory of (17.1) starting from $\phi$. Then it follows from (17.1) that

$$\frac{d[x_i(t) - x_i^*]}{dt} = \sum_{j=1}^{n} a_{ij} \left[ g_j(x_j(t)) - g_j(x_j^*) \right]$$
$$+ \sum_{j=1}^{n} b_{ij} \left[ g_j(x_j(t - \tau_{ij}(t))) - g_j(x_j^*) \right]$$
$$+ I_i(t) - I_i \qquad (16.5)$$

for $t \geq 0$ and $(i = 1, \cdots, n)$.

Using the fact that $M$ is a nonsingular matrix, there exist constants $\alpha_i > 0 (i = 1, \cdots, n)$ such that

$$a_{ii} + |b_{ii}| + \frac{1}{\alpha_i} \sum_{j \neq i, j=1}^{n} \alpha_j (|a_{ij}| + |b_{ij}|) < 0, (i = 1, \cdots, n). \tag{16.6}$$

Define

$$\begin{cases} z_i(t) = \dfrac{1}{\alpha_i} [x_i(t) - x_i^*] \\ h_i(t) = \dfrac{1}{\alpha_i} [g_i(x_i(t)) - g_i(x_i^*)] \end{cases}$$

for $t \geq 0$ and $(i = 1, \cdots, n)$. From (17.5) we have

$$D^+ |z_i(t)| \leq a_{ii} |h_i(t)| + \frac{1}{\alpha_i} \sum_{j=1, j\neq i}^{n} \alpha_j |a_{ij}| |h_j(t)|$$

$$+ \frac{1}{\alpha_i} \sum_{j=1}^{n} \alpha_j |b_{ij}| |g_j(x_j(t - \tau_{ij}(t))) - g_j(x_j^*)|$$

$$+ \frac{1}{\alpha_i} |I_i(t) - I_i|$$

$$\leq a_{ii} |h_i(t)| + \frac{1}{\alpha_i} |I_i(t) - I_i|$$

$$+ \frac{1}{\alpha_i} \sum_{j=1}^{n} \alpha_j \left[ |a_{ij}|(1 - \delta_{ij}) + |b_{ij}| \right] \cdot \sup_{t-\tau \leq \theta \leq t} |h_j(\theta)| \tag{16.7}$$

and

$$D^+ |h_i(t)| \leq D^+ g_i(x_i(t)) \left[ a_{ii} |h_i(t)| + \frac{1}{\alpha_i} |I_i(t) - I_i| \right.$$

$$\left. + \frac{1}{\alpha_i} \sum_{j=1}^{n} \alpha_j \left( |a_{ij}|(1 - \delta_{ij}) + |b_{ij}| \right) \cdot \sup_{t-\tau \leq \theta \leq t} |h_j(\theta)| \right] \tag{16.8}$$

for all $t \geq 0$ and $(i = 1, \cdots, n)$, where

$$\delta_{ij} = \begin{cases} 1, & \text{if } i = j \\ 0, & \text{if } i \neq j \end{cases}$$

By (17.4), $I_i(t) \to I_i (i = 1, \cdots, n)$ as $t \to +\infty$, then a constant $m > 0$ exists, such that $|I_i(t) - I_i|/\alpha_i \leq m (i = 1, \cdots, n)$ for all $t \geq 0$.

Denote

$$\eta_i = - \left[ a_{ii} + |b_{ii}| + \frac{1}{\alpha_i} \sum_{j=1, j\neq i}^{n} \alpha_j \left( |a_{ij}| + |b_{ij}| \right) \right], (i = 1, \cdots, n)$$

then, by (17.6), $\eta_i > 0 (i = 1, \cdots, n)$.

Given any constant $H > 0$ such that

$$H > \frac{m}{\min_{1 \leq i \leq n}(\eta_i)},$$

we will prove that $\sup_{-\tau \leq \theta \leq 0} |h_i(\theta)| \leq H (i = 1, \cdots, n)$ imply $|h_i(t)| \leq H (i = 1, \cdots, n)$ for all $t \geq 0$. If not, there must exist a $t_1 > 0$ and some $i$ such that

$$|h_i(t_1)| > H, D^+|h_i(t_1)| > 0, \text{ and } |h_j(t)| \leq |h_i(t_1)|, -\tau \leq t \leq t_1, (j = 1, \cdots, n).$$

However, from (17.8) we have

$$D^+|h_i(t_1)| \leq D^+ g_i(x_i(t_1)) \cdot (-\eta_i|h_i(t_1)| + m)$$
$$\leq D^+ g_i(x_i(t_1)) \cdot (-\eta_i H + m)$$
$$\leq 0.$$

This is a contradiction and it shows that $\sup_{-\tau \leq \theta \leq 0} |h_i(\theta)| \leq H (i = 1, \cdots, n)$ imply $|h_i(t)| \leq H (i = 1, \cdots, n)$ for all $t \geq 0$. Thus, $|h_i(t)| (i = 1, \cdots, n)$ are bounded for all $t \geq 0$.

Suppose that

$$\lim_{t \to +\infty} \sup |h_i(t)| = \sigma_i, (i = 1, \cdots, n).$$

Clearly, $0 \leq \sigma_i \leq H < +\infty, (i = 1, \cdots, n)$. Without loss of generality, assume that $\sigma_1 = \max_{1 \leq i \leq n} \{\sigma_i\}$.

Next, we will prove $\sigma_1 = 0$. If this is not true, i.e., $\sigma_1 > 0$, we will prove that it leads to a contradiction.

By $\lim_{t \to +\infty} I_i(t) = I_i (i = 1, \cdots, n)$, there must exist a $t_2 \geq 0$ such that

$$\frac{|I_i(t) - I_i|}{\alpha_i \eta_i} \leq \frac{\sigma_1}{4}, (i = 1, \cdots, n)$$

for all $t \geq t_2$. Choosing a constant $\epsilon > 0$ such that

$$0 < \epsilon \leq \frac{\alpha_i \eta_i \sigma_1}{4 \sum_{j=1}^n \alpha_j (|a_{ij}| + |b_{ij}|)}, (i = 1, \cdots, n).$$

By $\lim_{t \to +\infty} \sup |h_i(t)| = \sigma_i, (i = 1, \cdots, n)$, there exists a $t_3 \geq t_2$, such that $t \geq t_3$ implies

$$\sup_{t - \tau \leq \theta \leq t} |h_i(\theta)| \leq \sigma_i + \epsilon \leq \sigma_1 + \epsilon, 1 \leq i \leq n.$$

To complete the proof of $\sigma_1 = 0$, let us first prove that there exists a $t_4 \geq t_3$ such that

$$D^+|h_1(t)| \leq 0 \tag{16.9}$$

for all $t \geq t_4$. Otherwise, there must exist a $t_5 > t_3$ such that

$$D^+|h_1(t_5)| > 0 \quad \text{and} \quad |h_1(t_5)| \geq \sigma_1 - \epsilon.$$

However, on the other hand, from (17.8) we have

$$D^+|h_1(t_5)| \leq D^+g_1(x_1(t_5)) \cdot \left[ a_{11}(\sigma_1 - \epsilon) \right.$$
$$\left. + \frac{1}{\alpha_1}\sum_{j=1}^{n}\alpha_j\left(|a_{1j}|(1-\delta_{1j})+|b_{1j}|\right)(\sigma_1+\epsilon) + \frac{\eta_1\sigma_1}{4} \right]$$
$$= D^+g_1(x_1(t_5)) \cdot \left[ -\eta_1\sigma_1 \right.$$
$$\left. + \left(\frac{1}{\alpha_1}\sum_{j=1}^{n}\alpha_j(|a_{1j}|+|b_{1j}|)\right)\epsilon + \frac{\eta_1\sigma_1}{4} \right]$$
$$\leq -\frac{\eta_1\sigma_1}{2} \cdot D^+g_1(x_1(t_5))$$
$$\leq 0$$

which is a contradiction. This proves that (17.9) is true.

Next, we continue to prove $\sigma_1 = 0$. By (17.9), $|h_1(t)|$ must be monotonically decreasing. Thus, the limit of $|h_1(t)|$ exists, i.e.,

$$\lim_{t\to+\infty}|h_1(t)| = \lim_{t\to+\infty}\sup|h_1(t)| = \sigma_1.$$

Then, there must exist a $t_6 \geq t_4$ such that

$$\begin{cases} |h_1(t)| > \sigma_1 - \epsilon \\ \sup_{t-\tau\leq\theta\leq t}|h_i(\theta)| < \sigma_1 + \epsilon, (i=1,\cdots,n) \end{cases}$$

for all $t \geq t_6$.

From (17.7), it follows that

$$D^+|z_1(t)| \leq a_{11}(\sigma_1 - \epsilon) + \frac{1}{\alpha_1}\sum_{j=1}^{n}\alpha_j\left(|a_{1j}|(1-\delta_{1j})+|b_{1j}|\right)(\sigma_1+\epsilon)$$
$$+\frac{1}{\alpha_1}|I_1(t) - I_1|$$
$$\leq -\eta_1\sigma_1 + \left[\frac{1}{\alpha_1}\sum_{j=1}^{n}\alpha_j\left(|a_{1j}|+|b_{1j}|\right)\right]\epsilon + \frac{\eta_1\sigma_1}{4}$$
$$\leq -\frac{\eta_1\sigma_1}{2}$$

for all $t \geq t_6$, and so

$$|z_1(t)| \leq |z_1(t_6)| - \frac{\eta_1\sigma_1}{2}(t - t_6)$$
$$\to -\infty$$

as $t \to +\infty$, which is a contradiction to $|z_1(t)| \geq 0$. This contradiction implies that $\sigma_1 = 0$.

Since $\sigma_1 = \max_{1 \leq i \leq n}\{\sigma_i\}$, then $\sigma_i = 0 (i = 1, \cdots, n)$. It follows that
$$\lim_{t \to +\infty} |h_i(t)| = \lim_{t \to +\infty} \sup |h_i(t)| = 0, (i = 1, \cdots, n).$$

That is
$$\lim_{t \to +\infty} g_i(x_i(t)) = g_i(x_i^*), (i = 1, \cdots, n).$$

This shows the network (17.1) is GOC and the proof is completed.

Conditions given in the above Theorem 16.4 are quite simple and checkable. They are represented by simple algebra relations among the coefficients of the network model (17.1). Moreover, the conditions for GOC do not depend on the delays. Thus, the delays could be uncertain, and this gives robust for designing networks. If we let $\alpha_i = 1 (i = 1, \cdots, n)$ in Theorem 16.4, then it gives an especially simple results.

**Corollary 16.5.** *If*
$$a_{ii} + |b_{ii}| + \sum_{j \neq i, j=1}^{n} (|a_{ij}| + |b_{ij}|) < 0, (i = 1, \cdots, n),$$
*then the network (17.1) is GOC.*

From the conditions given in Theorem 16.4 or Corollary 16.5, it can be found that the negativity of the coefficients $a_{ii}(i = 1, \cdots, n)$ play a crucial role for holding the given conditions. Requiring $a_{ii}(i = 1, \cdots, n)$ to be negative seems quite reasonable. For example, let us consider a very simple one dimension network $\dot{x} = a \cdot g(x)$, where $g(x) = \max(0, x)$. It is easy to see that if $a > 0$, then the network output $g(x)$ will diverge.

Next, we give another theorem for GOC.

**Theorem 16.6.** *Suppose that*
$$\int_0^{+\infty} [I_i(t) - I_i]^2 dt < +\infty, (i = 1, \cdots, n),$$
*and*
$$\dot{\tau}_{ij}(t) \leq \gamma_{ij} < 1, t \geq 0, (i, j = 1, \cdots, n),$$
*where $\gamma_{ij}(i, j = 1, \cdots, n)$ are constants. Denote*

$$\Pi = \begin{pmatrix} \sum_{j=1}^{n}\left(|b_{1j}| + \frac{|b_{j1}|}{1-\gamma_{j1}}\right) & 0 & \cdots & 0 \\ 0 & \sum_{j=1}^{n}\left(|b_{2j}| + \frac{|b_{j2}|}{1-\gamma_{j2}}\right) & \cdots & 0 \\ \vdots & \vdots & \ddots & \vdots \\ 0 & 0 & \cdots & \sum_{j=1}^{n}\left(|b_{nj}| + \frac{|b_{jn}|}{1-\gamma_{jn}}\right) \end{pmatrix}.$$

If
$$A + A^T + \Pi$$
is a negative definite matrix, then the network (17.1) is GOC.

Proof. By $\Omega \neq \emptyset$, a constant vector $x^* \in \Omega$ exists such that

$$\sum_{j=1}^{n} (a_{ij} + b_{ij}) g_j(x_j^*) + I_i = 0, (i = 1, \cdots, n)$$

Rewrite (17.1) as

$$\begin{aligned}\dot{x}_i(t) &= \sum_{j=1}^{n} a_{ij} \left[g_j(x_j(t)) - g_j(x_j^*)\right] \\ &+ \sum_{j=1}^{n} b_{ij} \left[g_j(x_j(t - \tau_{ij}(t))) - g_j(x_j^*)\right] \\ &+ I_i(t) - I_i \end{aligned} \quad (16.10)$$

for $t \geq 0$.

Define a differentiable function

$$U(t) = \sum_{i=1}^{n} \int_{x_i^*}^{x_i(t)} [g_i(s) - g_i(x_i^*)] \, ds \quad (16.11)$$

for $t \geq 0$. Clearly, $U(t) \geq 0$ for all $t \geq 0$ by Lemma 16.3.

Let $-\lambda$ be the largest eigenvalue of the matrix $(A + A^T + \Pi)/2$. Clearly, $\lambda > 0$. Choosing a constant $\beta > 0$ such that $\beta < \lambda$. It follows from (17.11) and (17.10) that

$$\dot{U}(t) = \sum_{i=1}^{n} [g_i(x_i(t)) - g_i(x_i^*)] \dot{x}_i(t)$$

$$= \sum_{i=1}^{n} \sum_{j=1}^{n} a_{ij} [g_i(x_i(t)) - g_i(x_i^*)] \cdot [g_j(x_j(t)) - g_j(x_j^*)]$$

$$+ \sum_{i=1}^{n} \sum_{j=1}^{n} b_{ij} [g_i(x_i(t)) - g_i(x_i^*)] \cdot [g_j(x_j(t - \tau_{ij}(t))) - g_j(x_j^*)]$$

$$+ \sum_{i=1}^{n} [I_i(t) - I_i^*] \cdot [g_i(x_i(t)) - g_i(x_i^*)]$$

$$\leq \sum_{i=1}^{n} \sum_{j=1}^{n} a_{ij} [g_i(x_i(t)) - g_i(x_i^*)] \cdot [g_j(x_j(t)) - g_j(x_j^*)]$$

$$+ \frac{1}{2} \sum_{i=1}^{n} \sum_{j=1}^{n} |b_{ij}| [g_i(x_i(t)) - g_i(x_i^*)]^2$$

$$+ \frac{1}{2} \sum_{i=1}^{n} \sum_{j=1}^{n} |b_{ij}| [g_j(x_j(t - \tau_{ij}(t))) - g_j(x_j^*)]^2$$

$$+ \beta \sum_{i=1}^{n} [g_i(x_i(t)) - g_i(x_i^*)]^2 + \frac{1}{4\beta} \sum_{i=1}^{n} [I_i(t) - I_i]^2$$

$$= \sum_{i=1}^{n} \sum_{j=1}^{n} a_{ij} [g_i(x_i(t)) - g_i(x_i^*)] \cdot [g_j(x_j(t)) - g_j(x_j^*)]$$

$$+ \sum_{i=1}^{n} \left( \frac{1}{2} \sum_{j=1}^{n} |b_{ij}| + \beta \right) [g_i(x_i(t)) - g_i(x_i^*)]^2$$

$$+ \frac{1}{2} \sum_{i=1}^{n} \sum_{j=1}^{n} |b_{ij}| [g_j(x_j(t - \tau_{ij}(t))) - g_j(x_j^*)]^2$$

$$+ \frac{\|I(t) - I\|^2}{4\beta} \tag{16.12}$$

for all $t \geq 0$, where $\|\cdot\|$ is a *Euclidean* norm.

Define a second differentiable function

$$V(t) = U(t) + \frac{1}{2} \sum_{i=1}^{n} \sum_{j=1}^{n} \frac{|b_{ij}|}{1 - \gamma_{ij}} \cdot \int_{t-\tau_{ij}(t)}^{t} |g_j(x_j(s)) - g_j(x_j^*)|^2 ds \tag{16.13}$$

for $t \geq 0$. Clearly, $V(t) \geq U(t)$ for all $t \geq 0$. Then, using (17.13) and (17.12), it follows that

$$\dot{V}(t) = \dot{U}(t) + \frac{1}{2}\sum_{i=1}^{n}\sum_{j=1}^{n}\frac{|b_{ij}|}{1-\gamma_{ij}}\left[g_j(x_j(t)) - g_j(x_j^*)\right]^2$$

$$-\frac{1}{2}\sum_{i=1}^{n}\sum_{j=1}^{n}\frac{|b_{ij}|(1-\dot{\tau}_{ij}(t))}{1-\gamma_{ij}}\left[g_j(x_j(t-\tau_{ij}(t))) - g_j(x_j^*)\right]^2$$

$$\leq \dot{U}(t) + \frac{1}{2}\sum_{i=1}^{n}\sum_{j=1}^{n}\frac{|b_{ji}|}{1-\gamma_{ji}}\left[g_i(x_i(t)) - g_i(x_i^*)\right]^2$$

$$-\frac{1}{2}\sum_{i=1}^{n}\sum_{j=1}^{n}|b_{ij}|\left[g_j(x_j(t-\tau_{ij}(t))) - g_j(x_j^*)\right]^2$$

$$\leq \sum_{i=1}^{n}\sum_{j=1}^{n}a_{ij}\left[g_i(x_i(t)) - g_i(x_i^*)\right]\cdot\left[g_j(x_j(t)) - g_j(x_j^*)\right]$$

$$+\sum_{i=1}^{n}\left(\sum_{j=1}^{n}\frac{1}{2}\left(|b_{ij}| + \frac{|b_{ji}|}{1-\gamma_{ji}}\right) + \beta\right)\left[g_i(x_i(t)) - g_i(x_i^*)\right]^2$$

$$+\frac{\|I(t) - I\|^2}{4\beta}$$

$$= [G(x(t)) - G(x^*)]^T \left[\frac{A + A^T + \Pi}{2}\right][G(x(t)) - G(x^*)]$$

$$+\beta\|G(x(t)) - G(x^*)\|^2 + \frac{\|I(t) - I\|^2}{4\beta}$$

$$\leq -(\lambda - \beta)\cdot\|G(x(t)) - G(x^*)\|^2 + \frac{\|I(t) - I\|^2}{4\beta} \qquad (16.14)$$

for all $t \geq 0$, where $G(x^*) = (g_1(x_1^*), \cdots, g_n(x_n^*))^T$.

Define another differentiable function

$$W(t) = \left[V(t) + \frac{1}{4\beta}\right]\cdot e^{-\int_0^t \|I(s)-I\|^2 ds} \qquad (16.15)$$

for all $t \geq 0$. Since $V(t) \geq 0$ for $t \geq 0$, it is easy to see that

$$W(t) \geq \frac{1}{4\beta}\cdot e^{-\int_0^t \|I(s)-I\|^2 ds} > 0$$

for $t \geq 0$. Moreover,

$$\dot{W}(t) \leq -\|I(t)-I\|^2 W(t) + \dot{V}(t)e^{-\int_0^t \|I(s)-I\|^2 ds}$$

$$\leq -\frac{\|I(t)-I\|^2}{4\beta} e^{-\int_0^t \|I(s)-I\|^2 ds}$$

$$-\left[(\lambda-\beta)\|G(x(t))-G(x^*)\|^2 - \frac{\|I(t)-I\|^2}{4\beta}\right]$$

$$\times e^{-\int_0^t \|I(s)-I\|^2 ds}$$

$$\leq -(\lambda-\beta) \cdot \|G(x(t))-G(x^*)\|^2 \cdot e^{-\int_0^{+\infty} \|I(s)-I\|^2 ds}$$

$$\leq -(\lambda-\beta) \cdot |g_i(x_i(t))-g_i(x_i^*)|^2 \cdot e^{-\int_0^{+\infty} \|I(s)-I\|^2 ds} \quad (16.16)$$

$$\leq 0 \quad (16.17)$$

for all $t \geq 0$ and all $(i = 1, \cdots, n)$.

By (16.17), $\dot{W}(t) \leq 0$ for $t \geq 0$, we have $W(t) \leq W(0)$ for all $t \geq 0$, i.e.,

$$\left[V(t) + \frac{1}{4\beta}\right] \cdot e^{-\int_0^t \|I(s)-I\|^2 ds} \leq \left[V(0) + \frac{1}{4\beta}\right]$$

for $t \geq 0$. Thus,

$$\int_{x_i^*}^{x_i(t)} [g_i(s) - g_i(x_i^*)] ds \leq U(t) \leq V(t) \leq \left[V(0) + \frac{1}{4\beta}\right] e^{\int_0^{+\infty} \|I(s)-I\|^2 ds}$$

for all $t \geq 0$ and $i = 1, \cdots, n$. Using *Lemma 2*, we have

$$[g_i(x_i(t)) - g_i(x_i^*)]^2 \leq 2\omega \int_{x_i^*}^{x_i(t)} [g_i(s) - g_i(x_i^*)] ds$$

$$\leq 2\omega \left[V(0) + \frac{1}{4\beta}\right] e^{\int_0^{+\infty} \|I(s)-I\|^2 ds}$$

for all $t \geq 0$ and $(i = 1, \cdots, n)$. Clearly, it shows that $|g_i(x_i(t)) - g_i(x_i^*)|(i = 1, \cdots, n)$ are bounded. It is easy to see from (17.10) that $|\dot{x}_i(t)|(i = 1, \cdots, n)$ are also bounded, i.e., there exists a constant $d > 0$ such that $|\dot{x}_i(t)| \leq d(i = 1, \cdots, n)$ for all $t \geq 0$.

Next, we will prove that

$$\lim_{t \to +\infty} |g_i(x_i(t)) - g_i(x_i^*)| = 0, (i = 1, \cdots, n). \quad (16.18)$$

Suppose this is not true, then there must have some $i(1 \leq i \leq n)$, for some $\epsilon > 0$ there exists a divergent sequence $\{t_k\}$ for which $|g_i(x_i(t_k)) - g_i(x_i^*)| \geq 2\epsilon$. Then, on the intervals

$$t_k \leq t \leq t_k + \frac{\epsilon}{d\omega}, (k = 1, 2, \cdots)$$

it holds from (17.3) that

$$|g_i(x_i(t)) - g_i(x_i^*)| = |g_i(x_i(t)) - g_i(x_i(t_k)) + g_i(x_i(t_k)) - g_i(x_i^*)|$$
$$\geq |g_i(x_i(t_k)) - g_i(x_i^*)| - |g_i(x_i(t)) - g_i(x_i(t_k))|$$
$$\geq 2\epsilon - \omega|x_i(t) - x_i(t_k)|$$
$$\geq 2\epsilon - \omega|\dot{x}_i(\xi)| \cdot |t - t_k|$$

where $\xi$ is some value between $t$ and $t_k$, then,

$$|g_i(x_i(t)) - g_i(x_i^*)| \geq 2\epsilon - d\omega|t - t_k|$$
$$\geq \epsilon. \qquad (16.19)$$

The above intervals can be assumed to be disjoint by taking, if necessary, a subsequence of $t_k$.

From (16.16) and (16.19), we have

$$\dot{W}(t) \leq -(\lambda - \beta) \cdot \epsilon^2 \cdot e^{-\int_0^{+\infty} \|I(s) - I\|^2 ds}$$

for $t_k \leq t \leq t_k + \epsilon/d\omega$. By integrating, it follows that

$$W\left(t_k + \frac{\epsilon}{d\omega}\right) - W(t_k) \leq -\frac{(\lambda - \beta)\epsilon^3}{d\omega} \cdot e^{-\int_0^{+\infty} \|I(s) - I\|^2 ds} \qquad (16.20)$$

for all $k = 1, 2, \cdots$. Since by (16.17) $\dot{W}(t) \leq 0$ for all $t \geq 0$, and the intervals $[t_k, t_k + \epsilon/d\omega](k = 1, 2, \cdots)$ are disjoint of each other, then

$$W(t_k) \leq W\left(t_{k-1} + \frac{\epsilon}{d\omega}\right) \qquad (16.21)$$

for all $k = 2, 3, \cdots$. From (16.19) and (16.21),

$$W\left(t_k + \frac{\epsilon}{d\omega}\right) - W\left(t_{k-1} + \frac{\epsilon}{d\omega}\right) \leq -\frac{(\lambda - \beta)\epsilon^3}{d\omega} \cdot e^{-\int_0^{+\infty} \|I(s) - I\|^2 ds}$$

for all $k = 2, 3, \cdots$. Then,

$$W\left(t_k + \frac{\epsilon}{d\omega}\right) - W\left(t_1 + \frac{\epsilon}{d\omega}\right)$$
$$\leq -\frac{(\lambda - \beta)\epsilon^3}{d\omega} e^{-\int_0^{+\infty} \|I(s) - I\|^2 ds} \cdot (k-1)$$
$$\to -\infty$$

as $k \to +\infty$, which contradicts to $W(t) > 0$. This proves that (16.18) to be true, i.e., the network (17.1) is GOC. The proof is completed.

If the matrix $M$ is a nonsingular $M$-matrix, clearly, $(A+B)^{-1}$ exists. If $(A+A^T+\Pi)$ is negative definite, it also implies that $(A+B)^{-1}$ exists. In fact, given any $x \in R^n$, it holds that $x^T(A+B)x \leq x^T((A+A^T+\Pi)/2)x \leq -\lambda x^T x$. This shows that the matrix $(A+B)$ cannot have any zero eigenvalue and so it is invertible. It should be noted that the matrix $(A+B)$ is nonsingular implies that the set $G(\Omega)$ is actually a singleton. From these facts, then, by the *Lemma 16.1*, to check that $\Omega \neq \emptyset$ in Theorem 16.4 or Theorem 16.6, we only need to check

$$-\sum_{j=1}^{n} c_{ij} b_j \in \left(\underline{g}_i, \overline{g}_i\right), (i = 1, \cdots, n)$$

where $c_{ij}(i, j = 1, \cdots, n)$ are elements of $(A + B)^{-1}$.

Conditions given in Theorem 16.4 or Theorem 16.6 guarantee global output convergence of the network (17.1). A natural problem is that: can these conditions also guarantee state convergence? Generally speaking, this answer is no. Consider a simple one dimensional network

$$\dot{x}(t) = -\sigma(x(t)) - \frac{1}{t+1}$$

for $t \geq 0$, where $\sigma(s) = \max(0, s)$. Clearly, this network satisfies conditions of Theorem 16.4 or Theorem 16.6, thus, it is globally output convergent. However, some state trajectories of this network are not convergent. In fact, it is easy to check that $x(t) = -\ln(t+1)$ is a trajectory of this network starting from zero point. Clearly, $x(t) \to -\infty$, i.e., it diverges to negative infinite. Such properties cannot be observed in Hopfield neural network.

## 16.4 Simulation Results

In this section, some examples will be used to further illustrate the global output convergence results obtained in last section. For simulation convenience, constant delays are considered instead of variable ones.

*Example 16.7.* Consider the following neural network with time varying inputs

$$\begin{bmatrix} \dot{x}_1(t) \\ \dot{x}_2(t) \\ \dot{x}_3(t) \end{bmatrix} = \begin{bmatrix} -1.48 \cdot g(x_1(t)) \\ -2.88 \cdot g(x_2(t)) \\ -3.72 \cdot g(x_3(t)) \end{bmatrix} + \begin{bmatrix} 0.4 + \dfrac{1}{\sqrt{1+t}} \\ 0.2 + \dfrac{1}{\sqrt{1+t}} \\ 0.3 + \dfrac{1}{\sqrt{1+t}} \end{bmatrix}$$

$$+ \begin{bmatrix} 0 & 0.648 & -0.288 \\ 0.648 & 0 & 0.62 \\ 0.882 & -0.72 & 0 \end{bmatrix} \begin{bmatrix} g(x_1(t-0.1)) \\ g(x_2(t-0.2)) \\ g(x_3(t-0.3)) \end{bmatrix} \quad (16.22)$$

for $t \geq 0$, where

$$g(s) = \frac{e^s - e^{-s}}{e^s + e^{-s}}, s \in R.$$

It is easy to check that network (16.22) satisfies the conditions of Theorem 16.4, then it is GOC. Fig. 16.1 shows the convergence of the output of this network for four output trajectories corresponding to four state trajectories starting from $(3,3,3)^T$, $(3,-3,0)^T$, $(-3,0,-3)^T$ and $(-3,3,0)^T$, respectively. Clearly, the four output trajectories converge to $(-0.7641, -0.1502, 0.1167)^T$.

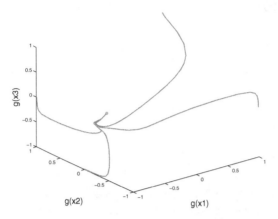

**Fig. 16.1.** Output convergence of network (16.22). The four output trajectories are generated by four state trajectories starting from $(3,3,3)^T$, $(3,-3,0)^T$, $(-3,0,-3)^T$ and $(-3,3,0)^T$, respectively. Clearly, the four output trajectories converge to $(-0.7641, -0.1502, 0.1167)^T$.

The GOC of network (16.22) cannot be checked by Theorem 16.6. In fact, the inputs of (16.22) do not satisfy the condition of Theorem 16.6.

Next, we consider a two dimensional network with time varying inputs and variable delays.

*Example 16.8.* Consider the following neural network

$$\begin{bmatrix} \dot{x}_1(t) \\ \dot{x}_2(t) \end{bmatrix} = \begin{bmatrix} -3 & 1 \\ -1 & -2 \end{bmatrix} \begin{bmatrix} g(x_1(t)) \\ g(x_2(t)) \end{bmatrix} + \begin{bmatrix} 2 + \frac{1}{1+t^2} \\ 0.5 + \frac{1}{1+t^2} \end{bmatrix}$$
$$+ \begin{bmatrix} -0.9 & -0.9 \\ 0.9 & 0.9 \end{bmatrix} \begin{bmatrix} g(x_1(t - 0.1\sin^2 t)) \\ g(x_2(t - 0.1\cos^2 t)) \end{bmatrix} \quad (16.23)$$

for $t \geq 0$, where

$$g(s) = \max(0, s), s \in R.$$

This network does not satisfy the conditions of Theorem 16.4. Next, we invoke Theorem 16.6 instead. We have

$$A = \begin{bmatrix} -3 & 1 \\ -1 & -2 \end{bmatrix}, \quad B = \begin{bmatrix} -0.9 & -0.9 \\ 0.9 & 0.9 \end{bmatrix}, \quad I(t) = \begin{bmatrix} 2 + \frac{1}{1+t^2} \\ 0.5 + \frac{1}{1+t^2} \end{bmatrix}.$$

Clearly,

$$\int_0^{+\infty} [I_i(t) - I_i]^2 \, dt = \int_0^{+\infty} \left(\frac{1}{1+t^2}\right)^2 dt = \frac{\pi}{4} < +\infty.$$

Since
$$\tau_{11}(t) = \tau_{21}(t) = 0.1 \sin^2 t, \quad \tau_{12}(t) = \tau_{22}(t) = 0.1 \cos^2(t)$$
then
$$\dot{\tau}_{11}(t) = \dot{\tau}_{21}(t) = -\dot{\tau}_{12}(t) = -\dot{\tau}_{22}(t) = 0.1 \sin 2t \leq 0.1 < 1,$$
for all $t \geq 0$, we can take $\gamma_{11} = \gamma_{21} = \gamma_{12} = \gamma_{22} = 0.1$. Then,
$$\Pi = \begin{bmatrix} 3.8 & 0 \\ 0 & 3.8 \end{bmatrix},$$
and
$$A + A^T + \Pi = \begin{bmatrix} -2.2 & 0 \\ 0 & -0.2 \end{bmatrix}$$
is a negative definite matrix. By using Theorem 16.6, the network (16.23) is GOC. Simulations for GOC of this network is shown in Fig. 16.2. In this figure, we simulated four outputs of four trajectories of the network starting from the points $(0, 0)$, $(1, 0)$, $(0, 1)$ and $(1, 1)$, respectively.

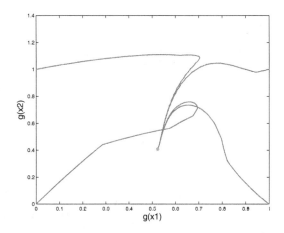

**Fig. 16.2.** Convergence of output of network (16.23)

## 16.5 Conclusion

In this chapter, we studied the global output convergence of a class of recurrent neural networks with time varying inputs and with delays. This class of neural networks possesses some interesting properties that can provide a powerful tool for solving some optimization problems. Some sufficient conditions were derived to guarantee the GOC of the networks. These GOC conditions could be used to assist the design of this model of recurrent neural networks with time varying inputs and time delays.

# 17

# Background Neural Networks with Uniform Firing Rate and Background Input

In this chapter, the dynamic properties of the background neural networks with the uniform firing rate and background input is investigated with a series of mathematical arguments including nondivergence, global attractivity and complete stability analysis. Moreover, it shows that shifting the background level affects the existence and stability of the equilibrium point. Depending on the increase or decrease in background input, the network can engender bifurcation and chaos. It may be have one or two different stable firing levels. That means the background neural network can exhibit not only monostability but also multistability.

## 17.1 Introduction

A bright red light may trigger a sudden motor action in a driver crossing an intersection: stepping at once on the brakes. The same red light, however, may be entirely inconsequential if it appears, say, inside a movie theater. Clearly, context determines whether a particular stimulus will trigger a motor response (see e.g., Ref. (Sommer and Wurtz, 2001)) or not. The background model of neural networks was proposed in Ref. (Salinas, 2003) in order to interpret the inner mechanism of this class phenomenon. It shows that networks of neurons have a built-in capacity to switch between two types of dynamic states: one in which activity is low and approximately equal for all neurons, and another in which different activity distributions are possible and may even change dynamically. This property allows whole circuits to be turned on or off by weak, unstructured background inputs. In other words, a uniform background input may determine whether a random network has one or two stable firing levels. This model of neural network can further explore the working of the cerebrum so that it is very significant to investigate the dynamical properties of it.

The dynamical properties of neural networks play important roles in their applications. Convergence analysis is one of the most important issues of dynamical analysis of neural networks. Convergence of neural networks has been extensively studied in recent years (see e.g., Refs. (Forti, 1994; Forti and Tesi, 1995; Liang and Si, 2001; Yi et al., 1999)). It can be roughly divided into two classes of convergent behaviors: monostability and multistability. Though monostability of neural networks has successful applications in certain optimization problems, monostable networks are computationally restrictive (see e.g., Ref. (Hahnloser, 1998)). On the other hand, in a multistable network, stable and unstable equilibria points may coexist in the network. In certain applications, such as Winner-take-all problems, the existence of unstable equilibria points are essential for digital constraints on selection. In addition, an unstable equilibrium point acts as a threshold that separates two stable equilibria points. In terms of the neural networks investigated in the paper, as mentioned above, depending on the direction of the movement of the bifurcation parameter, i.e. the background inputs, the saddle-node bifurcation explains the disappearance or the appearance of a pair of stable and unstable equilibria points. Consequently, the background inputs can determine if the network has one or two stable firing levels.

As mentioned above, so far, there have been only a few simulations and simple analysis instead of profound theoretical investigation. Therefore this paper studies the dynamical properties of the background neural networks presented in Ref. (Salinas, 2003) with a series of mathematical arguments. We will address three properties of the background network: nondivergence, global attractivity and complete stability. Nondivergence is an essential property that will allow multistability (see e.g., Refs. (Hahnloser, 1998; Wersing et al., 2001b)). Global attractivity is very useful for determine the final behavior of network's trajectories. Complete stability describes a kind of convergence characteristics of networks. A completely stable network may also possess the multistability property. Based on these analysis, further, it shows that the background input can indeed gate the transition between unique and multiple attractors of this class of neural networks.

The remaining parts of this chapter is organized as follows. In Section 2, preliminaries will be given. Nondivergence and global attractivity will be studied in Section 3. In Section 4, we will study the complete stability of the networks. Discussions and simulations will be given in Section 5 and Section 6 respectively. Finally, Section 7 presents the conclusion.

## 17.2 Preliminaries

Consider the following background network presented in (Salinas, 2003) in which the firing rates of all neurons are equal and refer to this uniform rate as $R(t)$.

$$\tau \frac{dR(t)}{dt} = -R(t) + \frac{\left(wR(t) + h\right)^2}{s + vNR^2(t)} \tag{17.1}$$

for $t \geq 0$, where $\tau > 0$ is a time constant, $s > 0$ is a saturation constant, and $N$ is the total number of neurons. $h \geq 0$ represents the background input whose value is independent of the network's activity. And the total synaptic input $w > 0$ to all neurons is same.

We assume the initial condition $R(0) \geq 0$ throughout the paper. Clearly, with this initial condition, we have $R(t) \geq 0$ for all $t \geq 0$ by the reason of

$$R(t) = R(0)e^{-\frac{t}{\tau}} + \frac{1}{\tau}\int_0^t e^{-\frac{t-\theta}{\tau}} \cdot \frac{(wR(\theta)+h)^2}{s+vNR^2(\theta)} d\theta \geq 0. \qquad (17.2)$$

This means the trajectories will remain in the positive region if the initial conditions are positive. This property will make the study latter more easy.

Next, a useful property of the network (17.1) is given.

**Lemma 17.1.** *For all $t \geq 0$, we have*

$$\frac{(wR(t)+h)^2}{s+vNR^2(t)} \leq \frac{w^2}{vN} + \frac{h^2}{s}. \qquad (17.3)$$

*Proof:* Clearly,

$$\frac{(wR(t)+h)^2}{s+vNR^2(t)} - \frac{w^2}{vN} - \frac{h^2}{s}$$
$$= \frac{svN(wR(t)+h)^2 - s(s+vNR^2(t))w^2 - vN(s+vNR^2(t))h^2}{svN(s+vNR^2(t))}$$
$$= -\frac{(sw - h\cdot vNR(t))^2}{svN(s+vNR^2(t))} \leq 0.$$

This implies that (17.3) is true. The proof is completed.

**Definition 17.2 ((Yi and Tan, 2002)).** *Let $S$ be a compact subset of $\Re$. Denote the $\epsilon$-neighborhood of $S$ by $S_\epsilon$. The compact set $S$ is called a global attractive set of a network if for any $\epsilon > 0$, all trajectories of that network ultimately enter and return in $S_\epsilon$.*

**Definition 17.3.** *The network (17.1) is said to be completely convergent if the equilibrium set $S^e$ is not empty and every trajectory $R(t)$ of that network converges to $S^e$, that is,*

$$dist(R(t), S^e) \triangleq \min_{R^* \in S^e} |R(t) - R^*| \to 0$$

*as $t \to +\infty$.*

## 17.3 Nondivergence and Global Attractivity

In this section, we will show the network (17.1) is always bounded and then further give the globally attractive set, which is computed explicitly.

**Theorem 17.4.** *The network (17.1) is always bounded. Moreover, the compact set*
$$S = \left\{ R \mid 0 \leq R \leq \frac{w^2}{vN} + \frac{h^2}{s} \right\}$$
*globally attracts the network (17.1).*

*Proof.* Denote
$$\Pi = R(0) + \frac{w^2}{vN} + \frac{h^2}{s} + 1.$$
We will prove that
$$R(t) < \Pi \tag{17.4}$$
for all $t \geq 0$. Otherwise, if (17.4) is not true, since $R(0) < \Pi$, there must exist a $t_1 > 0$ such that $R(t_1) = \Pi$ and $\dot{R}(t_1) \geq 0$. However, from (17.1) and Lemma 1, we have
$$\tau \frac{dR(t_1)}{dt} = -R(t_1) + \frac{(wR(t_1) + h)^2}{s + vNR^2(t_1)}$$
$$\leq -\Pi + \frac{w^2}{vN} + \frac{h^2}{s}$$
$$< 0. \tag{17.5}$$

This is a condition that proves (17.4) is true. Thus, the network (17.1) is bounded.

Next, we prove the global attractivity of the network.

Given any $\epsilon > 0$, there exists a $\eta = \frac{\epsilon}{R(0) + \frac{w^2}{vN} + \frac{h^2}{s}}$, by the (random) the choice of the $\epsilon$, $\eta$ is also random. Moreover, since $\lim_{t \to +\infty} e^{-\frac{t}{\tau}} = 0$, for the choice of $\eta$, there exists a $T$ such that
$$\left| e^{-\frac{t_1}{\tau}} - e^{-\frac{t_2}{\tau}} \right| < \eta$$
for all $t_1, t_2 > T$. Then from (17.2), it follows that
$$|R(t_1) - R(t_2)| \leq R(0) \left| e^{-\frac{t_1}{\tau}} - e^{-\frac{t_1}{\tau}} \right|$$
$$+ \frac{1}{\tau} \cdot \left( \frac{w^2}{vN} + \frac{h^2}{s} \right) \cdot \left| \int_0^{t_1} e^{-\frac{t_1 - \theta}{\tau}} d\theta - \int_0^{t_2} e^{-\frac{t_2 - \theta}{\tau}} d\theta \right|$$
$$= R(0) \left| e^{-\frac{t_1}{\tau}} - e^{-\frac{t_1}{\tau}} \right| + \left( \frac{w^2}{vN} + \frac{h^2}{s} \right) \cdot \left| e^{-\frac{t_1}{\tau}} - e^{-\frac{t_1}{\tau}} \right|$$
$$< \left( R(0) + \frac{w^2}{vN} + \frac{h^2}{s} \right) \cdot \eta$$
$$= \epsilon.$$

This means the limit of $R(t)$ exists, that is,
$$\lim_{t\to+\infty} R(t) = \lim_{t\to+\infty} \sup R(t).$$
Then from (17.2), it is easy to see that
$$\lim_{t\to+\infty} \sup R(t) = \lim_{t\to+\infty} R(t)$$
$$= \lim_{t\to+\infty} R(0)e^{-\frac{t}{\tau}} + \lim_{t\to+\infty} \frac{1}{\tau} \int_0^t e^{-\frac{t-\theta}{\tau}} \cdot \frac{(wR(\theta)+h)^2}{s+vNR^2(\theta)} d\theta$$
$$\leq \frac{w^2}{vN} + \frac{h^2}{s}.$$
This shows that $S$ is a global attractive set of the network (17.1). The proof is completed.

## 17.4 Complete Stability

Complete stability requires that every trajectory of a network converges to an equilibrium point. This property guarantees a network to work well without exhibiting any oscillations or chaotic behavior. In addition, a completely stable network may possess a multistable property. Stable and unstable equilibrium points may coexist in a completely stable network. This property has its important applications in certain networks (see e.g., Ref. (Wersing et al., 2001b)). Complete stability analysis for other models of neural networks can be found in Refs (Takahashi, 2000; Yi and Tan, 2002; Yi et al., 2003). In this section, we will show the network (17.1) is completely stable using the energy function method.

**Theorem 17.5.** *The network (17.1) is completely stable.*

*Proof:* Constructing an energy function
$$E(t) = \frac{R^2(t)}{2} - \frac{w^2}{vN} R(t) - \frac{wh}{vN} \cdot \ln\left(s + vNR^2(t)\right)$$
$$- \frac{vNh^2 - sw^2}{vN\sqrt{svN}} \cdot \left(\sqrt{\frac{vN}{s}} R(t)\right). \tag{17.6}$$

Clearly, since $R(t)$ is bounded, $E(t)$ is also bounded. It follows that
$$\dot{E}(t) = R(t)\dot{R}(t) - \frac{w^2}{vN}\dot{R}(t) - \frac{2whR(t)}{s+vNR^2(t)}\dot{R}(t)$$
$$- \frac{vNh^2 - sw^2}{vN(s+vNR^2(t))}\dot{R}(t)$$
$$= \dot{R}(t)\left[R(t) - \frac{(wR(t)+h)^2}{s+vNR^2(t)}\right]$$
$$= -\tau \dot{R}^2(t) \tag{17.7}$$

for all $t \geq 0$. Thus, $E(t)$ is monotonically decreasing. Since $E(t)$ is bounded, there must exist a constant $E_0$ such that

$$\lim_{t \to +\infty} = E_0 < +\infty.$$

From Eq. (17.7) we have

$$\begin{aligned}
\int_0^{+\infty} \dot{R}^2(\theta) d\theta &= \lim_{t \to +\infty} \int_0^t \dot{R}^2(\theta) d\theta \\
&= \lim_{t \to +\infty} \int_0^t \frac{-\dot{E}(\theta)}{\tau} d\theta \\
&= -\frac{1}{\tau} \lim_{t \to +\infty} E(t) \\
&= -\frac{E_0}{\tau} \\
&< +\infty.
\end{aligned} \quad (17.8)$$

Moreover, since $R(t)$ is bounded, from Eq. (17.1) it follows that $\dot{R}(t)$ is bounded. Then $R(t)$ is uniformly continuous on $[0, +\infty)$. Again, from Eq. (17.1), it follows that $\dot{R}(t)$ is also uniformly continuous on $[0, +\infty)$. Using (17.8), it must follows that

$$\lim_{t \to +\infty} \dot{R}(t) = 0.$$

Since $R(t)$ is bounded, every subsequence of $R(t)$ must contain convergent subsequence. Let $R(t_m)$ be any of such a convergent subsequence. There exists a $R^* \in \Re$ such that

$$\lim_{t_m \to +\infty} R(t_m) = R^*.$$

Then, from Eq. (17.1), we have

$$-R^* + \frac{(wR^* + h)^2}{s + vNR^{*2}} = \lim_{t_m \to +\infty} \dot{R}(t_m) = 0.$$

Thus, $R^* \in S^e$. This shows that $S^e$ is not empty, and any convergent subsequence of $R(t)$ converges to a point of $S^e$.

Next, we use the method in Ref. (Liang and Wang, 2000) to prove

$$\lim_{t \to +\infty} \text{dist}(R(t), S^e) = 0. \quad (17.9)$$

Suppose Eq. (17.9) is not true. Then there exists a constant $\epsilon_0 > 0$ such that for any $T \geq 0$, there exist a $\bar{t} \geq T$ that satisfies $\text{dist}(R(\bar{t}), S^e) \geq \epsilon_0$. From this and by the boundedness property of $R(t)$, we can choose a convergent subsequence $R(\bar{t}_m) = R^\ddagger \in S^e$, such that

$$\text{dist}(R(\bar{t}_m), S^e) \geq \epsilon_0, (m = 1, 2, \dots).$$

Letting $t_m \to +\infty$, we have

$$\text{dist}(R(t^{\ddagger}), S^e) \geq \epsilon_0 > 0,$$

which contradicts $\text{dist}(R(t^{\ddagger}), S^e) = 0$ since $R(t^{\ddagger}) \in S^e$. This proves Eq. (17.9) is true. The network (17.1) is thus completely convergent. The proof is completed.

## 17.5 Discussion

Theorems 1 and 2 show the network (17.1) is always convergent. In fact, if the equilibria points exist, they must hold

$$-R^* + \frac{(wR^* + h)^2}{s + vNR^{*2}} = 0, \tag{17.10}$$

where $R^*$ denotes the equilibrium point. It is equivalent to

$$R^{*3} - \frac{w^2}{vN} R^{*2} + \frac{s - 2wh}{vN} R^* - \frac{h^2}{vN} = 0. \tag{17.11}$$

The discriminant of the cubic equation (17.11) is

$$\Delta = \frac{1}{108(vN)^4} \Big[ 4vN(s - 2wh)^3 + 27h^4(vN)^2$$

$$+ 4w^6 h^2 - w^4(s - 2wh)^2 - 18w^2 h^2 vN(s - 2wh) \Big]. \tag{17.12}$$

In fact, $R^*$ in Eq. (17.11) will be stable if it holds that $\frac{d(\frac{dR}{dt})}{dR}|_{R^*} < 0$. That is,

$$\frac{1}{\tau}\left[-1 + \frac{2w(s + vNR^{*2}) - (wR^* + h)^2 \cdot 2vNR^*}{(s + vNR^{*2})^2}\right] < 0. \tag{17.13}$$

From Eq. (17.10), it follows that

$$s + vNR^{*2} = \frac{(wR^* + h)^2}{R^*}.$$

Then (17.13) is equivalent to

$$2w(wR^* + h) < s + 3vNR^{*2}. \tag{17.14}$$

That means that an equilibrium point of the network (17.1) must be stable, if it holds for (17.14). In addition, with the increase or decrease in the background input, the sign of $\Delta$ will be transition between positive and negative. It further affects the existence and stability of the equilibrium point of the network. So the background neural network can exhibit not only monostability but also multistability. Table 17.1 depicts their relations to each other.

**Table 17.1.** The relationship of the existence and stability of equilibrium point to $\Delta$.

| $\Delta$ | Roots of Eq. (17.10) | Dynamic properties of the network (17.1) |
|---|---|---|
| $<0$ | Three different real roots | Among the three equilibria points, the maximum one and the minimum one are both local attractive, the middle one is a saddle node of the system and unstable. |
| $=0$ | Three real roots, two of them are equivalent. | The saddle-node bifurcation. There is only one solution that is stable. |
| $>0$ | One real root and two imaginary roots | There is a unique solution that is global attractive. |

## 17.6 Simulation

In this section, we will employ example to further illustrate the analysis above.

*Example 17.6.* Consider the network (17.1) with $w = 1.12, vN = 0.0136$ and $\tau = 1, s = 40$.

Condition (a), where $h = 8.2$. From Eq. (17.12), here $\Delta = -2.1341 \times 10^7 < 0$, and
$$\frac{w^2}{vN} + \frac{h^2}{s} = 93.9163.$$

By Theorem 1, the network (17.1) is bounded, and there exists a compact set
$$S = \{R \mid 0 \leq R \leq 93.9163\},$$

which globally attracts all the trajectories of the network. In fact, it is easy to check that there are three equilibria points of the network. Clearly, they are all located in the set $S$. Moreover, by Theorem 2, $R = 3.9928$, and $R = 70.7376$ are stable while $R = 17.5049$ is unstable. Based on this condition, stable and unstable equilibria points can coexist in the network. The unstable equilibrium point acts as a threshold that separates two stable equilibria points. Fig. 1 shows the complete convergence of the network (17.1) in condition (a).

Condition (b), where $h = 14$. From Eq. (17.12), here $\Delta = 3.0772 \times 10^8 > 0$. We can see that only the high-rate solution $R = 86.8303$ is global attractive, while the low-rate disappear with the increase in the uniform background input $h$. Fig. 2 shows the global convergence of the network (17.1) in condition (b).

## 17.7 Conclusions

In this chapter, we have studied the background neural network with uniform firing rate and background input. Some of the dynamical properties of it, such

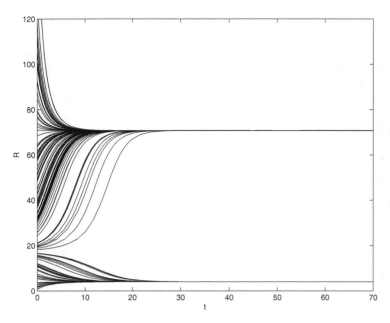

**Fig. 17.1.** Complete convergence of the network (17.1) in condition (a) in Example 1.

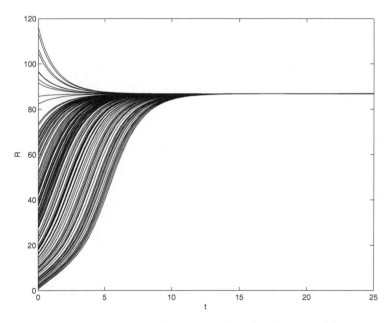

**Fig. 17.2.** Complete convergence of the network (17.1) in condition (a) in Example 1.

as boundedness, attractivity, and complete convergence have been investigated with theoretical arguments. The network considered in this paper, possesses the property that the trajectories will remain in the positive regime if the initial conditions are positive. In addition, the background input, as the bifurcation parameter, can determine if the network has one or two stable equilibria points, namely, one or two different stable firing levels. That means that an increase in background activity can eliminate the low-rate equilibrium point. These conclusions have been tested in the simulation examples successfully. However, this chapter presents a basic framework for the dynamical analysis of the background neural network. It remains more interesting problem to study. For example, the general investigation for this class of networks is expectable when the condition of uniform rate and total synaptic connection weight is removed.

# References

Abe, S. (1993). Global convergence and suppression of spurious states of the hopfield neural networks. *IEEE Trans. Circuits and Systems I* **40**(4), 246–257.

Abid, S., F. Fnaiech and M. Najim (2001). A fast feedforward training algorithm using a modified form of the standard backpropagation algorithm. *IEEE Trans. Neural Networks* **12**(2), 424–429.

Adorjan, P., J.B. Levitt, J.S. Lund and K. Obermayer (1999). A model for the intracortical orgin of orientation preference and tuning in macaque striate cortex. *Visual Neuroscience* **16**, 303–318.

Aiyer, S.V.B., M. Niranjan and F. Fallside (1990). A theoretical investigation into the performance of the hopfield model. *IEEE Trans. Neural Networks* **1**(2), 204–215.

Amari, S. and N. Ksabov (1997). *Brain-like computing and intelligent information systems*. Springer. New York.

Angel, R.D., W.L. Caudle, R. Noonan and A. Whinson (1972). Computer assisted school bus scheduling. *Management Science* **18**, 279–288.

Bai, Y., W. Zhang and Z. Jin (2006). An new self-organizing maps strategy for solving the traveling salesman problem. *Chaos, Solitons, and Fractals* **28**, 1082–1089.

Baldi, P. and A. Atiya (1994). How delays affect neural deynamics and learning. *IEEE Trans. Neural Networks* **5**, 612–621.

Bauer, U., M. Scholz, J.B. Levitt, K. Obermayer and J.S. Lund (1999). A biologically-based neural network model for geniculocortical information transfer in the primate visual system. *Vision Research* **39**, 613–629.

Bay, J.S. (1999). *Fundamentals of Linear State Space Systems*. McGraw-Hill.

Ben-Yishai, R., R. Lev Bar-Or and H. Sompolinsky (1995). Theory of orientation tuning in visual cortex. *Prof. Nat. Acad. Sci. USA* **92**, 3844–3848.

Berman, A. and R.J. Plemmons (1994). *Nonnegative Matrices in The Mathematical Science*. Academic Press. New York.

Bertels, K., L. Neuberg, S. Vassiliadis and D.G. Pechanek (2001). On chaos and neural networks: The backpropagation paradigm. *Artificial Intelligence Review* **15**(3), 165–187.

Bouzerdoum, A. and T. R. Pattison (1993). Neural network for quadratic optimization with bound constraints. *IEEE Trans. Neural Networks* **4**(2), 293–303.

Brandt, R.D., Y. Wang, A.J. Laub and S.K. Mitra (1988). Alternative networks for solving the traveling salesman problem and the list-matching problem. *Proceedings Int. Joint Conference on Neural Networks* **2**, 333–340.

Cao, J., J. Wang and X. Liao (2003). Novel stability criteria of delayed cellular neural networks. *Int. Journal of Neural Systems* **13**(5), 367–375.

Caulfield, H.J. and M. Kinser (1999). Finding the path in the shortest time using pcnn's. *IEEE Trans. Neural Networks* **10**(3), 604–606.

Chen, K. and D.L. Wang (2002). A dynamically coupled neural oscillator network for image segmentation. *Neural Networks* **15**, 423–439.

Chen, L. and K. Aihara (1995). Chaotic simulated annealing by a neural network model with transient chaos. *Neural Networks* **8**(6), 915–930.

Chen, T. and S.I. Amari (2001). Exponential convergence of delayed dynamical systems. *Neural Computation* **13**, 621–635.

Cheng, K.S., J.S. Lin and C.W. Mao (1996). The application of competitive hopfield neural network to medical image segmentation. *IEEE Trans. on Medical Imaging* **15**(4), 560–567.

Christofides, N. and S. Eilon (1969). An algorithm for the vehicle dispatching problem. *Research Quarterly* **20**, 309–318.

Chua, L.O. and L. Yang (1988). Cellular neural networks: Theory. *IEEE Trans. Circuits Systems* **35**, 1257–1272.

Civalleri, P.P., M. Gilli and L. Pandolfi (1993). On stability of cellular neural networks with delay. *IEEE Trans. Circuits and Systems-I* **40**, 157–165.

Cohen, M.A. and S. Grossberg (1983). Absolute stability of global pattern formation and parallel memory storage by competitive neural networks. *IEEE Trans. Systems, Man, Cybernetics* **13**, 815–826.

Cooper, B. (2002). Stability analysis of higher-order neural networks for combinatorial optimization. *Int. Journal of Neural Systems* **12**(3), 177–186.

Costantini, G., D. Casali and R. Perfetti (2003). Neural associative memory for storing gray-coded gray-scale images. *IEEE Trans. Neural Networks* **14**(3), 703–707.

Dayan, P. and L.F. Abbott (2001). *Theoretical Neuroscience*. MIT Press.

Deb, K. (2001). *Multiobjective Optimization Using Evolution Algorithms*. Wiley. Chichester.

Douglas, R., C. Koch, M. Mahowald, K. Martin and H. Suarez (1995). Recurrent excitation in neocortical circuits. *Science* **269**, 981–985.

Driessche, P.V.D. and X. Zou (1998). Global attractivity in delayed hopfield neural network models. *SIAM J. Appl. Math.* **58**(6), 1878–1890.

Duda, R.O., P.E. Hart and D.G. Stork (2001). *Pattern Classification*. 2nd Ed. John Wiley & Sons. New York.

Egmont-Pertersen, M., D. de Ridder and H. Handels (2002). Image processing with neural networks–a review. *Pattern Recognition* **35**, 2279–2301.

Ellias, S. A. and S. Grossberg (1975). Pattern formation, contrast control, and oscillations in the short term memory of shunting on-center off-surround networks. *Biol. Cybernetics* **20**, 69–98.

Ephremids, A. and S. Verdu (1989). Control and optimization methods in communication network problems. *IEEE Trans. Automat. Contr.* **34**, 930–942.

Ergezinger, S. and E. Thomsen (1995). An accelerated learning algorithm for multilayer perceptrons: Optimization layer by layer. *IEEE Trans. Neural Networks* **6**(1), 31–42.

Ermentrout, B. (1992). Complex dynamics in winner-take-all neural nets with slow inhibition. *Neural Networks* **5**, 415–431.
Farina, M., K. Deb and P. Amato (n.d.). Dynamic multiobjective optimization problems: test cases, approximations and applications. *Personal correspondance.*
Feldman, J.A. and D.H. Ballard (1982). Connectionist models and their properties. *Cognitive Science* **6**, 205–254.
Feng, J. (1997). Lyapunov functions for neural nets with nondifferentiable input-output characteristics. *Neural Computation* **9**, 43–49.
Feng, J. and K.P. Hadeler (1996). Qualitative behavior of some simple networks. *Journal of Physics A* **29**, 5019–5033.
Feng, J., H. Pan and V.P. Roychowdhury (1996). On neurodynamics with limiter function and linsker's developmental model. *Neural Computation* **8**, 1003–1019.
Forti, M. (1994). On global asymptotic stability of a class of nonlinear systems arising in neural network theory. *Journal of Differential Equations* **113**, 246–264.
Forti, M. and A. Tesi (1995). New conditions for global stability of neural networks with application to linear and quadratic programming problems. *IEEE Trans. Circuits and Systems-I* **42**(7), 354–366.
Franca, P.M., M. Gendreau, G. Laporte and F.M. Muller (1995). The m-traveling salesmanproblem with minmax object. *Transportation Science* **29**(3), 267–275.
Frederickson, G.N., M.S. Hecht and C.E. Kim (1978). Approximation algorithms for some routing problems. *SIAM Journal on Computing* **17**, 178–193.
Funahashi, K.I. (1989). On the approximate realization of continuous mapping by neural networks. *Neural Networks* **2**(3), 183–192.
G. Xiaodong, Y. Daoheng and Z. Liming (2005). Image shadow removal using pulse coupled neural network. *IEEE Trans. Neural Networks* **16**(3), 692–698.
Gallant, J.L., D.C. van Essen and H.C. Nothdurft (1995). Two-dimensional and three-dimensional texture processing in visual cortex of the macaque monkey. In: *Early Vision and Beyond* (T. Papathomas, C. Chubb, A. Gorea and E. Kowler, Eds.). pp. 89–98. MIT Press. Cambridge, MA.
Goldberg, J.L. (1992). *Matrix Theory With Applications.* McGraw-Hill.
Gonzalez, R.C. and R.E. Woods (2002). *Digital image processing.* Vol. 2nd Edition. Prentice Hall. New Jersey.
Gopal, S.S., B. Sahiner, H.P. Chan and N. Petrick (1997). Neural network based segmentation using a priori image models. *International Conference on Neural Networks* **4**, 2455–2459.
Gopalsamy, K. and X.Z. He (1994). Stability in asymmetric hopfield nets with transmission delays. *Physica D* **76**, 344–358.
Gray, C.M. and W. Singer (1989). Stimulus-specific neuronal oscillations in the orientation columns of cat visula cortex. *Proc. Nat. Academy Sci.* **86**, 1689–1702.
Grossberg, S. (1988). Nonlinear neural networks: Principles, mechanisms, and architectures. *Neural Networks* **1**, 17–61.
Grossberg, S. and E. Mingolla (1985). Neural dynamics of perceptual grouping: Textures, boundaries and emergent segmentation. *Percept. Psychophys.* **38**(2), 141–171.
Guckenheimer, J. and P. Holmes (1983). *Nonlinear oscillations, dynamical systems, and bifurcations of vector fields.* Vol. 42 of *Applied Mathematical Sciences.* Springer. New York.
Habib, Y., S.M. Sait and A. Hakim (2001). Evolutionary algorithms, simulated annealing and tabu search: a comparative study. *Artificial Intelligence* **14**, 167–181.

Hadeler, K.P. and D. Kuhn (1987). Stationary states of the hartline-ratliff model. *Biological Cybernetics* **56**, 411–417.

Hahnloser, R.H.R. (1998). On the piecewise analysis of networks of linear threshold neurons. *Neural Networks* **11**, 691–697.

Hahnloser, R.H.R., H.S. Seung and J.J. Slotine (2003). Permitted and forbidden sets in symmetric threshold-linear networks. *Neural Computation* **15**(3), 621–638.

Hahnloser, R.H.R., R. Sarpeshkar, M.A. Mahowald, R.J. Douglas and H.S. Seung (2000). Digital selection and analog amplification coexist in a cortex-inspired silicon circuit. *Nature* **405**, 947–951.

Haken, H. (2002). *Brain dynamics, synchronization and activity patterns in pulse-coupled neural nets with delays and noise.* Springer. New York.

Hale, J.K. (1977). *Theory of Functional Differential Equations.* Springer. New York.

Hartline, H.K. and F. Ratliff (1958). Spatial summation of inhibitory influence in the eye of limulus and the mutual interaction of receptor units. *Journal of General Physiology* **41**, 1049–1066.

Hasegawa, M., T. Ikeguchi and K. Aihara (2002a). Solving large scale traveling salesman problems by chaotic neurodynamics. *Neural Networks* **15**, 271–283.

Hasegawa, M., T. Lkeguchi and K. Aihara (2002b). Solving large scale traveling problems by choaotic neurodynamics. *Neural Networks* **15**, 271–385.

Haykin, S. (1999). *Neural Networks A Comprehensive Foundation.* 2nd Ed. Prentice Hall. New Jersey.

Hertz, J., A. Krogh and R.G. Palmer (1991). *Introduction to the Theory of Neural Computation.* Santa Fe Institute Studies in the Sciences of Complexity. Addison Wesley.

Hopfield, J.J. (1982). Neural networks and physical systems with emergent collective computational abilities. *Proc. Natl. Acad. Sci. USA* **79**, 2554–2558.

Hopfield, J.J. (1984). Neurons with grade response have collective computational properties like those of two-state neurons. *Proc. Natl. Acad. Sci. USA* **81**, 3088–3092.

Hopfield, J.J. and D.W. Tank (1985). Neural computation of decision in optimization problem. *Biol. Cybern.* **52**, 141–152.

Hu, S. and J. Wang (2002). Global exponential stability of continuous-time interval neural networks. *Physical Review E* **65**(3), 036133.1–9.

Hummel, J.E. and I. Biederman (1992). Dynamic binding in a neural network for shape recognition. *Psychol. Rev.* **9**, 480–517.

J. Cao, J. Wang (2003). Global asymptotic stability of a general class of recurrent neural networks with time-varying delays. *IEEE Trans. Circuits and Systems-I* **50**(1), 34–44.

Jankowski, S., A. Lozowski and J.M. Zurada (1996). Complex-valued multistate neural associative memory. *IEEE Trans. Neural Networks* **7**(6), 1491–1496.

Johnson, J.L. (1993). Waves in pulse coupled neural networks. *Proceedings of World Congress on Neural Networks.*

Johnson, J.L. (1994). Wpulse-coupled neural nets: Translation, rotation, scale, distortion, and intensity signal invariance for images. *Appl. Opt.* **33**(26), 6239–6253.

Johnson, J.L. and D. Ritter (1993). Observation of periodic waves in a pulse-coupled neural network. *Opt. Lett.* **18**(15), 1253–1255.

Junying, Z., W. Defeng, S.Meihong and W.J. Yue (2004). Output-threshold coupled networks for solving the shortest path problems. *Science in China* **47**(1), 20–33.

Kamgar-Parsi, B. and B. Kamgar-Parsi (1992). Dynamical stability and parameter selection in neural optimization. *Proceedings Int. Joint Conference on Neural Networks* **4**, 566–571.

Kapadia, M.K., M. Ito, C.D. Gilbert and G. Westheimer (1995). Improvement in visual sensitivity by changes in local context: Parallel studies in human observers and in v1 of alert monkeys. *Neuron* **15**(4), 843–856.

Kennedy, M. P. and L. O. Chua (1988). Neural networks for nonlinear programming. *IEEE Trans. Circuits and Systems* **35**(5), 554–562.

Knierim, J.J. and D.C. van Essen (1992). Neuronal responses to static texture patterns in area v1 of the alert macaque monkeys. *J. Neurophysiol.* **67**, 961–980.

Kohonen, T. (1989). *Self-organization and associative memory*. 3rd Edition. Springer-Verlag. Berlin.

Li, X., L. Huang and J. Wu (2004). A new method of lyapunov functionals for delayed cellular neural networks. *IEEE Trans. Circuits and Systems-I* **51**(11), 2263–2270.

Li, Z. (1998). A neural model of contour integration in the primary visual cortex. *Neural Computation* **10**, 903–940.

Li, Z. (2001). Computational design and nonlinear dynamics of a recurrent network model of the primary visual cortex. *Neural Computation* **13**, 1749–1780.

Li, Z. and P. Dayan (1999). Computational differences between asymmetrical and symmetrical networks. *Network* **10**, 59–77.

Liang, X. B. and J. Si (2001). Global exponential stability of neural networks with globally lipschitz continuous activations and its application to linear variational inequality problem. *IEEE Trans. Neural Networks* **12**(2), 349–359.

Liang, X.B. (2001). A recurrent neural network for nonlinear continuously differentiable optimization over a compact convex set. *IEEE Trans. Neural Networks* **12**(6), 1487–1490.

Liang, X.B. and J. Wang (2000). A recurrent neural network for nonlinear optimization with a continuously differentiable objective function and bound constraints. *IEEE Trans. Neural Networks* **11**(6), 1251–1262.

Liao, X. and J. Wang (2003). Global dissipativity of continuous-time recurrent neural networks with time delay. *Physical Review E* **68**(1), 016118. 1–7.

Likas, A. and V. Paschos (2002). A note on a new greedy-solution representation and a new greedy parallelizable heuristic for the traveling salesman problem. *Chaos, Solitons, and Fractals* **13**, 71–78.

Lin, W.C., E. Tsao and C.T. Chen (1992). Constraint satisfaction neural networks for image segmentation. *Pattern Recognition* **25**, 679–693.

Lippmann, R.P. (1987). An introduction to computing with neural nets. *IEEE ASSP* pp. 4–22.

Liu, D., S. Hu and J. Wang (2004). Global output convergence of a class of continuous-time recurrent neural networks with time-varying thretholds. *IEEE Trans. Circuits and Systems-II* **51**(4), 161–167.

Lu, H. (2001). Stability criteria for delayed neural networks. *Physical Review E* **64**, 051901.

Luo, F., R. Unbehauen and A. Cichocki (1997). A minor component analysis algorithm. *Neural Networks* **10**(2), 291–297.

Luo, F., R. Unbehauen and Y.D. Li (1995). A principal component analysis algorithm with invariant norm. *Neurocomputing* **8**, 213–221.

Maa, C.Y. and M. Shanblatt (1992). Linear and quadratic programming neural network analysis. *IEEE Trans. Neural Networks* **3**(4), 580–594.

Marcus, C.M. and R.M. Westervelt (1989). Stability of analog neural networks with delay. *Physical Review A* **39**, 347–359.

Martin-Valdivia, M., A. Ruiz-Sepulveda and F. Triguero-Ruiz (2000). Simproved local minima of hopfield networks with augmented lagrange multipliers for large scale tsps. *Neural Networks Letter* **13**, 283–285.

Matsuda, S. (1998). Optimal hopfield network for combinatorial optimization with linear cost function. *IEEE Trans. Neural Networks* **9**(6), 1319–1330.

Michalewicz, Z. (1994). *Genetic Algorithms + Data Structures = Evolution Programs*. second ed. Springer-Verlag. Berlin.

Michalewicz, Z. (n.d.). Genocop evolution program.

Mohan, R. and R. Nevatia (1992). Perceptual organization for scene segmentation and description. *IEEE Trans. Pattern Analysis and Machine Intelligence* **14**(6), 616–635.

Müezzinoğlu, M.K., C. Güzelis and J.M. Zurada (2003). A new design method for the complex-valued multistate hopfield associative memory. *IEEE Trans. Neural Networks* **14**(4), 891–899.

Muresan, R.C. (2003). Pattern recognition using pulse-coupled neural networks and discrete fourier transforms. *Neurocomputing* **51**, 487–493.

Nozawa, H. (1992). A neural network model as a globally coupled map and applications based on chaos. *Chaos* **2**(3), 377–386.

Ontrup, J. and H. Ritter (1998). Perceptual grouping in a neural model: Reproducing human texture perception. In: *Technical Report*. University of Bielefeld. Bielefeld.

Orloff, D.S. (1974). Routing a fleet of m vehicles to/from a central facility. *Networks* **4**, 147–162.

Pérez-Ilzarbe, M.J. (1998). Convergence analysis of a discrete-time recurrent neural networks to perform quadratic real optimization with bound constraints. *IEEE Trans. Neural Networks* **9**(6), 1344–1351.

Pakdaman, K., C. G. Ragazzo and C.P. Malta (1998). Transient reigme duration in continuous-time neuaral networks with delay. *Physical Review E* **58**(3), 3623–3627.

Papageorgiou, G., A. Likas and A. Stafylopatis (1998). Improved exploration in hopfield network state-space through parameter perturbation driven by simulated annealing. *European Journal of Operational Research* **108**, 283–292.

Park, S. (1989). Signal space interpretations for hopfield neural network for optimization. In: *Proceedings of IEEE Int. Symposium on Circuits Systems*. pp. 2181–2184.

Pavlidis, T. and Y.T. Liow (1990). Integrating region growing and edge detection. *IEEE Trans. Pattern Analysis and Machine Ingelligence* **12**, 225–233.

Peng, M., K. Narendra and A. Gupta (1993). An investigation into the improvement of local minimum of the hopfield network. *Neural Networks* **9**, 1241–1253.

Perko, L. (2001). *Differential equations and dynamical systems*. Vol. 3rd Edition of Santa Fe Institute Studies in the Sciences of Complexity. Springer. New York.

Phoha, V.V. and W.J.B. Oldham (1996). Image recovery and segmentation using competitive learning in a layered network. *IEEE Trans. Neural Networks* **7**(4), 843–856.

Pirlot, M. (1996). General local search methods. *European Journal of Operation Research* **92**, 493–511.

Poon, C.S. and J.V. Shah (1998). Hebbian learning in parallel and modular memories. *Biological Cybernetics* **78**(2), 79–86.

Potvin, J., G. Lapalme and J. Rousseau (1989). A generalized k-opt exchange procedure for mtsp. *INFOR* **27**(4), 474–481.

Protzel, P.W., D.L. Palumbo and M.K. Arras (1994). Perfomance and fault-tolerance of neural networks for optimization. *IEEE Trans. Neural Networks* **4**, 600–614.

Qiao, H., J. Peng, Z.B. Xu and B. Zhang (2003). A reference model approach to stability analysis of neural networks. *IEEE Trans. Systems, Man and Cybernetics B: Cybernetics* **33**(6), 925–936.

Qu, H., Z. Yi and H. Tang (n.d.). A columnar competitive model for solving multi-traveling salesman problem. *Chaos, Solitons, and Fractals* p. In Press.

R. Eckhorn, H.J. Reitboeck, M. Arndt and P.W. Dicke (1990). Feature linking via synchronous among ditributed assemblies: Simulations of results from cat visual cortex. *Neural Comput.* **2**, 1253–1255.

Ranganath, S.H. and G. Kuntimad (1999). Object detection using pulse coupled neural networks. *IEEE Trans. Neural Networks* **10**, 615–620.

Rego, C. (1998). Relaxed tours and path ejections for the traveling salesman problem. *European Journal of Operation Research* **106**, 522–538.

Reinelt, G. (1991). Tsplib - a traveling salesman problem library. *ORSA Journal on Computing* **3**, 376–384.

Ricca, F. and P. Tonella (2001). Understanding and restructuring web sites with reweb. *Web Engineering* **7**, 40–51.

Rumelhart, D.E. and D. Zipser (1985). Feature discovery by competitive learning. *Cognitive Science* **9**, 75–112.

Salinas, A. and L.F. Abbott (1996). A model of multiplicative responses in parietal cortex. *Prof. Nat. Acad. Sci. USA* **93**, 11956–11961.

Salinas, E. (2003). Background synaptic activity as a swich between dynamical states in network. *Neural Computation* **15**, 1439–1475.

Samerkae, S., M. Abdolhamid and N. Takao (1999). Competition-based neural network for the multiple traveling salesmen problem with minmax objective. *Computer Operations Research* **26**, 395–407.

Savelsbergh, M.W.P. (1995). The general pickup and delivery problem. *Transactions of Science* **29**(1), 17–29.

Scalero, R.S. and N. Tepedelenlioglu (1992). A fast new algorithm for training feed-forward neural networks. *IEEE Trans. Signal Processing* **40**(1), 202–210.

Scherk, J. (2000). *ALGEBRA: A Computational Introduction*. Chapman & Hall/CRC.

Seung, H.S. (1998). Continuous attractors and oculomotor control. *Neural Networks* **11**, 1253–1258.

Shah, J.V. and C.S. Poon (1999). Linear independence of internal representations in multilayer perceptrons. *IEEE Trans. Neural Networks* **10**(1), 10–18.

Smith, K.A. (1999). Neural networks for combinatorial optimization: A review of more than a decade of research. *INFORMS Journal on Computing* **11**(1), 15–34.

Sommer, M.A. and R.H. Wurtz (2001). Frontal eye field sends delay activity related to movement, memory, and vision to the superior colliculus. *J. Neurophysiol* **85**, 1673–1685.

Sudharsanan, S. and M. Sundareshan (1991). Exponential stability and systematic synthesis of a neural network for quadratic minimization. *Neural Networks* **4**(5), 599–613.

## References

Sum, J.P.F., C.S. Leung, P.K.S. Tam, G.H. Young, W.K. Kan and L.W. Chan (1999). Analysis for a class of winner-take-all model. *IEEE Trans. Neural Networks* **10**(1), 64–70.

Svestka, J.A. and V.E. Huckfeldt (1973). Computational experience with an m-salesman traveling salesman algorithm. *Management Science* **19**, 790–799.

Tachibana, Y., G.H. Lee, H. Ichihashi and T. Miyoshi (1995). A simple steepest descent method for minimizing hopfield energy to obtain optimal solution of the tsp with reasonable certainty. *Proceedings Int. Conference on Neural Networks* pp. 1871–1875.

Takahashi, N. (2000). A new sufficient condition for complete stability of cellular neural networks with delay. *IEEE Trans. Circuits Systems-I* **47**, 793–799.

Takefuji, Y., K.C. Lee and H. Aiso (1992). An artificial maximum neural network: a winner-take-all neuron model forcing the state of the system in a solution domain. *Biological Cybernetics* **67**(3), 243–251.

Talaván, P.M. and J. Yáñez (2002a). Parameter setting of the hopfield network applied to tsp. *Neural Networks* **15**, 363–373.

Talavan, P.M. and Yanez (2002b). Parameter setting of the hopfield network applied to tsp. *Neural Networks* **15**, 363–373.

Tan, K.C., H. Tang and S.S. Ge (2005). On parameter settings of hopfield networks applied to traveling salesman problems. *IEEE Trans. Circuits and Systems-I: Regular Papers* **52**(5), 994–1002.

Tan, K.C., H. Tang and Z. Yi (2004). Global exponential stability of discrete-time neural networks for constrained quadratic optimization. *Neurocomputing* **56**, 399–406.

Tan, K.C., T.H. Lee, D. Khoo and E.F. Khor (2001). A multiobjective evolutionary algorithm toolbox for computer-aided multiobjective optimization. *IEEE Trans. Systems, Man and Cybernetics-Part B* **31**(4), 537–556.

Tang, H., K.C. Tan and Z. Yi (2004). A columnar competitive model for solving combinatorial optimization problems. *IEEE Trans. Neural Networks* **15**(6), 1568–1574.

Tang, H.J., K.C. Tan and Z. Yi (2002). Convergence analysis of discrete time recurrent neural networks for linear variational inequality problem. In: *Proceedings of 2002 Int. Joint Conference on Neural Networks*. Hawaii, USA. pp. 2470–2475.

Tank, D.W. and J.J. Hopfield (1986). Simple neural optimization networks: An a/d converter, signal decision circuit, and a linear programming circuit. *IEEE Trans. Circuits Syst.* **33**(5), 533–541.

Veldhuizen, D.A.V. and G.B. Lamont (2000). Multiobjective evolutionary algorithms: analyzing the state-of-the-art. *Evolutionary Computation* **8**(2), 125–147.

Vemuri, V.R. (1995). *Artificial Neural Networks: Concepts and Control Applications*. IEEE Comput. Society Press.

Venetianer, P.L. and T. Roska (1998). Image compression by cellular neural networks. *IEEE Trans. Circuits and Systems-I* **45**(3), 205–215.

Vidyasagar, M. (1992). *Nonlinear Systems Analysis*. 2nd Edition. Prentice Hall. New Jersey.

von der Malsburg, C. (1973). Self-organization of orientation sensitive cells in the striate cortex. *Kybernetik* **14**, 85–100.

von der Malsburg, C. (1981). *The correlation theory of brain function*. Tech. Rep. No. 81-2. Max-Planck-Institute for Biophysical Chemistry. Gottingen.

von der Malsburg, C. and J. Buhmann (1992). Sensory segmentation with coupled neural oscillators. *Biol. Cybern.* **54**, 29–40.

Wang, C.H., H.L. Liu and C.T. Lin (2001). Dynamic optimal learning rates of a certain class of fuzzy neural networks and its applications with genetic algorithm. *IEEE Trans. Systems, Man, and Cybernetics-Part B: Cybernetics* **31**(3), 467–475.

Wang, D.L. (1999). Object selection based on oscillatory correlation. *Neural Networks* **12**, 579–592.

Wang, D.L. and D. Terman (1995). Locally excitatory globally inhibitory oscillator networks. *IEEE Trans. Neural Networks* **6**(1), 283–286.

Wang, D.L. and D. Terman (1997). Image segmentation based on oscillatory correlation. *Neural Computation* **9**, 805–836.

Wang, D.L. and X. Liu (2002). Scene analysis by integrating primitive segmentation and associative memory. *IEEE Trans. Systems, Man and Cybernectis: B* **32**(3), 254–268.

Wang, J. (1992). Analogue neural networks for solving the assignment problem. *Electronics Letters* **28**(11), 1047–1050.

Wang, J. (1993). Analysis and design of a recurrent neural network for linear programming. *IEEE Trans. Circuits and Systems-I* **40**(9), 613–618.

Wang, J. (1995). Analysis and design of an analog sorting network. *IEEE Trans. Neural Networks* **6**(4), 962–971.

Wang, J. (1996). A recurrent neural network for solving the shortest path problem. *IEEE Trans. Circuits and Systems-I* **43**(6), 482–486.

Wang, J. (1997a). Primal and dual assignment networks. *IEEE Trans. Neural Networks* **8**(3), 784–790.

Wang, L. (1997b). Discrete-time convergence theory and updating rules for neural networks with energy functions. *IEEE Trans. on Neural Networks* **8**(2), 445–447.

Wang, L. (1997c). On competitive learning. *IEEE Trans. Neural Networks* **8**(5), 1214–1216.

Wang, L. (1998). On the dynamics of discrete-time, continuous-state hopfield neural networks. *IEEE Trans. on Circuits and Systems-II: Analog and Digital Signal Processing* **45**(6), 747–749.

Wang, L. and F. Tian (2000). Noisy chaotic neural networks for solving combinatorial optimization problems. In: *Proceedings of Int. Joint Conference on Neural Networks (IJCNN)*. Combo, Italy.

Wang, L. and K. Smith (1998). On chaotic simulated annealing. *IEEE Trans. on Neural Networks* **9**(4), 716–718.

Weaver, S. and M. Polycarpou (2001). Using localizing learning to improve supervised learning algorithms. *IEEE Trans. Neural Networks* **12**(5), 1037–1045.

Wersing, H. and H. Ritter (1999). Feature binding and relaxation labeling with the competitive layer model. In: *Proc. Eur. Symp. on Art. Neur. Netw. ESANN*. pp. 295–300.

Wersing, H., J.J. Steil and H. Ritter (1997). A layered recurrent neural network for feature grouping. In: *Proceedings of Int. Conf. Art. Neur. Netw. ICANN*. Lausanne. pp. 439–444.

Wersing, H., J.J. Steil and H. Ritter (2001a). A competitive layer model for feature binding and sensory segmentation. *Neural Computation* **13**(2), 357–387.

Wersing, H., W.J. Beyn and H. Ritter (2001b). Dynamical stability conditions for recurrent neural networks with unsaturating piecewise linear transfer functions. *Neural Computation* **13**(8), 1811–1825.

Williamson, J.R. (1996). Neural network for dynamic binding with graph representation: form, linking and depth-from-occlusion. *Neural Computation* **8**, 1203–1225.

## References

Wilson, G.W. and G.S. Pawley (1988a). On the stability of the travelling salesman problem algorithm of hopfield and tank. *Biol. Cybern.* **58**, 63–70.

Wilson, H.R. (1999). *Spikes, Decisions and Actions Dynamical foundations of Neuroscience*. Oxford University Press. New York.

Wilson, V. and G.S. Pawley (1988b). On the stability of the tsp problem algorithm of hopfield and tank. *Bio. Cybern.* **58**, 63–70.

Wu, Y.Q., C.P. Yang and T.Z. Wang (2001). A new approach of color quantization of image based on neural network. *International Joint Conference on Neural Networks* pp. 973–977.

Xia, Y. and J. Wang (2000). Global exponential stability of recurrent neural networks for solving optimization and related problems. *IEEE Trans. Neural Networks* **11**(4), 1017–1022.

Xiaofu, Z. and A. Minai (2004). Temporally sequenced intelligent block-matching and motion-segmentation using locally coupled networks. *IEEE Trans. Neural Networks* **15**(5), 1202–1214.

Xie, X., R.H.R. Hahnloser and H.S. Seung (2002). Selectively grouping neurons in recurrent networks of lateral inhibition. *Neural Computation* **14**(11), 2627–2646.

Yam, J.Y.F. and T.W.S. Chow (2001). Feedforward networks training speed enhancement by optimal initialization of the synaptic coefficients. *IEEE Trans. Neural Networks* **12**(2), 430–434.

Yao, X. (1999). Evolving artificial neural networks. *Proceedings of IEEE* **87**(9), 1423–1447.

Yao, X. and Y. Liu (1997). A new evolutionary system for evolving artificial neural networks. *IEEE Trans. Neural Networks* **8**(3), 694–713.

Ye, H., A.N. Michel and K. Wang (1994). Global stability and local stability of hopfield neural networks with delays. *Physical Review E* **50**(5), 4206–4213.

Yen, S.C. and L.H. Finkel (1998). Extraction of perceptually salient contours by striate cortical networks. *Vision Research* **38**(5), 719–741.

Yeung, D.S. and X. Sun (2002). Using function approximation to analyze the sensitivity of mlp with antisymmetrix squashing activation function. *IEEE Trans. Neural Networks* **13**(1), 34–44.

Yi, Z. and K.K. Tan (2002). Dynamic stability conditions for lotka-volterra recurrent nerual networks with delays. *Physical Review E* **66**, 011910.

Yi, Z. and K.K. Tan (2004). *Convergence Analysis of Recurrent Neural Networks*. Vol. 13 of *Network Theory and Applications*. Kluwer Academic Publishers. Boston.

Yi, Z., K.K. Tan and T.H. Lee (2003). Multistability analysis for recurrent neural networks with unsaturating piecewise linear transfer functions. *Neural Computation* **15**(3), 639–662.

Yi, Z., P.A. Heng and A.W. Fu (1999). Estimate of exponential convergence rate and exponential stability for neural networks. *IEEE Trans. Neural Networks* **10**(6), 1487–1493.

Yi, Z., P.A. Heng and K.S. Leung (2001). Convergence analysis of cellular neural networks with unbounded delay. *IEEE Trans. Circuits and Systems-I* **48**(6), 680–687.

Yi, Z., P.A. Heng and P.F. Fung (2000). Winner-take-all discrete recurrent neural networks. *IEEE Trans. Circuits and Systems II* **47**(12), 1584–1589.

Yi, Z., Y. Fu and H. Tang (2004). Neural networks based approach for computing eigenvectors and eigenvalues of symmetric matrix. *Computer & Mathematics with Applications* **47**, 1155–1164.

Yu, X., M.O. Efe and O. Kaynak (2002). A general backpropagation algorithm for feedforward neural networks learning. *IEEE Trans. Neural Networks* **13**(1), 251–254.

Zak, S. H., V. Upatising and S. Hui (1995a). Solving linear programming problems with neural networks: a comparative study. *IEEE Trans. Neural Networks* **6**(1), 94–103.

Zak, S.H., V. Upatising and S. Hui (1995b). Solving linear programming problems with neural networks: A comparative study. *IEEE Trans. Neural Networks* **6**(1), 94–104.

Zhang, X.S. (2000). *Neural networks in optimization.* Kluwer Academic Publisher.

Zhang, Z., T. Ding, W. Huang and Z. Dong (1992). *Qualitative Theory of Differential Equations.* Translated by W.K. Leung. American Mathematical Society. Providence, USA.

Zitzler, E., K. Deb and L. Thiele (2000). Comparison of multiobjective evolutionary algorithms: empirical results. *Evolutionary Computation* **8**(2), 173–195.

Zucker, S.W., A. Dobbins and L. Iverson (1989). Two stages of curve detection suggest two styles of visual computation. *Neural Computation* **1**, 68–81.

Zuo, X.Q. and Y.S. Fan (n.d.). A chaos search immune algorithm with its application to neuro-fuzzy controller design. *Chaos, Solitons, and Fractals* p. In Press.

Zurada, J.M. (1995). *Introduction to Artificial Neural Systems.* PWS Publishing Company. Boston.

Zurada, J.M., I. Cloete and E. van der Poel (1996). Generalized hopfield networks for associative memories with multi-valued stable states. *Neurocomputing* **13**, 135–149.

CPSIA information can be obtained
at www.ICGtesting.com
Printed in the USA
LVHW082119251218
601613LV00011B/365/P